Plasticity and Signal Representation in the Auditory System

Plasticity and Signal Representation in the Auditory System

Edited by

Josef Syka
Institute of Experimental Medicine
Academy of Sciences of the Czech Republic
Prague, Czech Republic

and

Michael M. Merzenich
Keck Center for Integrative Neurosciences
University of California at San Francisco
San Francisco, California

Springer

Proceedings of the International Symposium on Plasticity of the Central Auditory System and Processing of Complex Acoustic Signals, held as a Satellite Symposium of the 6th IBRO World Congress of Neuroscience, July 7–10, 2003, in Prague, Czech Republic

ISBN 0-387-23154-4
eISBN 0-387-23181-1

Printed in the United States of America. (BS/DH)

9 8 7 6 5 4 3 2 1

springeronline.com

Preface

The symposium that has provided the basis for this book, "Plasticity of the Central Auditory System and Processing of Complex Acoustic Signals" was held in Prague on July 7-10, 2003. This is the fourth in a series of seminal meetings summarizing the state of development of auditory system neuroscience that has been organized in that great world city. Books that have resulted from these meetings represent important benchmarks for auditory neuroscience over the past 25 years. A 1980 meeting, "Neuronal Mechanisms of Hearing" hosted the most distinguished hearing researchers focusing on underlying brain processes from this era. It resulted in a highly influential and widely subscribed and cited proceedings co-edited by professor Lindsay Aitkin. The subject of the 1987 meeting was the "Auditory Pathway – Structure and Function". It again resulted in another important update of hearing science research in a widely referenced book co-edited by the late Bruce Masterton.

While the original plan was to hold a meeting summarizing the state of auditory system neuroscience every 7 years, historical events connected with the disintegration of the Soviet Empire and return of freedom to Czechoslovakia resulted in an unavoidable delay of what was planned to be a 1994 meeting. It wasn't until 1996 that we were able to meet for the third time in Prague, at that time to review "Acoustical Signal Processing in the Central Auditory System". To re-establish the 7-year period between meetings, this 4th meeting was organized as a satellite meeting to the 6th IBRO World Congress, which was held in Prague in July 2003. Participants included the majority of most distinguished research scientists in this subdiscipline, in the world.

While the topics of the 2003 symposium overlapped with those of the 1996 meeting, this field is in an era of remarkable progress, particularly in the field of plasticity of the auditory system. New advances in understanding auditory system plasticity, based substantially on a large and growing body of results from animal experiments, were related to innumerable new insights into the physiology and pathology of speech and music perception and production generated by behavioral studies, and from the application of modern brain imaging techniques. We are living in an especially exciting period of research, marked by an almost astounding rate of advance in the development of our understanding of the hearing brain. The extraordinary series of reports published in this book document this rapid, further advance.

The setting of the symposium was different from the previous ones – from the picturesque halls of ancient monasteries and the magnificent architecture of the Prague castle, the organizers transferred the symposium to the modern buildings of the largest clinical and research institute in the Czech Republic, the Institute for Clinical and Experimental Medicine. Its modern building is situated on the southern outskirts of Prague, next to a large forest. But contact with historical Prague was not lost completely: a concert consisting of famous pieces of Mozart and Dvořák was held in the famous Mirror Chapel, a concert hall of the former Jesuit baroque monastery (the Klementinum) and the symposium dinner was organized in the luxurious French Restaurant of the Art Noveau brilliant of Prague, the Municipal Hall.

Much water has flown through Vltava river since the first symposium in 1980 and many historical events happened since that time in Prague – the majority of them in the direction of improvement of life, enhancement of freedom and democracy and consequently also improvement of conditions for science in the Czech Republic. When preparing this volume for publication, the Czech Republic became member of the European Union. It is our hope that for science, this fact promises further rapid advance and bright possibilities.

Contents

Interaural Time Difference Processing

ALAN R. PALMER[*], and BENEDIKT GROTHE[#]
[*]MRC Institute of Hearing Research, University Park, Nottingham NG7 2RD, UK; and [#]Research Group Auditory Processing, Max Planck Institute of Neurobiology, Am Klopferspitz 18a, 82152 Martinsried, Germany

1. INTRODUCTION

Localisation of a sound source depends upon cues that result from having two ears separated in space by the head. The binaural cues consist of an interaural time delay (ITD) caused by a longer sound path to one ear and, depending on the frequency, an interaural level difference (ILD) resulting from the shadowing effect of the head. Head shadowing effects are minimal for low-frequencies, for which the wavelength is longer than the head width, whilst for higher frequencies the interaural phase differences (IPDs), that result from the ITD, present an ambiguous cue. For high-frequency or complex sounds, the different transmission delays to the ears also create a difference in the arrival times of the sounds at the ears, and an on-going delay of the envelope. In addition to these cues, interference between reflections of sound waves from the torso and within the outer ear results in frequency-dependent spectral colouring of broadband sounds. Low-frequency sounds can be localised using IPDs alone, high-frequency sounds by their ILDs, and for complex sounds the envelope delay may also be used. Spectral cues permit accurate monaural sound localisation of complex sounds and provide cues for sound elevation. In practice, in real listening environments with broadband signals ITDs, ILDs and spectral cues will all be simultaneously present and consistent with each other.

Here we are only concerned with the processing of the ITD cue. For humans this is the dominant localisation cue and the human ear is exquisitely sensitive to ITDs. For noise and clicks the just noticeable difference in ITD is 5-12 microseconds and for tones it is 20-30 microseconds (dependent on frequency and duration Durlach, 1972; Tobias and Zerlin, 1959).

In both mammals and birds ITD, ILD and pinna/torso cues are processed in anatomically-distinct auditory pathways (see Irvine, 1986; Konishi *et al.*, 1988; Yin and Chan, 1988) that begin with the projection of discrete neural populations in the ventral cochlear nucleus to the sub-nuclei of the superior olivary complex. Detailed reviews of the neural coding of interaural cues for localisation are available (Caird, 1991; Irvine, 1986; Kuwada *et al.*, 1997; Phillips *et al.*, 1991; Yin and Chan, 1988) and it is not the intention here to review this literature extensively, but rather to describe some recent findings that have shed new light on possible mechanistic and coding issues. The dominant view of the mechanism underlying ITD sensitivity is that of Jeffress (1948) in which neurons acting as detectors of coincident spike input from the two ears are connected via a series of delay lines to provide a transformation from ITD to a spatial code (see below for more detail and alternative mechanisms).

2. NEURONAL SENSITIVITY TO ITDS

Neurons sensitive to the ITDs of low-frequency signals have been recorded from auditory nuclei in a wide range of species (Caird and Klinke, 1983; Kuwada *et al.*, 1987; Palmer *et al.*, 1992; Spitzer and Semple, 1998; Yin and Kuwada, 1983; Yin and Chan, 1990). It is generally accepted that this sensitivity derives from comparison of activity in the medial superior olive, but it is stereotypical throughout the system at least for tonal stimulation. ITD sensitivity of a single neuron in the inferior colliculus (IC) is illustrated in Figure 1. The firing rate of the neuron varies periodically as a function of ITD with a period equal to that of the stimulus frequency. The non-sinusoidal nature of this function probably relates to the fact that it was measured in the IC where several other factors, such as convergence, also play a role. Whatever mechanism underlies this sensitivity, it involves a longer effective transmission delay from one ear than from the other. This imbues each ITD sensitive neuron with a "characteristic delay" at which it gives equal output at all stimulating frequencies. For simple coincidence detectors the characteristic delay corresponds to the peak of the curve which is also known as the best delay (BD).

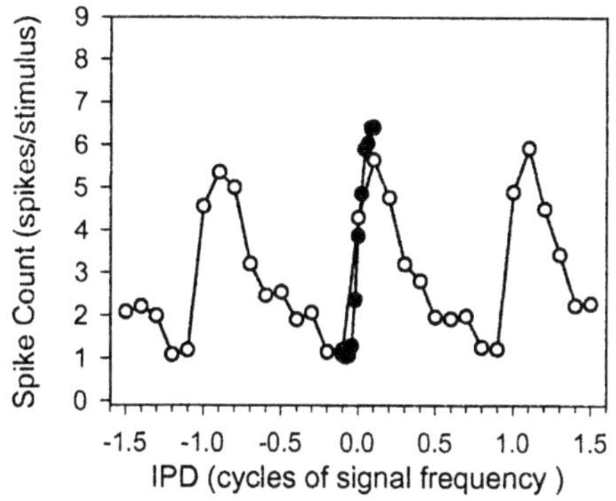

Figure 1. ITD sensitivity of a single inferior colliculus neuron to best frequency tones 50-ms duration at 20 dB above threshold. Black dots show higher resolution measurements. Modified from Shackleton *et al.* (2003)

2.1 Best Frequency Dependence of Best Delays

Following directly from Jeffress's delay line model there has been a focus on value of the ITD evoking the maximum output of the coincidence detectors. In response to best frequency tones or to broadband noise the distributions of neurons in the IC with the various BDs peaks at relatively long ITDs (150-400 microseconds) representing sensitivity preferentially to sounds in the contralateral hemifield and this appeared to be species independent (McAlpine *et al.*, 1996; Palmer *et al.*, 1990; Yin and Kuwada, 1983). However, recent studies have suggested that within such plots there was concealed a frequency dependence: BDs close to zero ITD were found in neurons with high best frequencies while the longer BDs were found in neurons with lower best frequencies. This was shown to be the case when BD to tones (Figure 2A McAlpine *et al.*, 1996) or BD in response to noise (Figure 2B McAlpine *et al.*, 2001) was measured as a function of best frequency. Peak ITD expressed in terms of phase was approximately 45°, or 1/8 of a cycle of interaural phase, independent of best frequency.

One consequence of the dependence of BD on BF is that the sharpest slope on the ITD function passes through the midline (ITD=0) and hence through the physiological range of ITDs. This is illustrated in Figure 2C, which shows the mean noise delay curves within frequency bands overplotted.

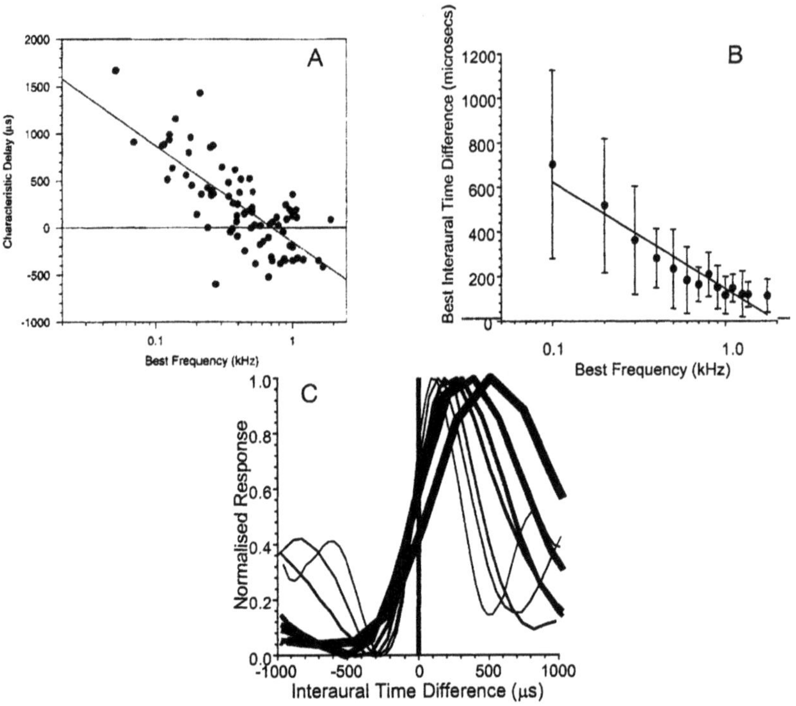

Figure 2. A. Dependence of characteristic delay (McAlpine *et al.*, 1996) on best frequency. The points represent individual neurons that had linear phase vs frequency plots and the line is a regression line fit. B. Dependence of noise best delay (McAlpine *et al.*, 2001) on best frequency in the guinea pig IC. The dots are the mean values in frequency bands and the error bars are standard deviations. The line is a regression line fit to the means. C. The mean noise delay curves for neurons in different frequency bands from 242 to 1400 Hz. The line thickness decreases with band centre frequency. All tones and noises 50-ms duration and 20 dB above threshold. Note the midline slopes all align.

2.2 ITD Resolution Achievable from the Activity of Individual Neurons

Slopes of functions generally provide the most sensitivity to change and human resolution is best on midline where these slopes occur. Shackleton and his colleagues (Shackleton *et al.*, 2003; Skottun, 1998) have measured the ability of neurons in the IC to signal changes in the ITD of a pure tone. To do this they measured high resolution ITD functions along the midline slope using multiple presentations (see Figure 1), so that they could employ signal detection theoretical methods to estimate the smallest just noticeable difference that a single neuron could signal given its discharge statistics. The results are shown in Figure 3. The best resolution measurable from single neurons was comparable to that of human subjects for the same stimulus conditions and showed similar frequency dependence. However, it is also clear that the neurons capable of such sensitivity will be swamped within a population of actively responding neurons that do not show such acuity. Thus it is highly likely that population codes of some kind are still necessary unless the system has some means of knowing exactly which sub population of neurons to use.

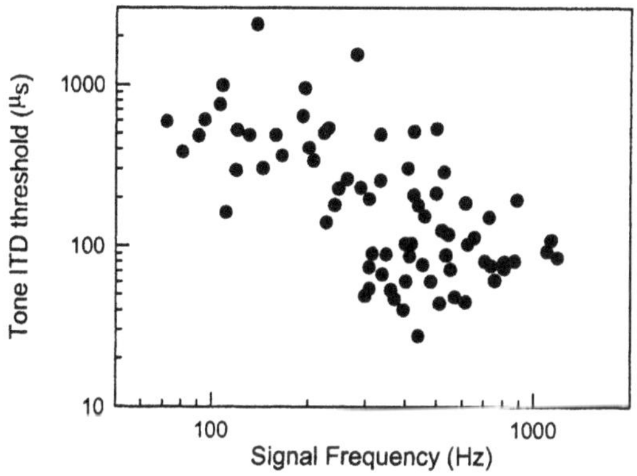

Figure 3. Single IC neurone just noticeable difference thresholds for 50-ms best frequency tones 20 dB above threshold. Each dot represents the best threshold from a single neuron. Modified from Shackleton *et al.* (2003)

3. SENSITIVITY TO INTERAURAL CORRELATION

Interaural correlation is central to nearly all theories of ITD processing and it has been shown that the cells in the MSO perform an interaural cross correlation (Yin and Chan, 1990). The process of coincidence detection is formally equivalent to the multiplicative step involved in cross correlation. Thus it is reasonable to expect that the mechanism underlying humans' exquisite sensitivity to decorrelation of the waveforms at the two ears is likely to also intimately involve the MSO coincidence detectors. However, there are two reasons why we might not expect single neurone correlation thresholds to match behavioural performance: the nature of the perception reported in psychophysical experiments, and stimulus variability.

Shackleton and Palmer (2003) measured directly the just noticeable difference in interaural correlation from cells in the IC using signal detection theoretical methods. Their results are shown in Figure 4. The lowest neural thresholds approach the human psychophysical data at a reference correlation of +1, and overlap it at a reference correlation of 0. Thresholds are much worse at a reference of -1. This failure at -1 is entirely consistent with the much reduced slopes of variation in firing rate with interaural correlations near -1 compared to the sharper slopes at +1.

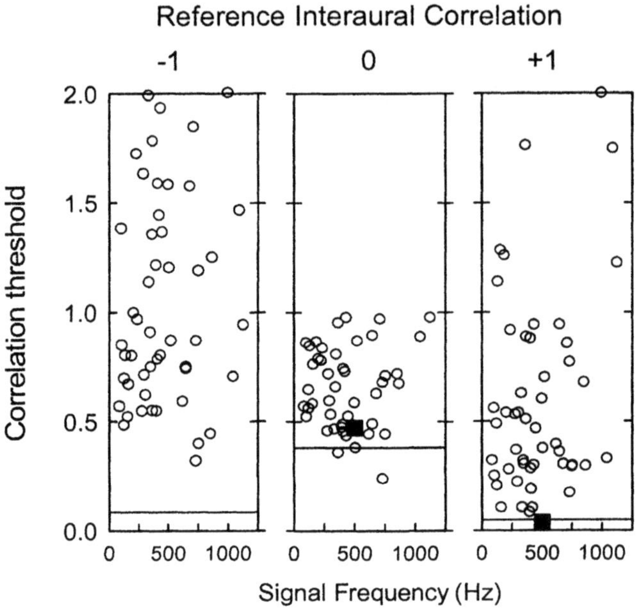

Figure 4. Just noticeable difference thresholds for interaural correlation in single IC neurones. Each circle represents the best threshold from a single neuron. Reference correlation is shown at the top. The squares show human interaural correlation thresholds for the same stimulus conditions estimated by interpolation from Bernstein and Trahiotis (1997). The lines show human thresholds for 400-ms duration broadband noises where the value at -1 was about twice that for +1 (Boehnke *et al.*, 2002). Modified from Shackleton and Palmer (2003).

Up to this point we have described the nature of ITD processing in the mammal and shown that the displacement of the peaks of ITD functions away from midline gives the best resolution for both tonal ITD and interaural correlation within the physiological range. Below we consider what mechanisms are responsible for the frequency dependent BD.

4. MECHANISMS PROPOSED FOR ITD PROCESSING

The dominant model for the neural processing of ITDs is that of Jeffress (1948). Jeffress proposed that somewhere in the brain, there existed a network of cells that acted as coincidence detectors receiving excitatory input from the two ears. Input to the coincidence detector from both sides occurring within a short time window leads to summation of the excitatory post synaptic potentials and generation of an action potential. Responses of the majority of ITD-sensitive neurons are consistent with Jeffress's notion of coincidence detectors. The ability of such cells to respond maximally to specific spatial locations was achieved in the Jeffress model by connections from each ear by axons of differing length to provide a delay line. Any specific coincidence detector would fire only when the internal delay compensates exactly for the later arrival of the sound at one ear. Thus the coincidence detector activated provides a code for spatial position and the network of coincidence detectors converts ITDs into a place code for position. Given the overlap in ITD functions, however, any sound source will activate a population of closely tuned cells.

The Jeffress model has been widely adopted because it accounts for many aspects of the psychophysics, and coincidence detection and characteristic delay are well supported by the physiological data. In the bird, there is good evidence for an anatomical substrate for a delay line (Carr and Konishi, 1990; Carr, 1993; Konishi *et al.*, 1988; Rubel and Parks, 1988). However, the evidence in the mammal for an actual delay line system is somewhat less compelling. Two studies that addressed this issue by reconstructing the paths of axons to the MSO (Bekius *et al.*, 1999; Smith *et al.*, 1993) showed axons that could provide the graded delays necessary and both concluded that the anatomical substrate for a delay line system is present in the mammal. However, the difficulty of these experiments means that even those few reports that exist are not comprehensive and unfortunately not entirely convincing.

A way of generating the interaural delays, not involving axons of different length, utilises the delays inherent in the conduction time of the activity along the basilar membrane to different frequencies, by mismatching the best frequencies of the inputs to the coincidence detectors from each ear (Bonham and Lewis, 1999; Schroeder, 1977; Shackleton *et al.*, 2000; Shamma, 1989). Recently, it has been shown that delays in auditory nerve fibres with different best frequencies vary in appropriate ways and over suitable ranges to account for ITD sensitivity measured in the IC (Joris *et al.*, 2003).

Another way to achieve the necessary apparent delays between the inputs to the coincidence detectors has recently been demonstrated by Brand *et al.* (2002). Using iontophoresis they blocked inhibitory inputs to MSO neurons. Their results are shown in Figure 5: blocking the glycinergic inhibitory input to the MSO cell results in a shift of the peak ITD sensitivity toward zero ITD. In Figure 5A is shown the data from a single MSO neuron before (control) and after (strychnine) blockade of the the glycinergic inputs. The increase in discharge rate caused by the blockade is accompanied by a shift in the BD from about 250 μs to zero. In Figure 5B the distribution of BDs in their sample before and during blockade are shown in histogram form. There is a statistically significant shift in the distribution toward zero ITD. On the basis of such data, Brand *et al.* proposed that phase-locked, but ITD insensitive inhibition operates to shift the peak of the ITD curve from zero to about 1/8 of the period of the best frequency tone, thus providing the dependence of BD on BF inhibition and accounting for the empirically observed BD.

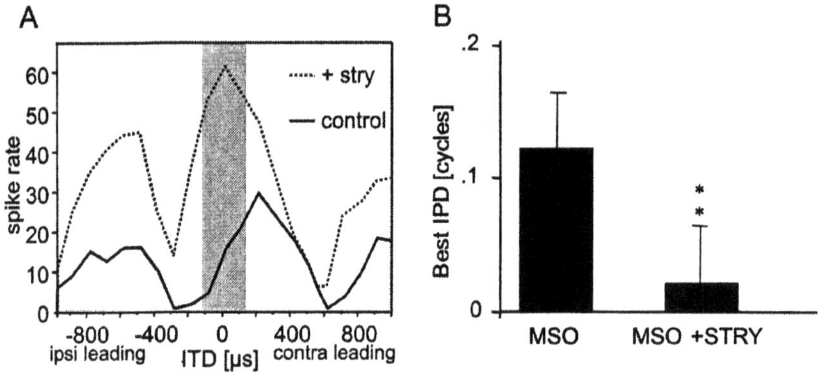

Figure 5. Effect of blocking glycinergic inhibition in the MSO using strychnine. A. Shift of peak ITD sensitivity toward zero ITD in a single neuron. B. BDs (in cycles) in MSO before (about 1/8 cycle) and during (near zero) the blockade of glycinergic inhibition. Asterisks show statistically significant reduction in BD. Modified from Brand *et al.* (2002).

4.1 Development of the ITD Sensitivity Circuit: A Structural Refinement of Inhibitory Inputs Is Correlated with Maturation of ITD Tuning

Kapfner and colleagues (Kapfner *et al.*, 2002) have shown that in gerbils the inhibitory inputs, which appear responsible for the tuning of MSO ITD

functions to about 1/8 the of the period of the best frequency, undergo an experience-dependent development within a critical period. During the first few days after hearing onset the glycinergic input to the MSO undergoes a specific structural refinement. The inhibitory synapses that are neonatally present on the soma and dendrites (see Figure 6: juvenile gerbil) are eliminated from the dendrites during development (Figure 6: adult gerbil). This refinement fails to occur unless the animal has appropriate auditory experience. When one cochlea is ablated preventing synchronous bilateral input to the MSO these synaptic rearrangements do not occur. Partial masking of directional cues by rearing the animals in omnidirectional noise substantially reduces the degree of synaptic refinement (Figure 6: noise reared gerbil, Kapfner *et al.*, 2002). The structural pattern with inhibitory inputs limited to the soma has also been found in other mammals that use ITDs (cat: Clark, 1969; chinchilla: Perkins, 1973), but not in animals that have only high-frequency hearing and thus do not use ITDs for localisation.
(rats, free-tailed bats, short-tailed opossum: Kapfner *et al.*, 2002).

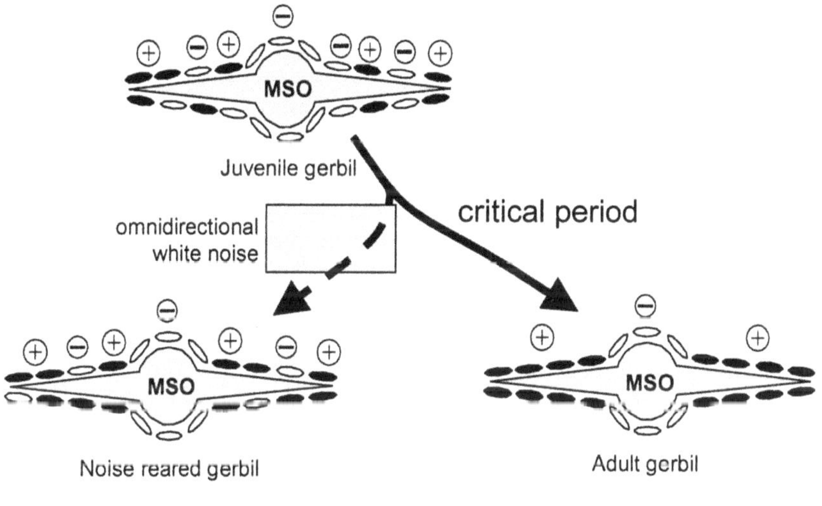

Figure 6. Developmental changes that occur in the distribution of synapses across the soma and dendrites of an MSO neuron (modified after Kapfner *et al.*, 2002). Filled ellipses represent excitatory synapses and unfilled ellipses inhibitory synapses. The changes in synaptic distribution that occur in development are reduced by presentation of omnidirectional noise during the critical period.

If this structural development of the inhibitory inputs has any functional significance, one should expect these inputs to have no effect or at least a different effect on ITD tuning in juvenile or noise-reared animals compared to normal adults. The ITD functions should be dominated by the excitatory inputs and, hence, should resemble those measured in adult animals during blockade of inhibition. Indeed, in young gerbils, three days after hearing onset, ITD tuning in the dorsal nucleus of the lateral lemniscus (DNLL: to obtain larger samples than possible recording in MSO) failed to show the adult-like pattern. That is, most neurons had BDs close to the midline (Seidl and Grothe, 2003). Consistent with the reduced refinement of the inhibitory inputs to the MSO when animals are raised in omnidirectional noise, their ITD sensitivity is also altered. When directional cues were partially masked by rearing gerbils in omnidirectional noise the ITD sensitivity looked more similar to the juvenile than the adult. More specifically the BDs in the DNLL of control adult gerbils showed the usual clustering at 1/8 cycle, whereas in noise reared animals the BDs clustered around zero ITD as they did in juveniles or after blockade of the glycinergic inputs.

5. SUMMARY AND CONCLUSION

Grothe (2003) has argued strongly that the mechanisms for processing ITD information are quite different in mammals and birds. The chick and the owl have anatomical arrangements that subserve the classical Jeffress model of coincidence detectors connected by delay lines. In the mammal the anatomical evidence is less clear and other mechanisms to account for the effective delay of signals from the two ears reaching the MSO coincidence detectors have been proposed. The evidence presented above presents a persuasive case that phase-locked ITD independent glycinergic input to the MSO from the medial nucleus of the trapezoid body and the lateral nucleus of the trapezoid body (see Grothe, 2000) may produce the effective delay by shifting the peak of the ITD sensitivity of MSO neurons. The fact that this inhibition changes during development as does the ITD sensitivity, and that removal of this inhibition results in juvenile-like ITD sensitivity, provide further circumstantial evidence of the involvement of the inhibition in tuning the ITD sensitivity.

The net result of the transmission delay that displaces the BD from zero ITD (from whatever cause) is to position the greatest slope of the ITD function through the physiological range of even the smallest animals. The utility of the discharge rate changes, embodied in this sharp slope, in detecting ITD and correlation changes has been demonstrated. Single neurons perform at levels that approximate human psychophysical acuity for tone and noise ITDs and for correlations at +1 and 0. However, unlike minimum thresholds, which also seem to depend on the most sensitive neurons, those neurons signalling very small ITD changes will be amongst a very active population. Further processing involving populations of cells will no doubt be necessary to exploit the neurons with highest sensitivity. The original Jeffress formulation suggested that ITD was signalled by the position of maximum activity within an array of neurons. An alternative suggestion, originally made by von Bekesy (1960), has recently been revived by McAlpine *et al.* (2001) following their description of the best frequency dependency of BD. This model evaluates the ITD values by a comparison of activity on the two sides of the brain.

From the different models being proposed for the generation of the delay at the input to the coincidence detectors, and for the way that ITD sensitive provides localisation information, it is clear that this area is dynamic and presently remains controversial.

ACKNOWLEDGEMENTS

We would like to acknowledge the huge contributions of our co-authors on the various papers on which this report is based and in particular Antje Brand, Christoph Kapfner, David McAlpine, Armin Seidl and Trevor Shackleton.

REFERENCES

Bekesy, G., 1960, *Experiments in hearing.* McGraw-Hill, New-York.

Beklus, G. E., Batra, R., and Oliver, D. L., 1999, Axons from anterventral cochlear nucleus that terminate in medial superior olive of the cat: observations related to delay lines. *J. Neurosci.* **19**: 3146-3161.

Bernstein, L. R., and Trahiotis, C., 1997, The effects of randomizing values of interaural disparities on binaural detection and on discrimination of interaural correlation. *J. Acoust. Soc. Am.* **102**: 1113-1120.

Boehnke, S. E., Hall, S. E., and Marquardt, T., 2002, Detection of static and dynamic changes in interaural correlation. *J. Acoust. Soc. Am.* **112**: 1617-1626.

Bonham, B. H., and Lewis, E. R., 1999, Localization by interaural time difference (ITD): effects of interaural frequency mismatch. *J. Acoust. Soc. Am.* **106**: 281-290.

Brand, A., Behrend, O., Marquardt, T., McAlpine, D., and Grothe, B., 2002, Precise inhibition is essential for microsecond interaural time difference coding. *Nature* **417**: 543-547.

Caird, D., 1991, Processing in the colliculi. In *Neurobiology of Hearing: The Central Auditory System* (R.A. Altschuler, R.P. Bobbin, B.M. Clopton and D.W. Hoffman, eds.), Raven Press, New York. pp. 141-162.

Caird, D. M., and Klinke, R., 1983, Processing of binaural stimuli by cat superior olivary complex neurons. *Exp. Brain Res.* **52**: 385-399.

Carr, C. E., and Konishi, M., 1990, A circuit for detection of interaural time differences in the brain stem of the barn owl. *J. Neurosci.* **10**: 3227-3246.

Carr, C. E., 1993, Delay line models of sound locaization in the barn owl. *Am. Zool.* **33**: 79-85.

Clark, G. M., 1969, The ultrastructure of nerve endings in the medial superior olive of the cat. *Brain Res.* **14**: 293-305.

Durlach, N. I., 1972, Binaural signal detection: equalization and cancellation theory. In *Foundations of Modern Auditory Theory* (J.V. Tobias, ed.), Academic Press, New York. pp. 369-462.

Grothe, B., 2003, New roles for synaptic inhibition in sound localization. *Nat. Rev.* **4**: 1-11.

Grothe, B., 2000, The evolution of temporal processing in the medial superior olive, an auditory brainstem structure. *Prog. Neurobiol.* **61**: 581-610.

Irvine, D.R.F., 1986, *The Auditory Brainstem.* Springer-Verlag, Berlin.

Jeffress, L.A., 1948, A place code theory of sound localization. *J. Comp. & Physiol. Psychol.* **41**: 35-39.

Joris, P. X., van der Heijden, M., Louage, D., van de Sande, B., and van Kerckhoven, C., 2003, Dependence of binarual and cochlear "best delays" on characteristic frequency. In *Auditory Signal Processing: Physiology, Psychophysics and Models* (D. Pressnitzer, A. de Cheveigne, S. McAdams and L. Collet, eds.), Springer-Verlag, New York. pp. 396-402.

Kapfner, C., Seidl, A.H., Schweizer, H., and Grothe, B., 2002, Experience-dependent refinement of inhibitory inputs to auditory coincidence-detector neurons. *Nat. Neurosci.* **5**: 247-253.

Konishi, M., Takahashi, T. T., Wagner, H., Sullivan, W. E., and Carr, C. E., 1988, Neurophysiological and anatomical substrates of sound localization in the owl. In *Auditory Function: Neurobiological Basis of Hearing* (E. M. Edelman, W. E. Gall and W. M. Cowans, eds.), J. Wiley, New York. pp. 721-746.

Kuwada, S., Batra, R., and Fitzpatrick, D. C., 1997, Neural processing of binaural temporal cues. In *Binaural and Spatial Hearing in Real and Virtual Environments* (R. H. Gilkey and T. R. Anderson, eds.), L. Erlbaum Associates, Mahwah, N.J. pp. 399-425.

Kuwada, S., Stanford, T. R. and Batra, R., 1987, Interaural phase-sensitive units in the inferior colliculus of the unanesthetized rabbit: Effects of changing frequency. *J. Neurophysiol.* **57**: 1338-1360.

McAlpine, D., Jiang, D., and Palmer, A. R., 1996, Monaural and binaural responses of low-best frequency neurones in the inferior colliculus of the guinea pig. *Hear. Res.* **97**: 136-152.

McAlpine, D., Jiang, D., and Palmer, A. R., 2001, A neural code for low-frequency sound localisation in mammals. *Nat. Neurosci.* **4**: 396-401.

Palmer, A. R., Rees, A., and Caird, D., 1992, Binaural masking and sensitivity to interaural delay in the inferior colliculus. In *Processing of Complex Sounds by the Auditory System* (R. P. Carlyon, C. J. Darwin and I. J. Russell, eds.), Clarendon Press, Oxford. pp. 415-422.

Palmer, A. R., Rees, A., and Carid, D., 1990, Interaural delay sensitivity to tones and broad band signals in the guinea-pig inferior colliculus. *Hear. Res.* **50**: 71-86.

Perkins, R. E., 1973, An electron microscopic study of synaptic organization in the medial superior olive of normal and experimental chinchillas. *J. Comp. Neurol.* **148**: 387-415.

Phillips, D. P., Reale, R. A., and Brugge, J. F., 1991, Stimulus processing in the auditory cortex. In *The Central Auditory System* (R. Altschuler, R. Bobbin, B. Clopton and D. Hoffman, eds.), Raven Press, New York. pp. 335-365.

Rubel, E. W., and Parks, T. N., 1988, Organization and development of the avian brainstem auditory system. In *Auditory Function: Neurobiological Basis of Hearing* (E. G. M. Edelman, W. E. Gall and W. M. Cowans, eds.), Wiley, New York. pp. 3-92.

Schroeder, M. R., 1977, New viewpoints in binaural interactions. In *Psychophysics and Physiology of Hearing* (E. F. Evans and J. P. Wilson, eds.), Academic Press, New York. pp. 455-467.

Seidl, A. H., and Grothe, B., 2003, Acoustic experience is essential for the development of normal ITD processing. *26th Midwinter Meeting of ARO*, Daytona Beach, Fla. pp. 930.

Shackleton, T. M., Skottun, B. C., Arnott, R. H., and Palmer, A. R., 2003, Interaural time difference discrimination thresholds for single neurons in the inferior colliculus of guinea pigs. *J. Neurosci.* **23**: 716-24.

Shackleton, T. M., McAlpine, D., and Palmer, A.R., 2000, Modelling convergent input onto interaural-delay sensitive inferior colliculus neurones. *Hear. Res.* **149**: 199-215.

Shamma, S. A., 1989. Stereausis: binaural processing without neural delays. *J. Acoust. Soc. Am.* **86**: 989-1006.

Skottun, B. C., 1998, Sound localization and neurons. *Nature* **393**: 531.

Smith, P. H., Joris, P. X., and Yin, T. C. T., 1993, Projections of physiologically characterized spherical bushy cell axons from the cochlear nucleus of the cat: evidence for delay lines to the medial superior olive. *J. Comp. Neurol.* **331**: 245-260.

Spitzer, M. W., and Semple, M. N., 1998, Transformation of binaural response properties in the ascending auditory pathway: Influence of time-varying interaural phase disparity. *J. Neurophysiol.* **80**: 3062-3076.

Tobias, J. V., and Zerlin, S., 1959, Lateralization threshold as a function of stimulus duration. *J. Acoust. Soc. Am.* **31**: 1591-1594.

Yin, T., and Kuwada, S., 1983, Binaural interaction in low-frequency neurons in inferior colliculus of the cat. III. Effects of changing frequency. *J. Neurophysiol.* **50**: 1020-1042.

Yin, T., and Chan, J., 1990, Interaural time sensitivity in medial superior olive of cat. *J. Neurophysiol.* **64**: 465-488.

Yin, T. C. T., and Chan, J. C. K., 1988, Neural mechanisms underlying interaural time sensitivity to tones and noise. In *Auditory Function: Neurobiological Basis of Hearing* (G.M. Edelman, W.E. Gall and W.M. Cowan, eds.), J. Wiley, New York. pp. 385-430.

Interplay of Excitation and Inhibition in Auditory Brainstem Processing at Endbulbs of Held of the MNTB and AVCN

SUSANNE DEHMEL, CORNELIA KOPP-SCHEINPFLUG, and RUDOLF RÜBSAMEN
Department of Neurobiology, University of Leipzig, Talstrasse 33, 04103 Leipzig, Germany

1. INTRODUCTION

Spherical bushy cells (SBC) of the anteroventral cochlear nucleus (AVCN) and principal cells of the medial nucleus of the trapezoid body (MNTB) receive their afferent excitatory input through large, excitatory synaptic terminals (endbulbs or calyces of Held) (Brawer and Morest, 1975; Ryugo and Sento, 1991; Smith *et al.*, 1991). Earlier, this finding was associated with the idea of a one-to-one signal transmission at SBCs resp. MNTB cells (Pfeiffer, 1966a,b; Rose *et al.*, 1974 ; Sommer *et al.*, 1993). In this chapter, we provide evidence that the SBCs and the principal cells in the MNTB also receive acoustically evoked inhibitory inputs, which effectively shape the strong excitatory input of the calyces (Kopp-Scheinpflug *et al.*, 2002 a, b). Inhibition thereby influences brainstem-mediated processing of interaural-time and intensity differences serving sound lateralization at a key position. Through its effect on MNTB neurons, the inhibition modifies the output of a nucleus which itself provides an important source of inhibition to several auditory brainstem nuclei besides the lateral superior olive, namely the medial superior olive (MSO), superior paraolivary nucleus and also cochlear nucleus (CN).

1.1 Inhibitory Input to SBCs and MNTB Principal Cells

Following the verification of inhibitory input to SBCs in immunocytochemical studies (Altschuler *et al.*, 1986; Altschuler *et al.*, 1993; Saint Marie *et al.*, 1989), light- and electron-microscopic investigations, revealed that the inhibitory synaptic terminals form synaptic nests on somata of SBC and also contact the dendrites of SBCs (Altschuler *et al.*, 1993; Caspary, 1986; Cant, 1991). Electron-microscopic studies, in addition, provided evidence that about 50% of the input to the SBCs are from non-cochlear origin (Saint Marie *et al.*, 1993). The ipsilateral dorsal CN (Osen *et al.*, 1990; Wickesberg and Oertel, 1990) and the contralateral ventral CN (Cant and Gaston, 1982; Pirsig *et al.*, 1968; Schofield and Cant, 1996; Wenthold, 1987) are sources of glycinergic input to the AVCN. Descending projections from the superior olivary complex to the CN originate in the medial, ventral and lateral nuclei of the trapezoid body and from the superior paraolivary nucleus (Covey *et al.*, 1984; Schofield, 1991, 1994; Warr and Beck, 1996; Winer *et al.*, 1995). Most of these descending projections carry GABA and/or glycine as transmitter (Ostapoff *et al.*, 1997; Saint Marie *et al.*, 1993). Colocalization of GABA and glycin was found in 42% of the superior olivary complex neurons projecting to the AVCN (Ostapoff *et al.*, 1997). Purely glycinergic projections come from the ipsilateral MNTB and from the contralateral, lateral nucleus of the trapezoid body (Benson and Potashner, 1990; Ostapoff *et al.*, 1990). In addition, small-sized GABAergic and glycinergic neurons are scattered throughout the ventral CN (Roberts and Ribak, 1987; Wenthold *et al.*, 1987; Wenthold *et al.*, 1986) possibly establishing intrinsic, inhibitory connections (Oertel *et al.*, 1990; Snyder and Leake, 1988; Wu and Oertel, 1986). Pharmacological studies in vivo and in vitro confirmed neurons in the AVCN to be sensitive to both GABA and glycine (Caspary *et al.*, 1994; Caspary *et al.*, 1993; Ebert and Ostwald, 1995; Walsh *et al.*, 1990; Wickesberg and Oertel, 1990; Wu and Oertel, 1986).

The MNTB is also the target of inhibitory GABAergic and glycinergic synapses (Adams and Mugnaini, 1990; Cant, 1991; Osen and Ottersen, 1996; Roberts and Ribak, 1987). GABAergic terminals may originate from neurons in the ventral nucleus of the trapezoid body (Ostapoff *et al.*, 1997) and from the superior paraolivary nucleus, in which most neurons label for GABA (Helfert *et al.*, 1989; Kulesza Jr. and Berrebi, 2000; Kuwabara and Zook, 1992). Glycinergic terminals arise, in part, from recurrent projections of MNTB neurons itself (Kuwabara and Zook, 1991; Smith *et al.*, 1998). In earlier studies, the physiological effect of inhibitory transmitters and of inhibitory input to the MNTB has only been investigated in vitro (Banks and Smith, 1992; Forsythe and Barnes-Davies, 1993; Turecek and Trussell, 2001; Wu and Kelly, 1995).

1.2 *In Vivo* Study of the Pre-to-Postsynaptic Signal Transmission

The extraordinarily large presynaptic terminals on the somata of MNTB principle cells and SBCs allow for simultaneous recording of the presynaptic calyceal discharge activity (prepotential, PP) and the evoked postsynaptic somatic action potential (AP) through a single recording electrode (Pfeiffer, 1966a; Shofner and Young, 1985; Winter and Palmer, 1990; Winter *et al.*, 1990; Young *et al.*, 1988). By comparing *presynaptic recordings* with *postsynaptic recordings* in an input–output (PP-AP) analysis, signal transmission at this synapse can be investigated (Kopp-Scheinpflug *et al.*, 2002 a, b). It is important to note that not all PPs are followed by an AP. The proportion of PPs not followed by an AP, which in the following are referred to as *isolated PPs*, varies with stimulus condition (Fig.2 C, D, E). Within sidebands flanking the excitatory response areas of AVCN units, isolated PPs could approach up to 90% (Kopp-Scheinpflug *et al.*, 2002b). The abundance of isolated PPs points to the effect of inhibition which prevents presynaptic signals to evoke a postsynaptic AP. Comparison between pre- and postsynaptic activity during two-tone stimulation allows to differentiate between the effects of inhibition and those of cochlear suppression on SBCs (Sachs and Kiang, 1968). Pharmacological experiments enables the differentiation between the effects of GABAergic and glycinergic inhibition and other cellular mechanisms affecting synaptic transmission at the AVCN calyx.

The details of the acoustic stimulation, the methods of single unit recording and drug application have been published elsewhere (Kopp-Scheinpflug *et al.*, 2002a,b). Here, only a brief outline of the methods will be presented. Mongolian gerbils were used in the experiments. Animals were anesthetized during surgery and experiments with ketamine hydrochlorid and xylazine hydrochloride (0.13 mg/g and 0.005 mg/g body weight). The body temperature was adjusted to 37.5°C. The animal was fixed in a stereotaxic device, and the recording electrodes were advanced through a drill hole in the skull above the cerebellum. After each experiment the recording sites were verified by HRP injections. Acoustic stimuli were delivered near field through earphone transducers (DT48 and DT770 Pro, Beyer Dynamics) coupled to the extension of an ear speculum which could be inserted into the pinna. Stimuli were generated with 16 bit accuracy, delivered at 250 k samples/sec through a 14 bit A/D converter including a custom-made low-pass resynthesis filter (50 kHz cutoff). Recording experiments were performed in a sound-attenuated chamber. Glass micropipettes filled with KCl (15-30 MΩ) were used for the single unit recordings, and neuropharmacological experiments were performed with piggy back electrodes (Havey and Caspary, 1980). Drugs (strychnine HCl: 7mM, pH 3, RBI; bicuculline methiodide: 5mM, pH 3, RBI/Sigma) were applied iontophoretically (+15-50nA). NaCl (pH 3) was used in control experiments to estimate the influence of artifacts, e.g. of the iontophoretic application, and spontaneous changes in the responses occurring during recording time. Differences e.g. in discharge rate between the pre- and postsynaptic recordings or recordings before and during drug application were only attributed to inhibitory influences when they were above changes seen in the control experiments (for details refer to Fig.7, 10, 11). Holding current for each barrel was –15nA. A balancing or summating channel was employed to alleviate current effects (barrel filled with 1M sodium acetate). Analog waveforms (Fig. 1) were recorded with an A/D conversion rate of 20,000/sec (50μsec sample interval). Spike times were acquired with 100 resp. 10 μsec resolution. During each single unit recording, the window discriminator was set to be triggered by either the PPs or by the postsynaptic APs, allowing for a differential analysis of the two types of signals. The units' excitatory and inhibitory response areas were acquired by presenting tone bursts (100 msec duration, 5 msec rise-fall time, 50 msec interstimulus interval) pseudorandomly varied in frequency and intensity (16 * 15 combinations), each combination being repeated five times (e.g. Fig. 5 A). The units' inhibitory areas were additionally evaluated by applying two-two-tone stimulation (Fig. 4), which was especially important in those units where spontaneous activity was too low to be lowered significantly by inhibition. For the first of the two signals, the stimulus setting was the same as described above, except that stimulus duration was shorter (40 ms, test tone). This test tone was paired with a probe tone (100 ms) preceding the test tone by 30 msec, presented at the unit's characteristic frequency (CF). The probe tone evoked a steady level of activation, the reduction of which indicated the effect of acoustically evoked inhibition. Responses under single and two-tone-stimulation established the unit's response area. Quality of phaselocking to the pure tone frequency was evaluated by the vector strenght (VS) analysis (Goldberg and Brown, 1968); values with $\alpha < 1\%$ were considered significant (Knipschild *et al.*, 1992). Difference plots of presynaptic- vs postsynaptic response areas and of response areas before and during drug application (Figs. 5 C, F) emphasize the pre-to-postsynaptic processing and the effect of elimination of the inhibitory input, respectively. In each single case, the responses to the five tone burst repetitions in the diverse stimulus conditions were compared by a permutation test, with $\alpha = 5\%$ for the significance of differences in the discharge rates or in the VS (Siegel and Castellan Jr., 1988).

The validity of the pre-to-postsynaptic comparison depends on the demonstration that the signals recorded from a unit are composed of only two basic constituents. For this, we applied the principal component analysis to the digitized signal waveforms (Abeles and Goldstein, 1977) in combination with a cluster analysis. Principal component analysis allows for describing each waveform from an ensemble as the weighted sum of a number of basic waveforms, its *principal components* (PC). Depending on the variance of the ensemble, only a few PCs and a correspondingly small number of weights may give a sufficient description of the great majority of the original waveforms. Goodness of fit is expressed as the quotient of the variance of the approximating set of waveforms divided by the variance of the original ensemble of waveforms, so that 100% indicates perfect approximation. In our application, all waveforms (2.5-msec time epochs) that met the user-defined trigger criterion comprising upper and lower thresholds, slope direction, and dead-time were selected from digitized waveform recordings. Figure 2 shows the results of a PC analysis obtained from one representative unit.

2. WAVEFORMS OF EXTRACELLULAR RECORDINGS

Out of a population of 109 low frequency units (CFs below 6.2 kHz, 83% below 2.5 kHz), displaying primarylike post-stimulus time histograms (PSTHs) and located in the most rostral tip of the AVCN, 68 % showed complex waveforms (Fig.1A,B). In recordings of 146 MNTB units with CFs between 1.0 ad 53.7 kHz, displaying primarylike or primarylike-with-notch PSTHs, 58 % showed complex waveforms (Fig.1C,D). These waveforms were comparable to those shown for units in the cat AVCN (Pfeiffer, 1966a) and MNTB (Guinan and Li, 1990; Guinan, 1972). In many recordings, a second type of waveform occured, which was identified as isolated PPs not followed by APs. Isolated PPs had previously been detected in SBC recordings and described as 'spike failures' (Pfeiffer, 1966a), but were never explored in more detail. At first, we were interested in the question, if the presynaptic endbulb that gives rise to PPs followed by APs also generates isolated PPs. Alternatively, the recordings could have sampled two independent units. We applied a principle component analysis to 21 AVCN units and 30 MNTB units showing complex signal waveforms (representative example in Fig. 2). For all 51 units, the analysis yielded a 85-95 % efficiency by which the first two principle components jointly described the recorded signals; the complex waveforms could be divided into just two distinctive waveforms: (i) spike preceding PPs plus postsynaptic spikes and (ii) isolated PPs. More than two signals would occur when two independent units with variable timing had been superimposed in extracellular recordings.

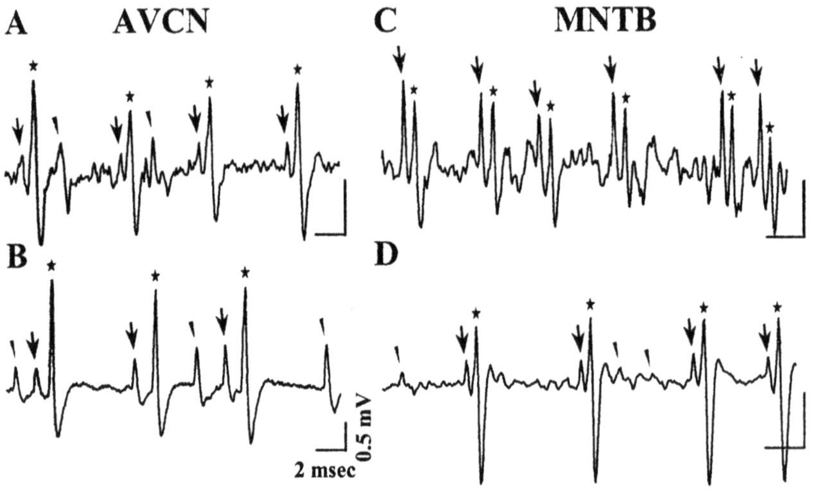

Figure 1. Waveforms of tone-evoked discharges of two SBCs (A,B) and two MNTB units (C,D). CFs [kHz] were (A)1.2 , (B) 0.8 , (C) 9.5, (D) 6.6. Isolated PPs (arrowheads), i.e. PPs which are not immediately followed by an AP (asterisks), occur in (A,B,D). PPs followed by an AP are marked by arrows.

A number of other observations corroborated the conclusion that isolated PPs and PPs preceding APs originate from the same source, namely the presynaptic endbulb terminal: (1) During recording, isolated PPs and PPs preceding APs always appeared and disappeared in concert, e.g. as the electrode was moved. Slight advancement of the recording electrode caused coherent changes in amplitudes of isolated PPs and PPs preceding APs. (2) Isolated PPs were only observed in recordings from PP units and never in recordings from units that lacked spike preceding PPs. (3) AP-preceding PPs and isolated PPs were comparable in size and shape, e.g., both signals had comparable onset slopes and thereby could be selected by the same slope criterion (0.15 mV/msec). These observations strongly suggest that the PPs preceding APs and the isolated PPs do, in fact, reflect the activity of the same endbulb.

The special features of extracellularly recorded waveforms of the SBC resp. MNTB units granted the unique possibility to simultaneously record and compare the units′ postsynaptic activity (APs) and their afferent input (all PPs, those preceding APs and isolated ones). Here, we used this approach to evaluate the pre-to-postsynaptic input-output functions, thereby considering the influence of neuronal inhibition.

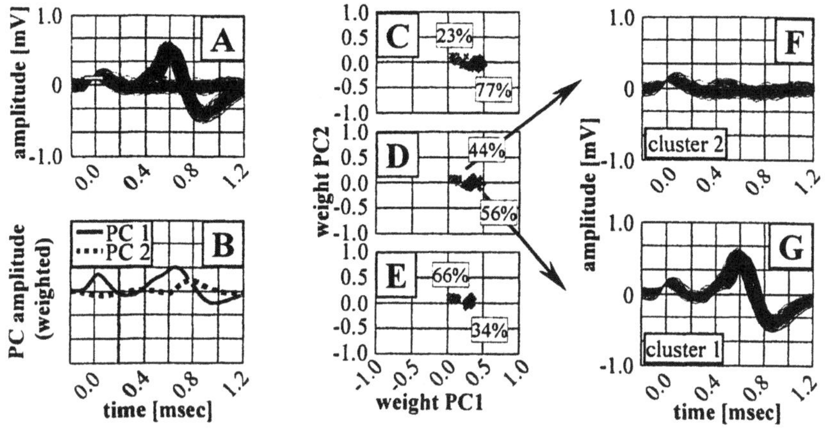

Figure 2. Separation of signal constituents of an AVCN unit using principal component (PC) analysis. (A) Superimposed waveforms of 301 discharges. Trigger level was set to the onset slope of PPs. The white bar (at 0.12 mV) indicates the upper trigger level of the combined threshold and slope criterion. (B) The first two PCs (PC1: solid, PC2: dotted) of the signals shown in (A) multiplied by their root-mean-square weight. The whole of the signals can be sufficiently well described by the first two PCs with an 87% efficiency, i.e., each of the 301 recorded signals can be adequately explained by the weighted sum of two functions, PC1 and PC2. (C-E) On the basis of PC1- and PC2 weights, the potentials are easily separated by a cluster algorithm into two clusters. In this scatter plot, the cluster with the higher PC1-weights (cluster 1, black) represents the complex waveforms (spike preceding PPs followed by postsynaptic spikes), the cluster with the lower PC1-weights (cluster 2, gray) identifies the isolated PPs. Relative proportions of both clusters vary with stimulus condition: (C) spontaneous activity, (D) excitatory area and (E) inhibitory sideband (F) and (G) show the original waveforms from (A) related to Cluster 1 (*spike preceding PPs followed by postsynaptic spikes*) and to Cluster 2 (*isolated PPs*), respectively.

3. FREQUENCY-INTENSITY RESPONSES ARE SHAPED AT CALYX SYNAPSES

In 38/42 AVCN units and in 38/54 MNTB units, the analysis of the response areas from presynaptic and postsynaptic recordings revealed stimulus conditions which induced postsynaptic inhibitory response areas not present in the presynaptic recordings. This inhibition could display different features (Fig. 3). Twenty seven units in the AVCN and 30 units in the MNTB already showed the effect of inhibition during single tone stimulation (Fig. 3 D,E). The remaining were units with spontaneous activity too low to be significantly reduced by inhibition (Fig. 3 C,F). In these cases, the two-tone stimulation revealed the inhibitory effects (Fig. 4 A-D darkly shaded areas). Furthermore, the comparison of the responses to two-tone stimulation recorded pre- and postsynaptically in the AVCN allowed for distinguishing postsynaptic inhibition from cochlear suppression, the latter of which appears in the PP recordings (showing auditory nerve activity; Fig. 4). Apart from the formation of inhibitory side bands, also some features of the purely excitatory response areas (dashed lines in Fig. 4 A-D) pointed to an influence of sound-evoked inhibition, e.g. differences in discharge rates, differences in width of the response area between the presynaptic and postsynaptic recordings (see also Fig. 3 C,F), postsynaptic non-monotonic rate-level functions (RLFs) and the lack of the low-frequency tails in the response areas (Fig. 4 C,D).

In our data sample, the prevalent effect of inhibition was the formation of inhibitory sidebands at the high and/or low-frequency sides of the excitatory response area and partly overlapping with it (AVCN 27/38 units, MNTB 33/38 units, Fig. 3 D). The remaining AVCN units displayed an on-CF inhibition, which is associated with non-monotonic excitatory RLFs. The inhibition could even cause the formation of closed excitatory response areas with upper thresholds (Fig. 3 E). The latter response characteristic was only found in SBCs but not in MNTB units. On-CF-inhibition and inhibitory sidebands could appear in combination with very broadband inhibition. Five MNTB units showed a low level inhibitory input around CF, resulting in a postsynaptic increase of threshold. This type of inhibitory input was never observed in SBCs.

Nineteen SBC units were subjected to pharmacological experiments, of which 12 showed various types of inhibitory response areas, e.g. inhibitory sidebands (8/12), on-CF inhibition (1/12), or broadband inhibition (3/12). In our pharmacological experiments, we were initially interested in differentiating between GABAergic and glycinergic inhibition, e.g. regarding the above defined types of inhibition (Fig. 3). Indeed, in four units (three broadband inhibition including on-CF inhibition, one on-CF inhibition), the on-CF inhibition restricted to the excitatory area could only be blocked by the application of strychnine (2/4 units) and not by bicuculline (2/4 units). This inhibition might originate from the glycinergic tuberculoventral cells in dorsal cochlear nucleus, which has previously been described as the source of on-CF inhibition (Oertel and Wickesberg, 1993; Wickesberg and Oertel, 1988; Young and Voigt, 1982).

However, in another four cases, in which this on-CF inhibition was not seen in the normal response as inhibitory areas, it still was blockable by bicuculline, i.e. it was detected by the comparison of the responses before and after the application of bicuculline (comparison as shown in Fig. 5). The data clearly show the effect of a GABAergic on-CF inhibitory input onto the SBCs. It might well be that GABA effects another population of SBCs as the glycinergic on-CF inhibition. Based on these results, it is tempting to assume functional differences in SBCs, which, however, will have to be explored in more detail in further experiments.

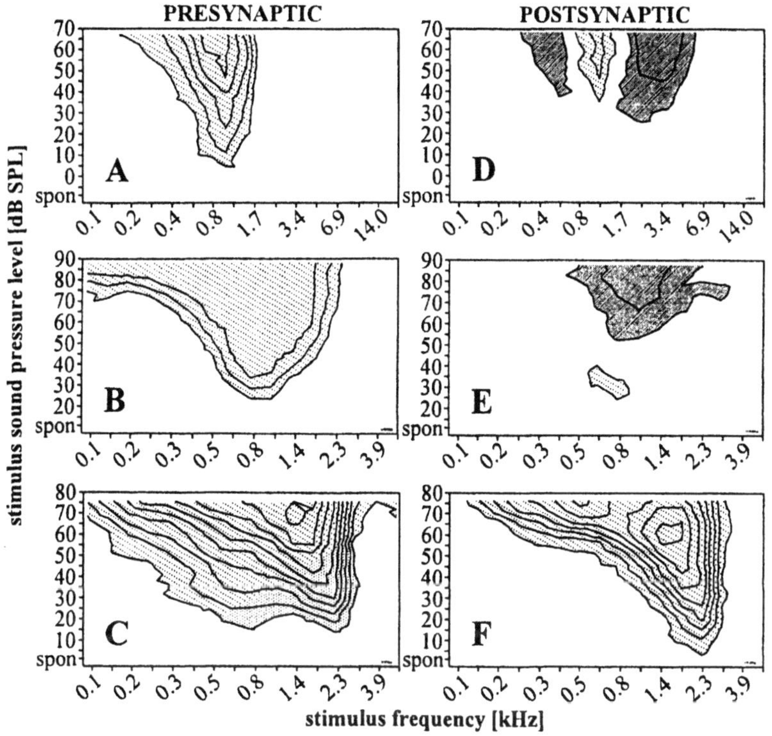

Figure 3. Response areas recorded pre- and postsynaptically show different features of inhibition. While presynaptic response areas reveal only excitatory input (dotted areas in (A-B), postsynaptically also the effect of inhibitory inputs is seen (D,E,F). Coherent inhibitory areas (darkly striped areas) can be formed either as high and low frequency sidebands (D) or they can widely overlap with the (presynaptic) excitatory response area (E). If inhibitory sidebands are formed, the bandwidth of the excitatory response areas are reduced pre-to-postsynaptically. Some units show this reduction of bandwidth, although they do not display postsynaptic inhibitory areas (discharge rate significantly below spontaneous rate) (F).

Only in one single case, it was possible to block the formation of inhibitory sidebands. In this unit inhibitory sidebands were blocked together with on-CF inhibition by application of strychnine. Such broadband inhibition might be caused by a broadband input transmitted through stellate cells located within the ventral CN itself (Ferragamo *et al.*, 1998; Nelken and Young, 1994; Smith and Rhode, 1989; Wickesberg and Oertel, 1990). Given the abundance of inhibitory sidebands in SBC response areas and the conspicuous effect they have on postsynaptic frequency selectivity (Fig. 6), one noticeable fact was that, in all but one case, this sideband inhibition seems to be neither GABA- nor glycinergic. This shows the need for detailed investigations of the effectiveness of other transmitters, synaptic mechanisms or intrinsic membrane properties in shaping the frequency response of SBCs (Caspary *et al.*, 1983; Manis and Marx, 1991; Rothman *et al.*, 1993).

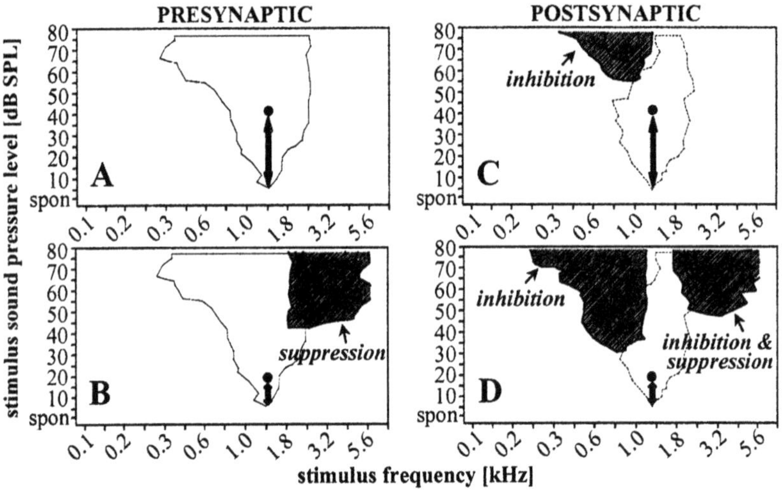

Figure 4. Distinction between presynaptic cochlear suppression and postsynaptic neuronal inhibition in an AVCN unit. The unit was recorded pre- and postsynaptically using two-tone stimulation. Dashed lines show response areas for single-tone stimulation. A constant probe tone (filled circle) was presented at two intensities at the CF (1.3 kHz): (A, C) 30 dB above threshold and (B, D) 10 dB above threshold. Shaded areas show frequency/intensity domains in which the test tone causes a reduction of probe-tone evoked discharge rate.

Since the pharmacological experiments showed the strongest effects of inhibition on-CF (9/19), these were analyzed in more detail: the RLFs at CF (Fig. 7) and the PSTHs (Fig. 8) were compared pre- and postsynaptically as well as before and during drug application in order to evaluate the level dependence and the time course of the inhibition, respectively.

It is important to keep in mind that for a single unit, lower postsynaptic than presynaptic discharge rates mean larger numbers of prepotentials than postsynaptic action potentials due to isolated prepotentials not followed by an action potential.

In the difference RLFs (Fig. 7 A,B) positive differences either indicate a pre-to-postsynaptic reduction or an increase in discharge rate during drug application. Both the pre-/postsynaptic recordings (Fig. 7 A) and the pharmacological experiments (Fig. 7 B) show pronounced inhibitory effects in comparison to the control units. The difference curves from pre-/postsynaptic recordings showed larger effects of inhibition than did those from the pharmacologically manipulated units.

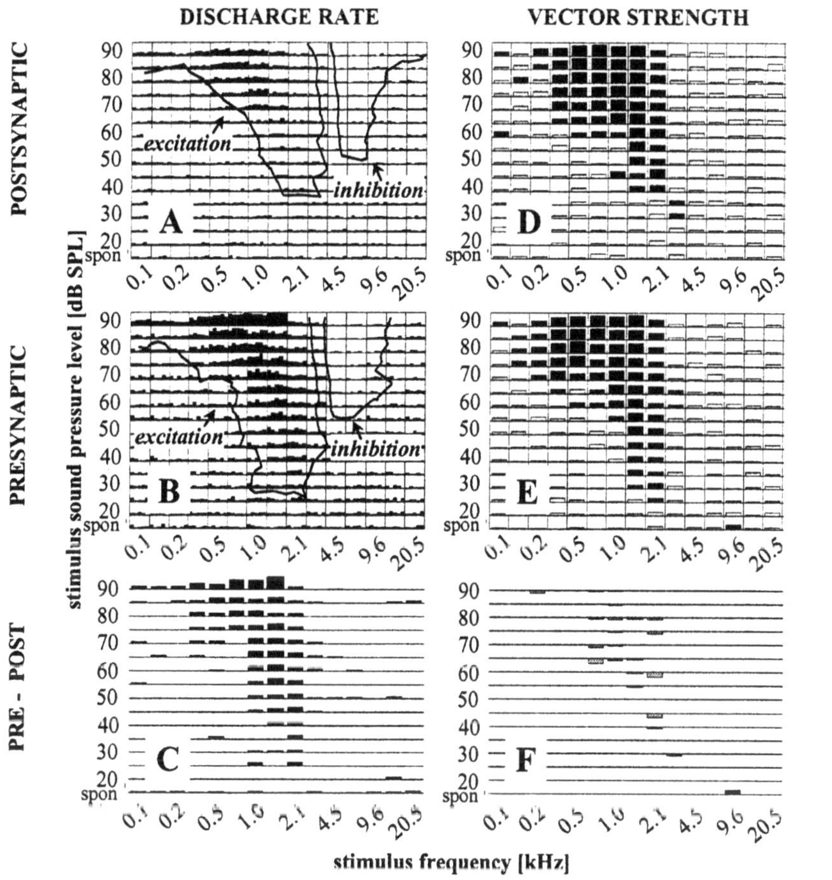

Figure 5. Acoustically evoked responses in a MNTB unit recorded pre- and postsynaptically. Response areas showing the discharge rate (A-C) and vector strenght values (D-F). (A, D) Upper row: postsynaptic response, (B, E) middle row: presynaptic response, bottom row: difference of both (presynaptic – postsynaptic). Height of the black bars in (A, B) symbolize the number of counted potentials per stimulus presentation, excitatory and inhibitory response areas calculated from the discharge rate are superimposed. Filled, black bars in (D, E) indicate stimulus parameter evoking significant phaselocking, height of bars corresponds to size of VS. The difference plot (bottommost row (C,F)) exclusively shows the significant values (permutation test $\alpha<5\%$), black bars above midline: positive values (presynaptic > postsynaptic and drug application > predrug condition), gray bars below midline: negative values (presynaptic < postsynaptic and drug application < predrug condition).

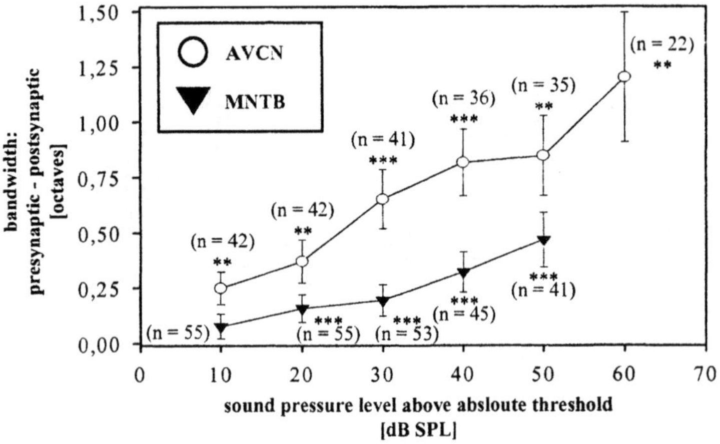

Figure 6. Difference in bandwidth between pre- and postsynaptic response areas increases with stimulus level. All values (presynaptic vs. postsynaptic bandwidth) of MNTB units (triangle) and AVCN (dot) differed significantly (t-test, *** p = 0.001, ** p = 0.005). Due to the units' different threshold values, the number of tested pairs (n) decreased towards higher levels.

This could mean that (1) other transmitters besides GABA and/or glycine or other cellular mechanisms affect the synaptic transmission, (2) in the pharmacological experiments, the block of the inhibition is not exhaustive, or (3) the smaller difference is simply the result of the lower discharge rate in the normal state of the manipulated AVCN units compared to that of the postsynaptic recordings (see below, Fig. 8 A,B). The latter could possibly be explained by a sampling bias. All presynaptic AVCN and MNTB recordings showed monotonic RLFs, steadily increasing or forming a constant plateau of discharge rate (Fig. 7 C-F). When the full range of stimulus levels was regarded, the RLF sometimes switched from a presynaptically monotonic to a postsynaptically non-monotonic course; this was the case in 9 out of 22 units in the AVCN, and in 4 out of 13 units in the MNTB (Fig. 7 C,E). The remaining units exhibited a postsynaptically preserved monotonic course (Fig. 7 D,F). In the AVCN the postsynaptic dynamic range was increased by about 9 dB above the presynaptic range ($p<0.05$, paired t test; Fig. 7 D). Of all pharmacologically manipulated units which showed significant increases in discharge rates during the block of inhibition, three had non-monotonic RLFs under normal conditions. When inhibition was blocked, in two out of the three cases the units increase in discharge rate was accompanied by a change from a non-monotonic to a monotonic RLF (Fig. 7 G).

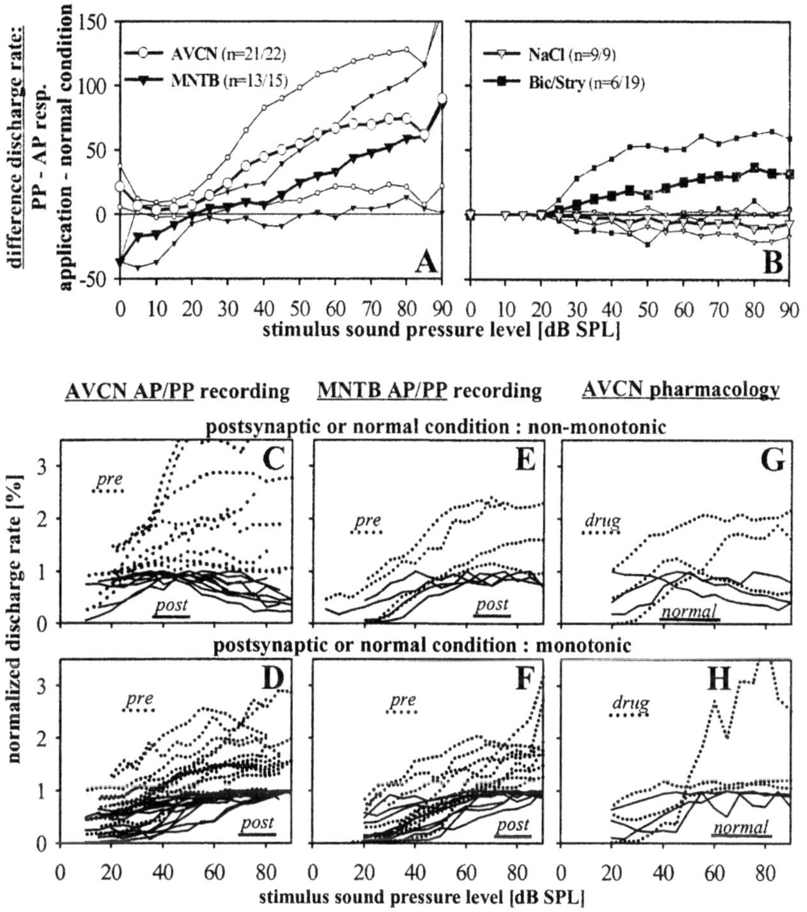

Figure 7. Significant changes of discharge rates seen from rate-level functions (RLFs). Shown are the difference RLFs of pre- and postsynaptically recorded MNTB and AVCN units (A) and of NaCl control AVCN units and AVCN units with pharmacological manipulation (Bic/Stry = bicuculline/strychnine) (B). (A): white circles: pre-post AVCN units, n=22; black triangles: pre-post MNTB units, n=13. (B): black squares: pharmacologically manipulated AVCN units, n=6; white triangles: all AVCN control units, n=9. Data for the difference RLFs were extracted from the respective response areas (e.g. Fig. 5 C) and averaged for CF and the two neighbouring frequencies, summed up for five repetitions of a 110 ms stimulus. Only significant differences were considered. Shown are mean ± stddev of all units of which 50% of the data points were above the standard deviation of the control units (gray triangles in (B)). Normalized RLFs of the units considered in (A,B) are shown in (C-H): pre- and postsynaptic recordings (AVCN: C, D, MNTB: E, F) and recordings under normal condition and drug application (G,H). Maximum discharge of postsynaptic or normal condition equals 100%. Units were sorted by their appearance of the postsynaptic or normal condition RLF respectively: (C,E,G) non monotonic (20% decrease under maximum) and (D,F,H) monotonic.

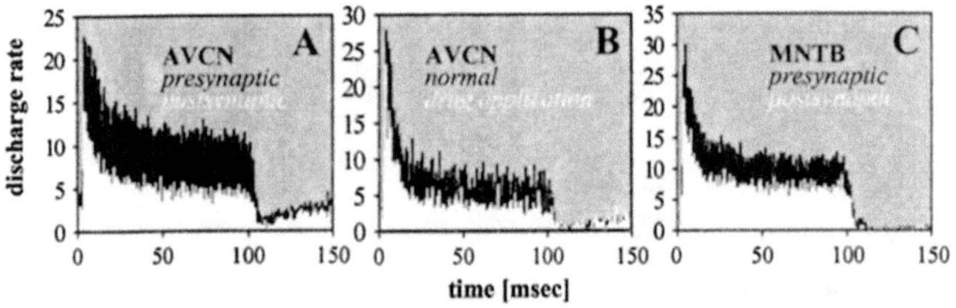

Figure 8. Averaged PSTHs of pre- and postsynaptic responses of AVCN and MNTB units and AVCN units under pharmacological manipulation. PSTHs were calculated from discharges at CF and the two neighboring test frequencies at the highest three intensities (same units as in Fig.7). Shown are pre-/postsynaptic PSTHs of AVCN units (n=22) (A), MNTB units (n=13) **(C)** and AVCN units before/after bicuculline or strychnine application (n=6) (B).

In response to high-intensity tone bursts at CF (90-85 dB SPL), all presynaptic recordings in the SBCs showed the typical primary like PSTHs (Fig. 8 A) (Pfeiffer, 1966b; Shofner and Young, 1985; Young *et al.*, 1988). This was also the case at CF in most of the postsynaptic recordings. In units with sideband inhibition, postsynaptic PSTHs could vary towards the low- and high frequency borders of the excitatory response area, or towards high stimulus levels at CF (in units with strong on-CF inhibition), from a primary-like pattern to an on- or on/off-pattern. Pre- and postsynaptic responses of the MNTB units showed primary-like or primary-like-with-notch patterns previously described for MNTB cells and also for globular bushy cells in the ventral CN (Fig. 8 C) (Smith *et al.*, 1991; Smith *et al.*, 1998; Tsuchitani, 1997). Both MNTB and AVCN units showed characteristic differences between pre- and postsynaptic PSTHs (Fig. 8 A,C). The timing of the postsynaptic onset peak was sharpened by reduction of its width, its height was increased relative to the unit's tonic activity. The pharmacological experiments revealed that the reduction of the steady state response (30 – 90 msec) in the AVCN was due to inhibiton, as activity was increased by 49% on average (86 sps predrug to 168 sps during drug application, n = 6). In the pre-/postsynaptically recorded AVCN units, the respective average reduction was 57% (261 sps presynaptically to 112 sps postsynaptically, n = 22, numbers averaged per unit and one stimulus repetition). The high correspondence between the results of the pharmacological experiments and those of the pre-/postsynaptic recordings in the AVCN suggests that the pre-to-postsynaptic reduction is indeed due to GABAergic and glycinergic inhibition. In the MNTB, the steady state discharge rate was on average reduced by 38% (265 sps pre- to 164 sps postsynaptically, n=15), pointing to a weaker effect of inhibition on the steady state response than in the AVCN.

4. CONTRIBUTION OF INHIBITION TO PHASELOCKING ACCURACY

According to the current models of low-frequency processing of interaural-time differences in the MSO, the phaselocked activity of SBCs provides a precisely timed excitatory input to the MSO and thus establishes the basis for encoding sound source laterality in the MSO. In this respect, the fidelity of phaselocking of the SBCs is of great importance.

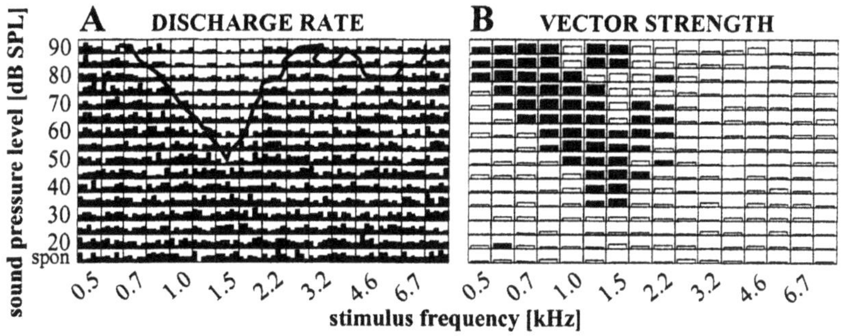

Figure 9. Response areas of discharge rate and rate synchronization of an AVCN SBC unit. (A) shows the number of APs for each of the five stimulus repetitions as columns. Areas with discharge rates significantly different from the spontaneous rate (bottommost row) are circumlined (here only inhibitory areas with stimulus evoked discharge rate below spontaneous rate) (B) VS of the responses shown in (A), filled black bars: significant VS (dependent on the number of actionpotentials), maximal bar height is 100%.

In many cases, SBC responses to tone bursts at threshold are not adequately characterized when only the increase in discharge rates is regarded (Fig. 9 A). Still, the typical shape of response areas of CN units becomes evident, if the VS of phaselocking is taken as an indicator of the response (Fig. 9 B). Apparently, the phase of low-frequency stimuli is one important feature coded by SBCs. Previous studies based on the comparison of unit samples, reported that the phaselocking ability of primarylike units exceeded that of auditory nerve fibers (Joris *et al.*, 1994a,b; Rothman, 1996). We investigated this issue by a direct input-output comparison for the individual SBCs, and we further explored whether GABAergic or glycinergic inhibition influences the fidelity of postsynaptic phaselocking.

In the pre-/postsynaptic recordings 9 out of 20 units, and in the pharmacological experiments 3 out of 18 units with phaselocking at CF showed a lower phaselocking accuracy presynaptically than postsynaptically or postsynaptically following the application of bicuculine or strychnine (Fig. 10).

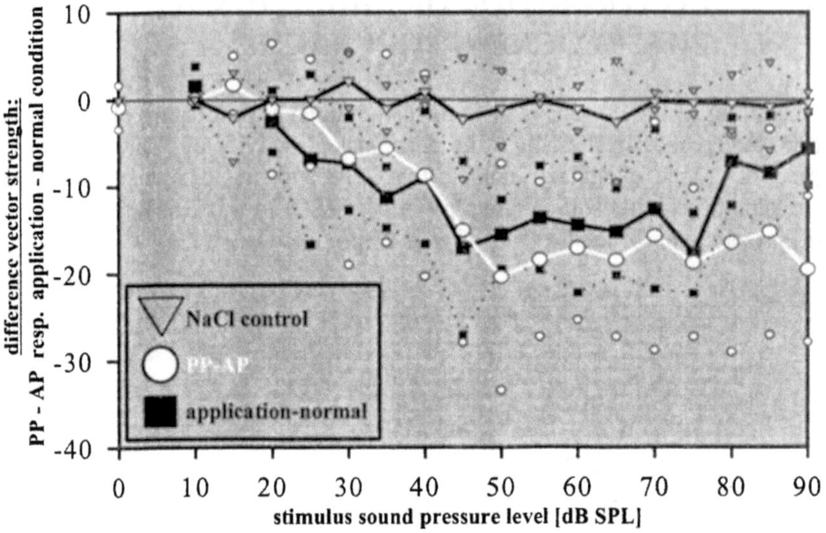

Figure 10. Significant changes in vector strength at CF. Curves were calculated in the same way as RLFs in Fig. 7 A,B. Shown are mean (solid lines) and stddev (dotted lines) of units with 50% of measured values below the control units' standard deviation (black squares: pre-/postsynaptic AVCN units (n=9), white circles: pharmacologically manipulated units (n=3), grey triangles: control AVCN units (n=7)). Positive differences indicate a pre- to postsynaptic reduction or increase in VS during drug application.

Thus, in half of the units, in which inhibition reduced the discharge rate from the presynaptic to the postsynaptic side, the phaselocking accuracy was increased (pre- and postsynaptically recorded units: 9/19, pharmacologically manipulated units: 3/5). In the other half of the units, phaselocking remained stable. Changes in phaselocking accuracy were also seen in period histograms, which were presynaptically broader than postsynaptically, and which also broadened postsynaptically during drug application (Fig. 11 A,B). The difference in width of the period histograms (Fig. 11) and the changes of VS seen at CF (Fig. 10) were approximately equal in pre-/postsynaptic recordings and during the block of inhibition. At the AVCN-MNTB synapse, an increase in phaselocking accuracy was observed in the CF region of low-frequency units (CF < 1.2 kHz; Fig. 5F), and in the area of the low-frequency tail of high-frequency units (Kopp-Scheinpflug *et al.*, 2002a).

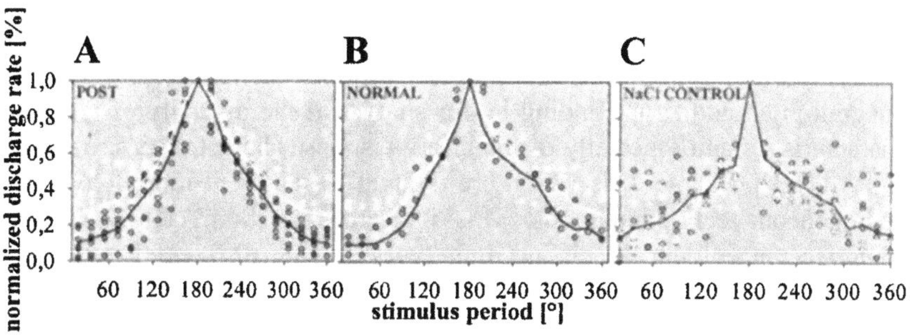

Figure 11. Changes of phaselocking seen in period histograms. Spike discharges were averaged for the CF and two neighbouring frequencies at the three highest intensities and normalized to the maximum bin centered at 180°. Units were the same as in Fig. 9: (A) pre- and postsynaptically recorded AVCN units (n=9); (B) pharmacologically manipulated AVCN units (n=3); (C) all AVCN control units (n=7).

5. CONCLUSIONS

The present study aimed at the integration of acoustically evoked excitation and inhibition in SBCs of the AVCN and in principal cells of the MNTB. In *in vivo* recording experiments, we were able to show that afferent excitatory synaptic transmission can be suppressed by acoustically evoked inhibition, which again can be pharmacologically blocked. This inhibition can sharpen the frequency selectivity, alter the RLFs (SBC and MNTB) and widen the dynamic range of the level dependent responses (SBC). Furthermore, inhibition can elevate the precision of onset activity and increase phaselocking accuracy for low-frequency pure tones (SBC and MNTB). The influence of inhibition on frequency selectivity and the RLFs was stronger in the SBCs than in the MNTB. Postsynaptic on-CF inhibition, strong enough to overwrite stimulus-induced activation at CF and thereby causing upper thresholds in the response areas, were only found in AVCN units and not in the MNTB. Such strong inhibition might cause a rate reduction abating monaural coincidence of afferent spike activity in the MSO (Yang *et al.*, 1999). Low level inhibition on-CF and a correspondingly postsynaptic increase in postsynaptic threshold was only seen in the MNTB but not in the SBCs and might be of importance in the context of processing of interaural intensity differences (Tsuchitani, 1997). The reduction of activity goes along with a sharpening of phaselocking in SBCs and MNTB principle cells, indicating an increase in temporal accuracy of the units' discharge activity. According to recently published models on processing of interaural-phase differences, such high precision phaselocking is one basic feature of excitatory as well as inhibitory inputs to the MSO (Brand *et al.*, 2002; Grothe, 2003). Aside from the insights in inhibitory-excitatory integration an additional finding in our study was the diversity of response characteristics and especially of inhibitory response characteristics, of SBCs and MNTB principle cells which are thought to represent morphologically homogeneous cell populations. This indicates that both cell types may comprise physiologically distinct unit types serving different functions in auditory brainstem processing.

ACKNOWLEDGEMENTS

This research was founded by the Deutsche Forschungsgemeinschaft (DFG) Graduiertenkolleg 250/1-96, 250/3 and DFG Ru 390-15/1-3. We thank G.J. Dörrscheidt and M.Weick for providing the software for data aquisition and analysis.

REFERENCES

Abeles, M., and Goldstein, M. H., 1977, Multispike train analysis. *Proc. IEEE* **65**: 762-773.

Adams, J. C., and Mugnaini, E., 1990, Immunocytochemical evidence for inhibitory and disinhibitory circuits in the superior olive. *Hear. Res.* **49**: 281-298.

Altschuler, R. A., Betz, H., Parakkal, M. H., Reeks, K. A., and Wenthold, R. J., 1986, Identification of glycinergic synapses in the cochlear nucleus through immunocytochemical localization of the postsynaptic receptor. *Brain Res.* **369**: 316-320.

Altschuler, R. A., Juiz, J. M., Shore, S. E., Bledsoe, S. C., Helfert, R. H., and Wenthold, R. J., 1993, Inhibitory amino acid synapses and pathways in the ventral cochlear nucleus. In *The mammalian cochlear nuclei: organization and function* (M. A. Merchan, ed.), Plenum Press, New York.

Banks, M. I., and Smith, P. H., 1992, Intracellular recordings from neurobiotin-labeled cells in brain slices of the rat medial nucleus of the trapezoid body. *J. Neurosci.* **12**: 2819-2837.

Benson, C. G., and Potashner, S. J., 1990, Retrograde transport of [3H]glycine from the cochlear nucleus to the superior olive in the guinea pig. *J. Comp. Neurol.* **296**: 415-426.

Brand, A., Behrend, O., Marquardt, T., McAlpine, D., and Grothe, B., 2002, Precise inhibition is essential for microsecond interaural time difference coding. *letters to nature* **417**: 543-547.

Brawer, J. R., and Morest, D. K., 1975, Relations between auditory nerve endings and cell types in the cat's anteroventral cochlear nucleus seen with the Golgi method and Nomarski optics. *J. Comp. Neurol.* **160**: 491-506.

Cant, N. B., 1991, Projections to the lateral and medial olivary nuclei from the spherical and globular bushy cells of the anteroventral cochlear nucleus. In *The central auditory system* (R. A. Altschuler, B. M. Clopton, R. P. Bobbin and D. W. Hoffman, eds.), Raven Press, New York, pp. 99-119.

Cant, N. B., and Gaston, K. C., 1982, Pathways connecting the right and left cochlear nuclei. *J. Comp. Neurol.* **212**: 313-326.

Caspary, D. M., (1986), Cochlear Nuclei: Functional neuropharmacology of the principal cell types. In *Neurobiology of hearing: The cochlea* (R. A. Altschuler, D. W. Hoffman and R. P. Bobbin, eds.), Raven Press, New York, pp.303-332.

Caspary, D. M., Backoff, P. M., Finlayson, P. G., and Palombi, P. S., 1994, Inhibitory inputs modulate discharge rate within frequency receptive fields of anteroventral cochlear nucleus neurons. *J. Neurophysiol.* **72**: 2124-2133.

Caspary, D. M., Palombi, S. P., Backoff, P. M., Helfert, R. H., and Finlayson, P. G., 1993, GABA and Glycine inputs control discharge rate within the excitatory response area of primarylike and phaselocked AVCN neurons. In *The mammalian cochlear nuclei: organization and function* (M. A. Merchan, ed.), Plenum Press, New York.

Caspary, D. M., Havey, D. C., and Faingold, C. L., 1983, Effects of acetylcholine on cochlear nucleus neurons. *Exp. Neurol.* **82**: 491-498.

Covey, E., Jones, D. R., and Casseday, J. H., 1984, Projections from the superior olivary complex to the cochlear nucleus in the tree shrew. *J. Comp. Neurol.* **226**: 289-305.

Ebert, U., and Ostwald, J., 1995, GABA can improve acoustic contrast in the rat ventral cochlear nucleus. *Exp. Brain. Res.* **104**: 310-322.

Ferragamo, M. J., Golding, N. L., and Oertel, D., 1998, Synaptic inputs to stellate cells in the ventral cochlear nucleus. *J. Neurophysiol.* **79**: 51-63.

Forsythe, I. D., and Barnes-Davies, M., 1993, The binaural auditory pathway: excitatory amino acid receptors mediate dual timecourse excitatory postsynaptic currents in the rat medial nucleus of the trapezoid body. *Proc. R. Soc. Lond. B. Biol. Sci.* **251**: 151-157.

Goldberg, J. M., and Brown, P. B., 1968, Functional organization of the dog superior olivary complex: an anatomical and electrophysiological study. *J. Neurophysiol.* **31**: 639-656.

Grothe, B., 2003, New roles for synaptic inhibition in sound localization. *Nat. Rev. Neurosci.* **4**: 540-550.

Guinan, J. J., Jr., and Li, R. Y., 1990, Signal processing in brainstem auditory neurons which receive giant endings (calyces of Held) in the medial nucleus of the trapezoid body of the cat. *Hear. Res.* **49**: 321-334.

Guinan, J. J., Jr., Guinan, S. S. and Norris, B. E., 1972, Single auditory units in the superior olivary complex I: responses to sounds and classification based on physiological properties. *Intern. J. Neurosci.* **4**: 101-120.

Havey, D. C., and Caspary, D. M., 1980, A simple technique for constructing 'piggy-back' multibarrel microelectrodes. *Electroencephalogr. Clin. Neurophysiol.* **48**: 249-251.

Helfert, R. H., Bonneau, J. M., Wenthold, R. J., and Altschuler, R. A., 1989, GABA and glycine immunoreactivity in the guinea pig superior olivary complex (published erratum appears in *Brain. Res.* 1990 Feb 12;**509**:180). *Brain. Res.* **501**: 269-286.

Joris, P. X., Carney, L. H., Smith, P. H., and Yin, T. C., 1994a, Enhancement of neural synchronization in the anteroventral cochlear nucleus. I. Responses to tones at the characteristic frequency. *J. Neurophysiol.* **71**: 1022-1036.

Joris, P. X., Smith, P. H., and Yin, T. C., 1994b, Enhancement of neural synchronization in the anteroventral cochlear nucleus. II. Responses in the tuning curve tail. *J. Neurophysiol.* **71**: 1037-1051.

Knipschild, M., Dörrscheidt, G. J., and Rübsamen, R., 1992, Setting complex tasks to single units in the avian auditory forebrain. I. Processing of complex artificial stimuli. *Hear. Res.* **57**: 216-230.

Kössl, M., and Vater, M., 1989, Noradrenaline enhances temporal auditory contrast and neuronal timing precision in the cochlear nucleus of the mustached bat. *J. Neurosci.* **9**: 4169-4178.

Kopp-Scheinpflug, C., Lippe, W. R., Dörrscheidt, G. J., and Rübsamen, R., 2002a, The medial nucleus of the trapezoid body in the gerbil is more than a relay: comparison of pre- and postsynaptic activity. *J. Assoc. Res. Otolaryngol.* **04**: 1-23.

Kopp-Scheinpflug, C., Dehmel, S., Dörrscheidt, G. J., and Rübsamen, R., 2002b, Interaction of excitation and inhibition in anteroventral cochlear nucleus neurons that receive large endbulb synaptic endings. *J. Neurosci.* **22**: 11004-11018.

Kulesza Jr., R. J., and Berrebi, A. S., 2000, Superior paraolivary nucleus of the rat is a GABAergic nucleus. *J. Assoc. Res. Otolaryngol.* **01**: 255-269.

Kuwabara, N., and Zook, J. M., 1992, Projections to the medial superior olive from the medial and lateral nuclei of the trapezoid body in rodents and bats. *J. Comp. Neurol.* **324**: 522-538.

Kuwabara, N., and Zook, J. M., 1991, Classification of the principal cells of the medial nucleus of the trapezoid body. *J. Comp. Neurol.* **314**: 707-720.

Manis, P. B., and Marx, S. O., 1991, Outward currents in isolated ventral cochlear nucleus neurons. *J. Neurosci.* **11**: 2865-2880.

Nelken, I., and Young, E. D., 1994, Two separate inhibitory mechanisms shape the responses of dorsal cochlear nucleus type IV units to narrowband and wideband stimuli. *J. Neurophysiol.* **71**: 2446-2462.

Oertel, D., and Wickesberg, R. E., 1993, Glycinergic inhibition in the cochlear nuclei: Evidence for tuberculoventral neurons being glycinergic. In *The mammalian cochlear nuclei: organization and function* (M. A. Merchan, ed.), Plenum Press, New York, pp. 225-237.

Oertel, D., Wu, S. H., Garb, M. W., and Dizack, C., 1990, Morphology and physiology of cells in slice preparations of the posteroventral cochlear nucleus of mice. *Journal of Comp. Neurol.* **295**: 136-154.

Osen, K. K., and Ottersen, O. P., 1996, Glutamat receptors in the medial nucleus of the trapezoid body. *International Symposium on Acoustical Signal Processing in the Central Nervous System*, Prauge, Abstr. 70.

Osen, K. K., Ottersen, O. P., and Storm-Mathisen, J., 1990, Colocalization of glycine-like and GABA-like immunoreactivities: a semiquantitative study of individual neurons in the dorsal cochlear nucleus of cat. In *Glycine neurotransmission* (O. P. Ottersen and J. Storm-Mathisen, eds.), Wiley and Sons.

Ostapoff, E. M., Benson, C. G., and Saint Marie, R. L., 1997, GABA- and glycine-immunoreactive projections from the superior olivary complex to the cochlear nucleus in guinea pig. *J. Comp. Neurol.* **381**: 500-512.

Ostapoff, E. M., Morest, D. K., and Potashner, S. J., 1990, Uptake and retrograde transport of [3H]GABA from the cochlear nucleus to the superior olive in the guinea pig. *J. Chem. Neuroanat.* **3**: 285-295.

Pfeiffer, R. R., 1966a, Anteroventral cochlear nucleus: wave forms of extracellularly recorded spike potentials. *Science* **154**: 667-668.

Pfeiffer, R. R., 1966b, Classification of response patterns of spike discharges for units in the cochlear nucleus: tone-burst stimulation. *Exp. Brain. Res.* **1**: 220-235.

Pirsig, W., Pfalz, R., and Sadanaga, M., 1968, Postsynaptic auditory crossed efferent inhibition in the ventral cochlear nucleus and the blocking of it by strychnine nitrate (guinea pig). *Kumamoto Med. J.* **21**: 75-82.

Roberts, R. C., and Ribak, C. E., 1987, GABAergic neurons and axon terminals in the brainstem auditory nuclei of the gerbil. *J. Comp. Neurol.* **258**: 267-280.

Rose, J. E., Kitzes, L. M., Gibson, M. M., and Hind, J. E., 1974, Observations on phase-sensitive neurons of anteroventral cochlear nucleus of the cat: nonlinearity of cochlear output. *J. Neurophysiol.* **37**: 218-253.

Rothman, J. S., and Young, E. D., 1996, Enhancement of neural synchronization in computational models of ventral cochlear nucleus bushy cells. *Audit. Neurosci.* **2**: 47-62.

Rothman, J. S., Young, E. D., and Manis, P. B., 1993, Convergence of auditory nerve fibers onto bushy cells in the ventral cochlear nucleus: implications of a computational model. *J. Neurophysiol.*, **70**: 2562-2583.

Ryugo, D. K., and Sento, S., 1991, Synaptic connections of the auditory nerve in cats: relationship between endbulbs of held and spherical bushy cells. *J. Comp. Neurol.* **305**: 35-48.

Sachs, M. B., and Kiang, N. Y., 1968, Two-tone inhibition in auditory- nerve fibers. *J. Acoust. Soc. Am.* **43**: 1120-1128.

Saint Marie, R. L., Morest, D. K., and Brandon, C. J., 1989, The form and distribution of GABAergic synapses on the principal cell types of the ventral cochlear nucleus of the cat. *Hear. Res.* **42**: 97-112.

Saint Marie, R. L., Ostapoff, E. M., Benson, C. G., Morest, D. K., and Potashner, S. J., 1993, Non-cochlear projections to the ventral cochlear nucleus: are they mainly inhibitory ? In *The mammalian cochlear nuclei: organization and function* (M. A. Merchan, ed.), Plenum Press, New York, pp.121-131.

Schofield, B. R., 1991, Superior paraolivary nucleus in the pigmented guinea pig: separate classes of neurons project to the inferior colliculus and the cochlear nucleus. *J. Comp. Neurol.* **312**: 68-76.

Schofield, B. R., 1994, Projections to the cochlear nuclei from principal cells in the medial nucleus of the trapezoid body in guinea pigs. *J. Comp. Neurol.* **344**: 83-100.

Schofield, B. R., and Cant, N. B., 1996, Origins and targets of commissural connections between the cochlear nuclei in guinea pigs. *J. Comp. Neurol.* **375**: 128-146.

Shofner, W. P., and Young, E. D., 1985, Excitatory/inhibitory response types in the cochlear nucleus: relationships to discharge patterns and responses to electrical stimulation of the auditory nerve. *J. Neurophysiol.* **54**: 917-939.

Siegel, S., and Castellan Jr., N. J., 1988, *Nonparametric statistics for the behavioral sciences.* McGraw-Hill Book Company.

Smith, P. H., Joris, P. X., Carney, L. H., and Yin, T. C., 1991, Projections of physiologically characterized globular bushy cell axons from the cochlear nucleus of the cat. *J. Comp. Neurol.* **304**: 387-407.

Smith, P. H., Joris, P. X., and Yin, T. C., 1998, Anatomy and physiology of principal cells of the medial nucleus of the trapezoid body (MNTB) of the cat. *J. Neurophysiol.* **79**: 3127-3142.

Smith, P. H., and Rhode, W. S., 1989, Structural and functional properties distinguish two types of multipolar cells in the ventral cochlear nucleus. *J. Comp. Neurol.* **282**: 595-616.

Snyder, R. L., and Leake, P. A., 1988, Intrinsic connections within and between cochlear nucleus subdivisions in cat. *J. Comp. Neurol.* **278**: 209-225.

Sommer, I., Lingenhohl, K., and Friauf, E., 1993, Principal cells of the rat medial nucleus of the trapezoid body: an intracellular in vivo study of their physiology and morphology. *Exp. Brain Res.* **95**: 223-239.

Tsuchitani, C., 1997, Input from the medial nucleus of trapezoid body to an interaural level detector. *Hear. Res.* **105**: 211-224.

Turecek, R., and Trussell, L. O., 2001, Presynaptic glycine receptors enhance transmitter release at a mammalian central synapse. *Nature* **411**: 587-590.

Walsh, E. J., McGee, J., and Fitzakerley, J. L., 1990, GABA actions within the caudal cochlear nucleus of developing kittens. *J. Neurophysiol.* **64**: 961-977.

Warr, W. B., and Beck, J. E., 1996, Multiple projections from the ventral nucleus of the trapezoid body in the rat. *Hear. Res.* **93**: 83-101.

Wenthold, R. J., 1987, Evidence for a glycinergic pathway connecting the two cochlear nuclei: an immunocytochemical and retrograde transport study. *Brain Res.* **415**: 183-187.

Wenthold, R. J., Huie, D., Altschuler, R. A., and Reeks, K. A., 1987, Glycine immunoreactivity localized in the cochlear nucleus and superior olivary complex. *Neurosci.* **22**: 897-912.

Wenthold, R. J., Zempel, J. M., Parakkal, M.H., Reeks, K. A., and Altschuler, R. A., 1986, Immunocytochemical localization of GABA in the cochlear nucleus of the guinea pig. *Brain Res.* **380**: 7-18.

Wickesberg, R. E., and Oertel, D., 1990, Delayed, frequency-specific inhibition in the cochlear nuclei of mice: a mechanism for monaural echo suppression. *J. Neurosci.* **10**: 1762-1768.

Wickesberg, R. E., and Oertel, D., 1988, Tonotopic projection from the dorsal to the anteroventral cochlear nucleus of mice. *J. Comp. Neurol.* **268**: 389-399.

Winer, J. A., Larue, D. T., and Pollak, G. D., 1995, GABA and glycine in the central auditory system of the mustache bat: structural substrates for inhibitory neuronal organization. *J. Comp. Neurol.* **355**: 317-353.

Winter, I. M., and Palmer, A. R., 1990, Responses of single units in the anteroventral cochlear nucleus of the guinea pig. *Hear. Res.* **44**: 161-178.

Winter, I. M., Robertson, D., and Yates, G. K., 1990, Diversity of characteristic frequency rate-intensity functions in guinea pig auditory nerve fibres. *Hear. Res.* **45**: 191-202.

Wu, S. H., and Kelly, J. B., 1995, Inhibition in the superior olivary complex: pharmacological evidence from mouse brain slice. *J. Neurophysiol.* **73**: 256-269.

Wu, S. H., and Oertel, D., 1986, Inhibitory circuitry in the ventral cochlear nucleus is probably mediated by glycine. *J. Neurosci.* **6**: 2691-2706.

Yang, L., Monsivais, P., and Rubel, E. W., 1999, The superior olivary nucleus and its influence on nucleus laminaris: a source of inhibitory feedback for coincidence detection in the avian auditory brainstem. *J. Neurosci.* **19**: 2313-2325.

Young, E. D., Robert, J. M., and Shofner, W. P., 1988, Regularity and latency of units in ventral cochlear nucleus: implications for unit classification and generation of response properties. *J. Neurophysiol.* **60**: 1-29.

Young, E. D., and Voigt, H. F., 1982, Response properties of type II and type III units in dorsal cochlear nucleus. *Hear. Res.* **6**: 153-169.

Topographic Representation of Periodicity Information: The 2nd Neural Axis of the Auditory System

GERALD LANGNER
Neuroacoustics, Institute for Zoology, TU-Darmstadt, Schnittspahnstr. 3, 64287 Darmstadt, Germany

1. INTRODUCTION

For acoustic communication the auditory system has to deal with two typical signal properties: first, many signals are embedded in noise and second, many signals are broadband harmonic sounds. Harmonic sounds, particularly voiced speech sounds and many animal communication signals, are characterized by a periodic envelope or amplitude modulation (AM). Signals with the same periodicity have the same pitch, while their frequency content defines their timbre. Note, that the terms 'period, periodic, or periodicity' are used here for the temporal envelope of a signal which may or may not correspond to the period of a pure tone, as in AM where each of its three frequency components may have a period differing from the period of the envelope. For the auditory system periodicity is a very useful property, because in periodic sounds the same feature, for example the peak of the envelope, appears repeatedly thereby facilitating signal detection in noise.

Acoustic signals may be described in the temporal as well as in the frequency domain. In order to understand the processing of complex signals by the auditory system one has to take into account that the first processing stage, the cochlea, performs a frequency analysis. From the beginning of modern hearing research this analysis was compared to a Fourier analysis (von Helmholtz, 1863). However, it turned out that the bandwidths of single cochlear filters, as may be measured in auditory nerve fibres, are broad enough to transmit several harmonics of a broadband sound, provided the fundamental frequency of the sound is lower than about ¼ of the centre frequency (CF) of the filters. For such low fundamental frequencies the travelling wave at the location of such a filter beats with a frequency equal to the fundamental frequency, because all harmonics of a periodic broadband sound are integer multiples of the fundamental frequency (Schouten *et al.*, 1962). As a result, spike trains in the auditory nerve show phase coupling to these beats and within to the fundamental frequency (Palmer, 1982) – even when the fundamental frequency component is 'missing'.

1.1 A Model for Processing Periodicity Information

As a consequence of periodicity information coded by spike intervals in the auditory nerve, neurons in the nucleus cochlearis also code periodicity in different ways (Møller, 1974; Frisina, *et al.,* 1990; Langner, 1992; Rhode and Greenberg, 1994). Based on electrophysiological recordings in Guinea fowl (Langner, 1981, 1983), cat (Langner and Schreiner, 1988) and more recently chinchilla (Langner *et al.*, 2002; Biebel and Langner, 2002) a neuronal model has been suggested that utilizes this temporal information for a correlation analysis of signal periodicity in the midbrain (Langner, 1992, 1997). Similar to the correlation model suggested by Licklider (1951, 1954), the model includes delayed responses and coincidence of delayed and undelayed neuronal activity as basic processing elements. In contrast to Licklider's model, which requires unrealistic long axonal or synaptic delays, it is suggested that delayed responses to amplitude modulations result from neuronal integration by neurons in the dorsal cochlear nucleus (DCN). A constant delay requires that these neurons receive their input from auditory nerve fibres with activity synchronized and thereby correlated to resolved harmonics. This is possible only in the saturation range of these fibres which is about 30 dB above their thresholds. In line with psychophysical findings (Bernstein and Oxenham, 2002) it is this theoretical condition which restricts optimal periodicity processing. As a result pitch percept is strongest in the range of lower harmonics of broad band signals where some nerve fibres resolve harmonics. Note, however, that the same neuronal mechanisms would work also in the range of unresolved higher harmonics of a broad band sound, although with less precision because of unreliable integrational delays in this range.

The undelayed response in this model comes from the responses of chopper neurons in the ventral cochlear nucleus (VCN) which respond and synchronize with short latency to modulated sounds. Their input is supposed to originate from fibres with low threshold which favour synchronization to the modulation which results from overlapping harmonics on the basilar membrane. As long as the chopper intervals are short, in comparison to the modulation period, their repetitive firing will only broaden the coincidence window of the correlation analysis or will result in additional side peaks close to the correlational maximum (Langner, 1983).

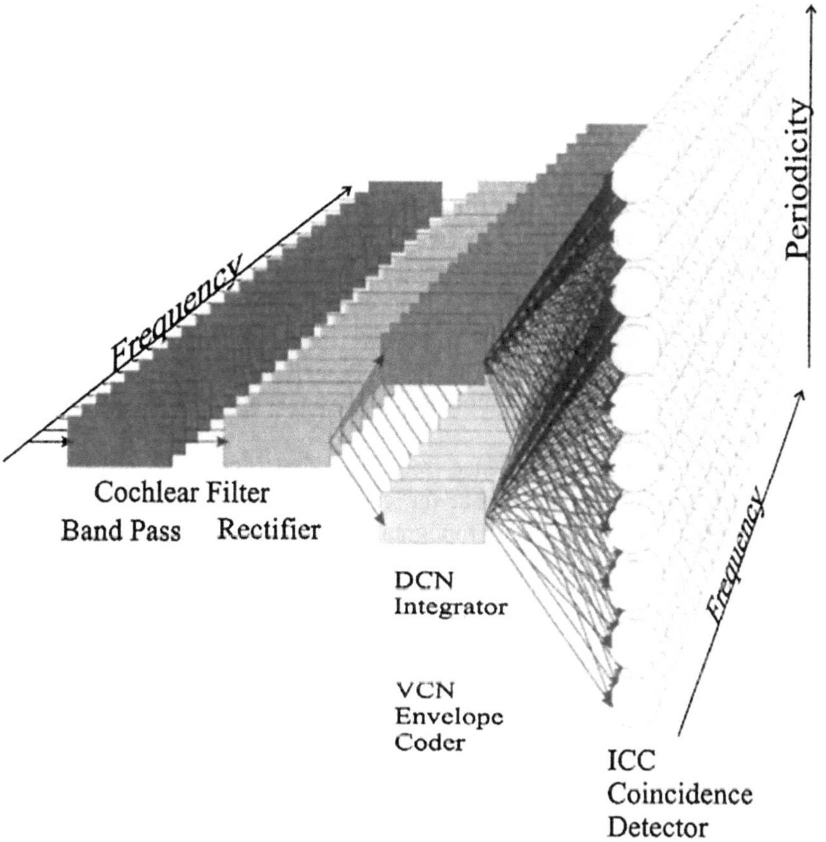

Figure 1. Delayed responses of integrator neurons in the DCN and undelayed responses by envelope coding of neurons in the VCN converge on coincidence neurons in the ICC along a 2nd axis of a frequency band lamina.

Furthermore, the model suggests that the projections from DCN and VCN converge on single neurons in the central nucleus of the inferior colliculus (ICC). Consequently, a coincidence neuron in the ICC will respond best when the signal is modulated with a period which just compensates for the delay of its DCN-input, because the delayed response to a modulation cycle will then coincide with the undelayed response to a previous cycle. Note that some kind of periodicity coding will already result when this coincidence increases the firing probability of the neuron, i.e. it is not necessary to assume that the 'coincidence neuron' fires exclusively when a coincidence occurred.

Under appropriate conditions such a coincidence neuron may also be activated by a harmonic of its best modulation frequency (BMF), resulting in a modulation transfer function with comb-filter instead of bandpass-filter characteristics. Our model therefore includes synchronized inhibition originating from the nucleus of the lateral lemniscus (not shown in Fig. 1).

It is assumed that the described temporal analysis creates a second frequency axis in the auditory system which would represent periodicity information orthogonal to the tonotopic axis (Fig. 1). The layer of coincidence neurons shown in Fig. 1 represents one of about 30 neuronal laminae in the ICC covering only a small range of cochlear frequency analysis. In contrast, it covers the full range of periodicity information which, for sinusoidal amplitude modulations, extends up to a certain BMF of about CF/4 (see above).

This paper summarizes evidence from different experiments and species which supports the hypothesis that the central auditory system up to the primary auditory cortex is indeed topographically organized with orthogonal representations of frequency and periodicity.

2. PERIODOTOPY IN THE INFERIOR COLLICULUS

At the level of the midbrain the inferior colliculus is the major nucleus for integration of auditory information. Information from both ears, distributed over different frequency channels, and processed differently in several brainstem nuclei converge on the layered network of its central nucleus (ICC).

In many neurons of the IC, response rate is maximal for a particular BMF. Examples of rate tuned neurons in the ICC of chinchilla are given in Fig. 2. Note, the broadening of curves for higher modulation frequency on a linear axis which disappears on a logarithmic axis.

In an investigation of the IC in cat most of the units were rate tuned, occasionally even up to 1000 Hz, while only a minority was also tuned in their synchronization (Langner and Schreiner, 1988). This suggests a synchronization-rate transformation at this auditory level, since units at more central stations in the central auditory pathway show progressively poorer responses to high modulation frequencies. Because the main input neurons of the ICC in the cochlear nucleus show high synchronization but nearly constant average rates when modulation frequency is varied (Kim *et al.*, 1990; Krishna and Semple, 2000), the auditory midbrain may be considered as a decoder for temporal information, transforming periodicity into rate/space information (Langner, 1992; Pinheiro *et al.*, 1991). Accordingly, although periodicity up to several hundred Hertz may be represented in the temporal response patterns of units in the ICC, neuronal processing in the brainstem results in a transformation of temporal into spatial information for periodicity pitch. Naturally, this synchronization-rate transformation does not exclude temporal representation and also transmission of low modulation frequencies to even higher processing centres.

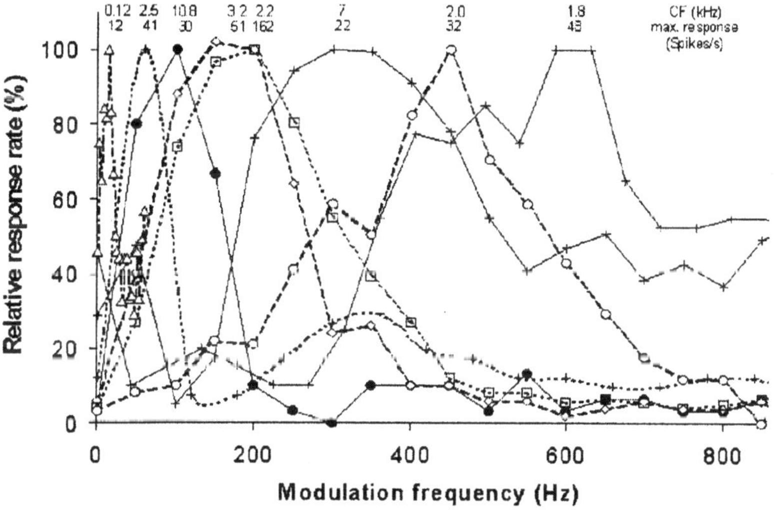

Fiure 2. Examples of normalized, band-pass tuned modulation transfer functions from neurons in the inferior colliculus of chinchilla (Langner *et al.*, 2002).

In line with the idea of the ICC as a temporal decoder, at different locations in the ICC units were found to be tuned not only to a certain frequency, but also to different modulation frequencies with a spatial distribution resulting in topographic maps for a wide range of modulation frequencies (Schreiner and Langner, 1988; Langner, 1992, 1997; Heil, *et al.*, 1995; Langner *et al.*, 2002).

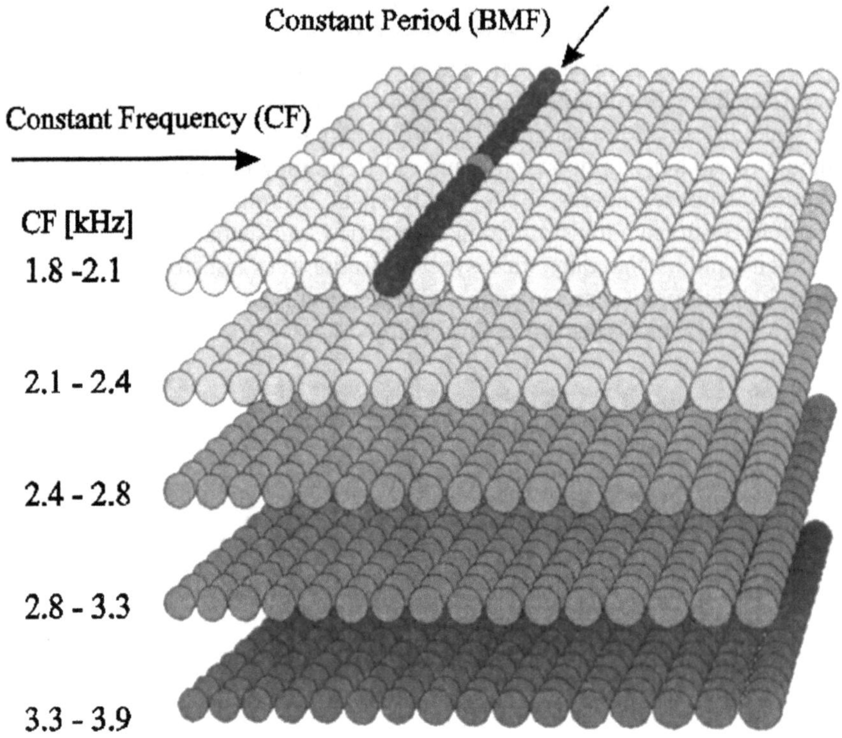

Figure 3. Scheme of tonotopy and periodotopy in layers of ICC.

The ICC of cat has a well-defined tonotopic or cochleotopic organization. Within the ICC, neuronal laminae, oriented approximately orthogonal to the main frequency gradient, may be anatomically defined. Some authors have identified these laminae with isofrequency planes. However, the ICC has only 30 - 40 laminae. Therefore, each lamina has to cover a certain frequency range and thus the neurons in each lamina should have slightly different CF values. Accordingly, a three-dimensional reconstruction of CF values obtained in many parallel electrode tracks revealed a fine-structure of tonotopic organization in these "frequency-band laminae". In each functionally-defined lamina, CF seems to increase over a small range of frequencies orthogonal to the main frequency gradient of the ICC (Schreiner and Langner, 1997).

The spatial distribution of BMFs was examined also within single "frequency band laminae". BMF increased along an isofrequency-line from about 10 Hz to CF/4 (<1000 Hz) with the highest BMFs located at the lateral border of the ICC. These results are summarized in the scheme of a model depicted in Fig. 3.

An example for a periodicity map in the ICC of chinchilla is given in Fig. 4. CFs of about 52 neurons were in a close frequency range around 6 kHz, however the BMF for sinusoidal amplitude modulations cover more than 5 octaves from 19 to 645 Hz. As in cat the gradient of periodicity is orientated roughly from medial to lateral.

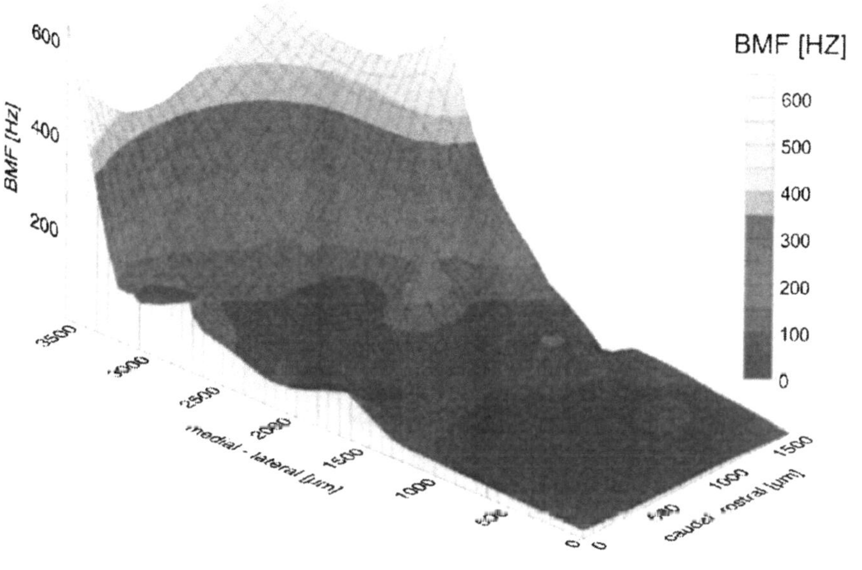

Figure 4. Electrophysiological mapping of best modulation frequencies in ‚iso-frequency' areas of the inferior colliculus of chinchilla (Langner *et al.*, 2002).

Recently we were able to obtain a direct visualization of periodotopy and its relation to tonotopy in the ICC of the Mongolian gerbil by using the ^{14}C-2-Deoxyglucose (2-DG) technique. Stimuli were pure tones (Fig. 5) and harmonic complexes (Fig. 6) with fundamental frequencies ranging from 40 to 1200 Hz and cut-off frequencies at the low (0.4 - 5 kHz) and the high frequency border (2 - 8 kHz) of their broad band spectra. The stimuli were presented either separately or consecutively in the same experiment. To avoid distortions an electrostatic headphone (STAX) and low sound pressure levels (0-55 dB SPL) were used. Animals were placed singly in a sound-proofed chamber and were stimulated for 90 min. Afterwards their brain were removed, frozen, and cut in a cryostat (Leica) into 20µm thick, transvers, serial sections. The sections were exposed to X-ray-films (Kodak) for 2 weeks.

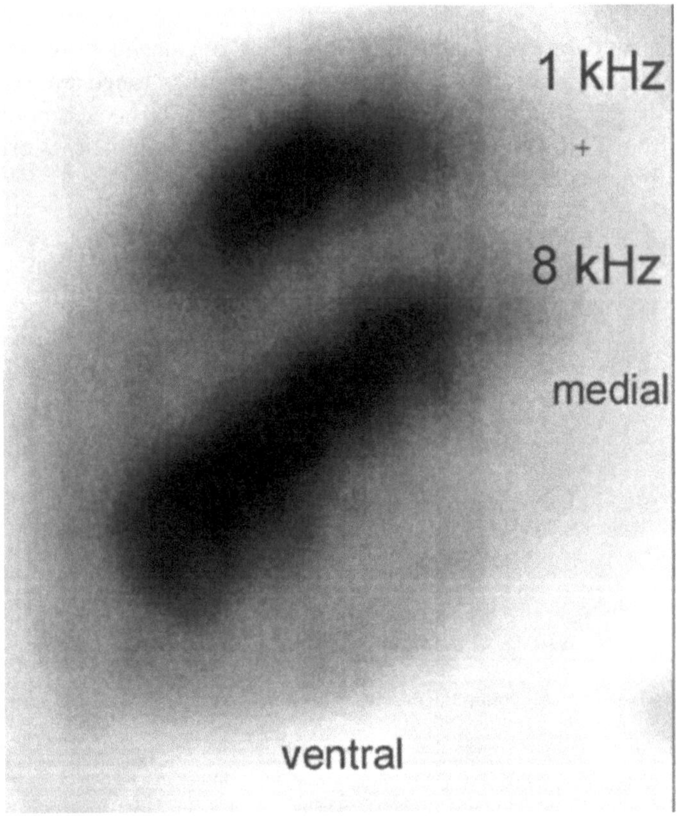

Figure 5. 2-DG mapping of responses to pure tones (1 and 8 kHz) in the inferior colliculus of gerbil.

The results were analyzed in three dimensions and visualized by 3D-computer reconstructions. ‚Iso-frequency' bands in the ICC of gerbil extend from dorso-medial to ventro-lateral with low frequencies dorsal and high frequencies ventral (Fig. 5). Fig. 6 shows the result of an experiment where 3 broadband harmonic sounds (40 dB SPL) with periodicities of 50, 400, and 800 Hz were presented alternatively. Different cut-off frequencies (see inset) allowed identification of the responses to different stimuli. Periodicity resulted in label which was roughly orthogonal to the label due to frequency content of the stimuli. Obviously, periodicity is represented in the ICC approximately orthogonal to the frequency representation. Orthogonality holds clearly at least for the middle third of the caudo-rostral extent of the ICC (not shown) with periodicity labels extending from dorso- lateral to ventro-medial.

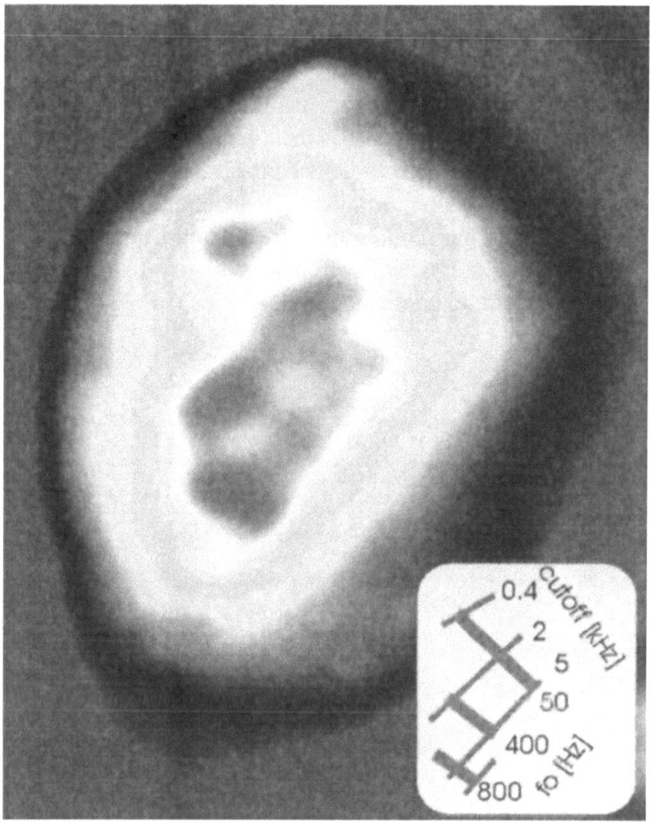

Figure 6. Mapping of 3 broadband harmonic sounds in inferior colliculus of gerbils.

3. PERIODOTOPY IN THE AUDITORY CORTEX

The 2-dimensional surface of the sensory cortex allows for a continuous mapping of two sensory dimensions corresponding, for example, to body surface or retina. In the auditory system it is well known that the one-dimensional receptor array in the cochlea leads to the tonotopic axis representing frequency (Merzenich *et al.*, 1976). Various other parameters like amplitude, signal bandwidth (Schreiner *et al.*, 2000)., and, in bat, echo delays may be mapped (e.g. Suga *et al.*, 1979), However, it is still an open question whether there is a single major parameter mapped along the 2nd axis of the primary auditory cortex. Evidence that such a parameter may be the periodicity of signal envelopes have been found in the auditory forebrain of mynah birds (Hose, *et al.*, 1987) and in the cortex of gerbils (Schulze and Langner, 1999; Schulze *et al.*, 2002).

Moreover, to study frequency and periodicity organization in the same individual, spatially highly resolved activity patterns were recorded in the primary auditory field (AI) of 5 cats using an optical imaging technique (Langner *et al.*, 1997b). Response maps for single stimulus conditions confirm the well-known cochleotopic organization of AI characterized by a anterior to posterior frequency gradient. In general, with the fundamental frequency of broadband harmonic sounds increasing from 25 Hz to 1200 Hz, response areas shift systematically from dorsal to ventral, approximately orthogonal to the tonotopic gradient.

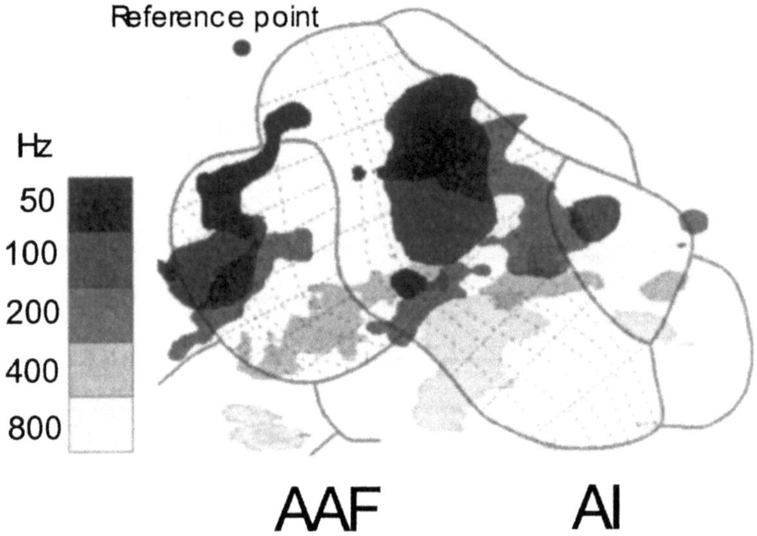

Figure 7. Periodicity maps in AI and AAF of gerbils demonstrated with 2-DG.

For further support of the hypothesis that the 2nd axis in the cortex is devoted to periodicity information, the representation of periodicity in gerbils was investigated using the 2-DG technique. The methods are the same as described above for the IC. The 20 μm thick, transvers, serial sections of the cortex were used for pseudo-3D reconstructions to survey the primary auditory cortex fields AI and AAF. Stimuli were the same broad-band periodic signals with a certain pitch as in the IC experiments (see above). Due to the frequency content of the stimuli, most of the label extends along the tonotopic gradient. Adjusting label with relation to the rostral hippocampus-tip and combining the single stimuli maps from 5 animals resulted in a composite map showing the periodotopy of AI and also AAF (Fig. 7). Note that a certain gray level indicates where the 2-DG labeling exceeds an arbitrarily selected threshold. The underlying scheme of a cortical map with roughly vertical lines for isofrequency-lines of the tonotopic maps in AI and AAF was obtained by electrophysiological recording (Thomas *et al.*, 1993). The lines plotted roughly orthogonal to the isofrequency-lines are based on an approximately logarithmic arrangement of periodicity from about 20 to 4000 Hz. The results indicate that in the auditory cortex periodicity is mapped dorso-ventrally, roughly orthogonally to the tonotopic gradient. Note, that the most dorsal labels are due to a pitch of 50 Hz (black dots) and the most ventral labels to 800 Hz (lightest gray) leaving enough space for the representation of even lower and higher pitches, respectively, in AI. This is not true for AAF, which may be due to an underestimation of its size in the underlying electrophysiological map. A reconstruction error in the 2-DG data is unlikely, because of the simultaneously obtained precise match of the expected periodotopy in AI.

Finally, in the auditory cortex of humans an orthogonal map for frequency and periodicity pitch has been demonstrated using magnetoencephalography (Langner *et al.*, 1997a).

In these experiments magnetic fields evoked in the human auditory cortex by harmonic sounds and pure tones were used to compute the corresponding equivalent current dipoles (ECD) for each hemisphere of the brain. ECD strength was varying in time and space. Around 60, 100, and 200 ms after stimulus onset, and even thereafter, local maxima in the brain signals can be observed.

These deflections are referred to as M60, M100, M150, M200 and sustained field (SF). Each local maximum may result from the dominant response of a certain cortical area. The peaks of M60 and M100 are quite sharp. Therefore, it is assumed that for their maxima the spatial position of the ECDs may correspond very well with the positions of their generators in the corresponding auditory areas.

In order to maximize the spatial resolution, for a topographic analysis, only those time segments of 10 ms around the maxima were evaluated during which the goodness of fit was maximal (about 90%).

In Fig. 8A a roughly orthogonal representation of periodicity and frequency in an auditory field related to M60 is demonstrated for the left hemisphere of one subject. Note, that harmonic stimuli with a low cut-off frequency of 400 Hz (p50 - p400) are located near the ECD position for pure tones of 400 Hz, while those with a cut-off frequency of 800 Hz (h100 and h200) are located near the ECD positions for pure tones of 800 Hz. The geometric arrangement of all stimuli, as well as the nearly equal spacing of 4 harmonic sounds representing octaves, defy the explanation that the different response positions may result from interactions of magnetic sources from different cortical fields.

In addition, the line connecting the ECD positions resulting from *'unmodulated'* pure tones runs approximately orthogonal to the corresponding lines for harmonic sounds, indicating orthogonality of tonotopy and periodicity also for the human auditory cortex. This result was confirmed by a statistical analysis of mapping results from 5 subjects.

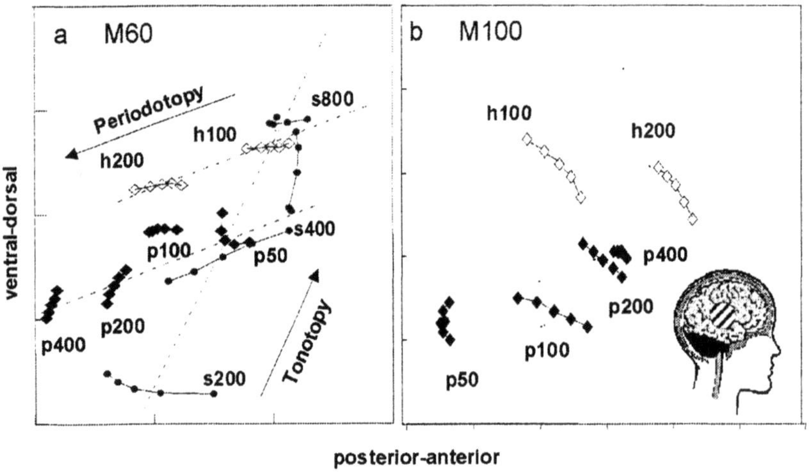

Figure 8. : MEG-mapping of periodicity pitch and frequency in human cortex.

Our results were recently verified by experiments using the same experimental paradigms and evaluation as ours (Fig. 8B, by courtesy of W. Smith, Centre for Brain And Cognitive Development, Birkbeck College, University of London). Although the results shown are from the M100 and the right cortex, the similarity with those shown in Fig. 8A is striking. They confirm that it is possible to establish pitch maps in the human auditory cortex using appropriate stimulation and evaluation.

4. THE 2ND NEURAL AXIS OF THE AUDITORY SYSTEM

In conclusion one may say that several different experiments in different species show that periodicity in the ICC is represented orthogonal to frequency. This orthogonality is preserved at the level of the auditory cortex at least in cats, gerbils, and humans. In gerbils it could be shown that orthogonality is present in AAF as well as in AI. Orthogonality may relate to the fact that timbre, a percept largely defined by the spectral envelope of an acoustic signal, and pitch, which correlates with temporal envelopes of signals, are to a certain extent orthogonal or independent variables (Plomp and Steeneken, 1971, Müller-Linow, 2003). Similarly to frequency, which is mapped along the tonotopic axis in all primary auditory areas, periodicity is mapped from midbrain to primary auditory cortex. Because in each case tonotopic and periodotopic organizations have approximately constant logarithmic gradients of similar size which are orthogonal to each other, the periodotopic axis may be considered as the 2nd neural axis of the auditory system.

ACKNOWLEDGEMENTS

Supported by grants from DFG, BMBF, Human-Frontier-Science-Program, and the VW-Stiftung. I would also like to thank Monika Albert, Ulrich Biebel, Susanne Braun, Thorsten Briede, Hubert Dinse, Ben Godde, Peter Heil, Michael Ochse, Mikko Sams, Christoph Schreiner, Holger Schulze, and Warwick Smith for their cooperation in the presented studies.

REFERENCES

Bernstein, J. G., and Oxenham, A. J., 2003, Pitch discrimination of diotic and dichotic tone complexes: Harmonic resolvability or harmonic number? *J. Acoust. Soc. Am.* **113**: 3323-3334.

Biebel, U. W., and Langner, G., 2002, Evidence for interactions across frequency channels in the inferior colliculus of awake chinchilla. *Hear. Res.* **169**:126-144.

Heil, P., Schulze, H., and Langner, G., 1995, Ontogenetic development of periodicity coding in the inferior colliculus of the mongolian gerbil. *Audit. Neurosci.* **1**:363-383.

Helmholtz, von, H. L. F., 1863, *Die Lehre von den Tonempfindungen*. F. Vieweg & Sohn, Braunschweig.

Frisina, R. D., Smith, R. L., and Chamberlain, S. C., 1990, Encoding of amplitude modulation of the gerbil cochlear nucleus: I. A hierarchy of enhancement. *Hear. Res.* **44**: 99-122.

Hose, B., Langner, G., and Scheich, H., 1987, Topographic representation of periodicities in the forebrain of the Mynah bird: one map for pitch and rhythm? *Brain Res.* **422**: 367-373.

Kim, D. O., Sirianni, J. G., and Chang, S. O., 1990, Responses of DCN-PVCN neurons and auditory nerve fibers in unanesthetized decerebrate cats to AM and pure tones: Analysis with autocorrelation/power-spectrum. *Hear. Res.* **45**: 95-113.

Krishna, B. S., and Semple, M. N., 2000, Auditory temporal processing: responses to sinusoidally amplitude-modulated tones in the inferior colliculus. *J. Neurophysiol.* **84**: 255-273.

Langner, G., 1981, Neuronal mechanisms for pitch analysis in the time domain. *Exp. Brain Res.* **44**: 450-454.

Langner, G., 1983, Evidence for neuronal periodicity detection in the auditory system of the guinea fowl: implications for pitch analysis in the time domain. *Exp. Brain Res.* **52**: 333-355.

Langner, G., 1992, Periodicity coding in the auditory system. *Hear. Res.* **60**: 115-142.

Langner, G., Schreiner, C. E., 1988, Periodicity coding in the inferior colliculus of the cat. I. Neuronal mechanisms. *J. Neurophysiol.* **60**: 1799-1822.

Langner, G., 1997, Neural processing and representation of periodicity pitch. *Acta Oto-Laryngol. Suppl.* **53**: 68-76.

Langner, G., and Schreiner, C. E., 1989, Orthogonal topographical representation of characteristic and best modulation frequency in the inferior colliculus of cat, *Soc. Neurosci. Abstr.*, **15**, Part 1, 1116.

Langner, G., Sams, M., Heil, P., and Schulze, H., 1997a, Frequency and periodicity are represented in orthogonal maps in the human auditory cortex: Evidence from Magnetoencephalography. *J. Comp. Physiol.* **181**: 665-676.

Langner, G., Dinse, H. R. B., Godde, B., and Langner, T., 1997b, Periodotopic organization of the cat auditory cortex demonstrated by optical recordings. *Soc. Neurosci. Abstr.* **23**, Part 1: 1034.

Langner, G., Albert, M., and Briede, T., 2002, Temporal and spatial coding of periodicity information in the inferior colliculus of awake chinchilla (Chinchilla laniger). *Hear. Res.* **168**: 110-130, 2002.

Licklider, J. C. R., 1951, A duplex theory of pitch perception. *Experientia* **7**: 128-134.

Licklider, J. C. R., 1954, 'Periodicity' pitch and 'place' pitch. *J. Acoust. Soc. Am.* **26**: 945 (abstract).

Merzenich, M. M., Knight, P. L., and Roth, G. L., 1976, Representation of the cochlea within primary auditory cortex in the cat. *J. Neurophysiol.* **38**: 231-249.

Müller-Linow, M., 2003, Master thesis *Psychophysikalische Studie zur Tonhöhen- und Klangwahrnehmung anhand reiner und harmonischer Stimuli.* Darmstadt University of Technology.

Møller, A.R., 1974, Coding of amplitude and frequency modulated sounds in the cochlear nucleus. *Acustica* **31**: 292-299.

Palmer, A. R., 1982, Encoding of rapid amplitude fluctuations by cochlear nerve fibres in the guinea pig. *Arch. Otorhinolaryngol.* **236**: 197-202.

Pinheiro, A. D., Jen, P. H. S., and Wu, M., 1991, Encoding repetition rate and duration in the inferior colliculus of the big brown bat, Eptesicus fuscus. *J. Comp. Physiol. [A.]* **169**: 69-85.

Plomp, R., and Steeneken, H. J. M., 1971, Pitch versus timbre. *7th Int. Congr. on Acoust,* Abstract 1602.

Rhode, W.S., and Greenberg, S., 1994, Encoding of amplitude modulation in the cochlear nucleus of the cat. *J. Neurophysiol.* **71**: 1797-1825.

Schreiner, C. E., and Langner, G., 1988, Periodicity coding in the inferior colliculus of the cat. II. Topographical organization. *J. Neurophysiol.* **60**: 1823-1840.

Schreiner, C. E., and Langner, G., 1997, Laminar fine structure of frequency organization in auditory midbrain. *Nature* **388**: 383-386.

Schreiner, C. E., Read, H. L., and Sutter, M. L., 2000, Modular organization of frequency integration in primary auditory cortex. *Annu. Rev. Neurosci.* **23**: 501-529.

Schulze, H., and Langner, G., 1999, Auditory cortical responses to amplitude modulations with spectra above frequency receptive fields: evidence for wide spectral integration. *J. Comp. Physiol [A];* **185**: 493-508.

Schulze, H., Hess, A., Ohl, F. W., and Scheich, H., 2002, Superposition of horseshoe-like periodicity and linear tonotopic maps in auditory cortex of the Mongolian gerbil. *Eur. J. Neurosci.* **15**: 1077-1084.

Suga, N., and O'Neill, W.E., 1979, Neural axis representing target range in the auditory cortex of the mustache bat. *Science* **206**: 351-353.

Thomas, H., Tillein, J., Heil, P., and Scheich H., 1993, Functional organization of auditory cortex in the Mongolian gerbil (*Meriones unguiculatus*). I. Electrophysiological mapping of frequency representation and distinction of fields. *Eur. J. Neurosci.* **5**: 882-897.

Complex Frequency Tuning of Neurons in the Mouse Inferior Colliculus

Responses to single tones and combinations of tones

CHRISTINE V. PORTFORS

School of Biological Sciences, Washington State University, Vancouver, WA 98686, USA

1. INTRODUCTION

Tonotopy is a fundamental organizing principle in the auditory system. This basic characteristic of ordering of frequency information leads to the impression that most neurons (at least in the brainstem and midbrain) have single excitatory frequency tuning curves. However, recent evidence suggests that spectral integration is a common feature of neurons in the inferior colliculus (IC) of the mustached bat. When tested with more complex stimuli, over fifty percent of single units in the mustached bat IC show multiple excitatory frequency tuning curves, facilitatory responses to combinations of tones and/or inhibitory response to combinations of tones (Portfors and Wenstrup, 1999, 2002). Multiple tuning and combination-sensitivity are features common to neurons in frequency representations important for echolocation behaviors, but they also occur in frequency representations thought to be important for social vocalizations (Portfors and Wenstrup 2002). Because many animals emit spectrally and temporally complex social vocalizations, multiply tuned and combination-sensitive neurons may be a mechanism by which other mammals process complex vocalizations. Thus far there are few reports of combination sensitivity in non-bat mammals. To determine whether spectral integration occurs in the IC of other mammals, we examined frequency-tuning characteristics of neurons in the IC of CBA/CaJ mice using single tones and combinations of tones. We found that the mouse IC has a high percentage of multiply tuned neurons and some combination-sensitive neurons. These findings suggest that frequency-tuning properties of the neurons in the mouse IC are more complex than previously thought and are similar to those found in the IC of the mustached bat.

2. METHODS

We recorded responses of single neurons in the IC of awake CBA/CaJ mice to single and combinations of tones. All procedures were approved by the Washington State University Institutional Care and Use Committee. Animals were restrained in a foam holder, and a pin cemented onto the skull was secured into a bar on a stereotaxic apparatus to immobilize the head. A small hole was cut in the skull to expose the left IC. Sound stimuli were synthesized using custom-designed software and presented through a speaker located 10 cm away and oriented 45 deg to the right ear. Single unit responses were recorded with micropipette electrodes. Characteristic frequency (CF) and minimum threshold (MT) were obtained audiovisually for each single unit. CF was defined as the frequency requiring the lowest intensity to elicit stimulus-locked spikes to 50% of the presentations. MT was defined as the lowest intensity to elicit spikes to 50% of the stimulus presentations. To obtain excitatory frequency-tuning curves, we measured the number of spikes elicited from single tone bursts (50 ms duration, 0.5 ms rise/fall, 4/s, 5-10 presentations per stimulus) that varied in frequency (6-100 kHz, 2 kHz steps) and intensity (10, 30 and 50 dB above MT). We tested for combination-sensitive facilitation and inhibition using a two-tone paradigm. A CF tone was presented at 10 dB above MT while the frequency and intensity of a second tone was varied (simultaneous onset, 6-100 kHz, 2 kHz steps). The timing between the two tones was varied in some tests.

3. RESULTS

We recorded 94 single units from the ICs of 10 mice. CFs ranged from 6 to 64 kHz (median CF = 25 kHz). Fifty-eight of the units had CFs in the ultrasonic range (> 20 kHz) and 36 had CFs in the audible range (< 20 kHz). Thirty-five (37%) units had multi-peaked excitatory frequency-tuning curves (Figure 1A). We classified units as multiply tuned if the tuning curves remained distinctly separate at the highest intensity tested (typically 80-90 dB SPL). Figure 1B shows that 26 of the 35 multiply tuned neurons had CFs in the ultrasonic range. Of these 26 ultrasonic neurons, 23 had their second excitatory frequency peak in the audible range. Only eight multiply tuned units had a low frequency CF and an ultrasonic second excitatory peak. Two units had three peaks of excitation.

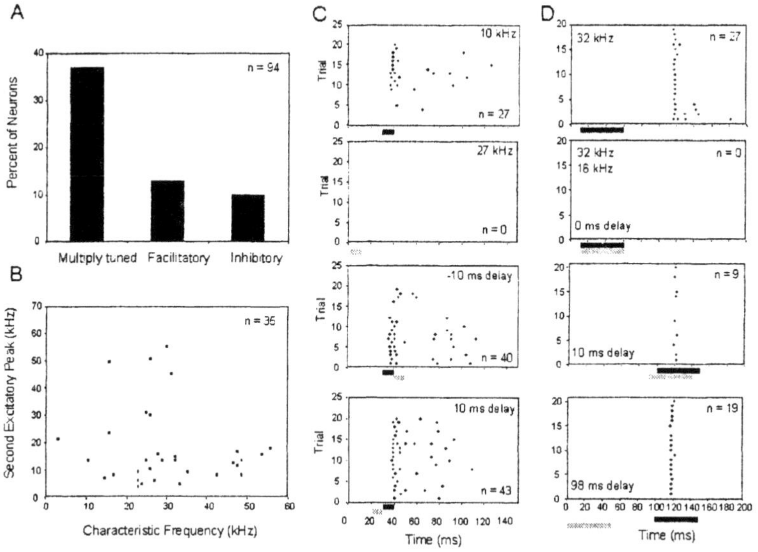

Figure 1. Frequency tuning characteristics of IC units. A: Percentage of neurons that showed multiple tuning or combination sensitivity. B: Distribution of CFs and second excitatory peaks of multiply tuned units. C: Raster plots showing facilitatory combination sensitivity. The unit had a CF of 10 kHz and did not respond to single tones at higher frequencies (plot 2). A second tone of 27 kHz presented 10 ms after (plot 3) or before (plot 4) the CF tone elicited facilitation. D: Raster plots showing inhibitory combination sensitivity. This unit's response to CF (32 kHz, plot 1) was completely eliminated with simultaneous onset of a second tone at 16 kHz (plot 2). Plots 3 and 4 show that at different delays, the response was not inhibited.

A small number of neurons (n=12, 13%) showed facilitation when presented with combinations of two tones of different frequencies. Figure 1C shows a facilitatory unit that responded with a mean spike count of 1.35 to a CF tone (10 kHz at 10 dB above MT). When presented with a second tone of 27 kHz that occurred 10 ms after the onset of the CF tone, the mean spike count increased to 2 spikes/stimulus. When the 27 kHz tone occurred 10 ms before the onset of the 10 kHz tone, the mean spike count was 2.15. Thus, the neuron had complex temporal tuning as well as frequency tuning. The majority of facilitatory responses occurred at non-zero delays between the two frequency tones. Further, there was no simple relation between the frequencies of tones eliciting facilitation; some combinations were harmonically related and others were not.

Ten percent of the units (n=9) showed combination-sensitive inhibition. In these neurons the excitatory response to CF was suppressed 20% or more by the addition of a second tone of a different frequency. Figure 1D shows a unit in which the response to CF (32 kHz) was completely eliminated with the presentation of a simultaneously presented 16 kHz tone. This inhibition was tuned in frequency and time. In all the units, the inhibition was strongest when the two tones overlapped in time. This type of inhibition is different than sideband inhibition in that the inhibiting frequency was not close to CF, but usually an octave or more away.

4. DISCUSSION

In this study we show that frequency-tuning properties of neurons in the IC of CBA mice are more complex than previously thought. Sixty percent of the neurons had sensitivity to two or more different frequencies or had combination-sensitive responses to sounds in different frequency bands. In many respects the characteristics of these complex responses in the IC of mouse are similar to those found in the IC of the mustached bat. This suggests that the IC of different mammals may contain similar mechanisms for processing complex sounds, including multiple tuning and combination sensitivity.

We found that almost 40% of neurons in the IC of mouse had tuning curves in more than one frequency band. This is a similar percentage to that reported in the non-sonar frequency representations of the mustached bat IC (Portfors and Wenstrup 2002) but far greater than has previously been reported in other species, including mouse (Egorova *et al.*, 2001; LeBeau *et al.*, 2001). We may have found a higher percentage of multiply tuned neurons because we consistently tested excitatory responses to tones across a large extent of the mouse hearing range (6-100 kHz). Other studies have focused their tone stimuli closer to CF, thereby potentially missing second excitatory peaks.

The function of multiply tuned neurons is not clear. One possibility is they are involved in analyzing social vocalizations. However, a neuron with many excitatory frequency response areas would likely respond to a multitude of social vocalizations and thereby be considered non-selective compared to a neuron with a single, sharply tuned frequency response area. Multiply tuned neurons may encode different types of sounds under different behavioral contexts. For example, the majority of social vocalizations emitted by mice are ultrasonic but sounds in the environment (ex. from predators) are more likely to be lower frequency. Multiply tuned neurons may respond to both types of sounds. If this is the case, a mechanism to transmit information about different types of sounds to higher processing areas may be necessary. One possibility may be differences in temporal firing patterns to different frequencies (unpubl. observations). Hypotheses regarding the function of multiply tuned neurons remain to be tested.

Combination sensitivity may be a mechanism for encoding complex sounds. Previous to our work, there has been little evidence for combination-sensitive neurons in the IC of non-bat mammals. This is most likely due to previous studies not specifically testing for this type of response rather than a lack of these neurons in non-bat mammals. Combination-sensitive neurons have been described in the IC of the mustached bat in frequency regions representing echolocation signals and in "non-sonar" regions, and consequently are thought to be important for analyses of echolocation and communication signals (Portfors and Wenstrup 1999, 2002). Combination-sensitive neurons in the mouse IC may also be important for analyzing social vocalizations, but may also be important for analyses of other types of complex sounds in the environment such as predator-generated sounds. Further studies are necessary to elucidate the importance of combination-sensitive neurons in the mouse IC.

ACKNOWLEDGEMENTS

This work was supported by National Institute of Health grant R01 DC 04733 from the National Institute on Deafness and Other Communication Disorders.

REFERENCES

Egorova, M., Ehret, G., Vartanian, I., and Esser, K. H., 2001, Frequency response areas of neurons in the mouse inferior colliculus. *Exp. Brain. Res.* **140**: 145-161.

LeBeau, F. E. N., Malmierca, M., and Rees, A., 2001, Iontophoresis in vivo demonstrates a key role for GABAA and glycinergic inhibition in shaping frequency response areas in the inferior colliculus of the guinea pig. *J. Neurosci.* **21**: 7303-7312.

Portfors, C. V., and Wenstrup, J. J., 1999, Delay-tuned neurons in the inferior colliculus of the mustached bat: implications for analyses of target distance. *J. Neurophysiol.* **82**: 1326-1338.

Portfors, C. V., and Wenstrup, J. J., 2002, Excitatory and facilitatory frequency response areas in the inferior colliculus of the mustached bat. *Hear. Res.* **168**: 131-138.

Role of KCC2 in Auditory Processing of the Brainstem

MOTOI KUDO, TAKAAKI NAKAMURA, and KIYOSHI KUROKAWA
Department of Anatomy, Shiga University of Medical Sciences, Seta, Otsu 520-2192, Japan

1. INTRODUCTION

GABA and/or Glycine have been thought to be inhibitory neurotransmitter. However, depending on the chloride-ion (Cl^-) concentration in post-synaptic neurons, GABA and/or Glycine do act as excitatory. In such chloride-ion regulation, potassium chloride co-transporter (KCC2) does play a main role in the central nervous system (Rivera *et al.*, 1999). Considering the fact that the central auditory pathways contains Glycine acting channels far much more than other sensory pathways, it is implicated that the Glycine channels have anything else than simple and persistent inhibitory role particularly in the auditory brain stem.

In the lateral superior olive (LSO), it has been shown that the cell property changes from inhibitory to excitatory with the developmental expression of KCC2 (Kakazu *et al.*, 1999; Balakrishnan *et al.*, 2003). These data show that developmental changes of the co-transporter alter intracellular Cl^- concentration and are responsible for the switch from the neonatal Cl^- to the mature Cl^- influx in LSO neurons. Such maturational changes in Cl^- co-transporters might have the important functional roles for Glycinergic and GABAnergic synaptic transmission and the broader implications for auditory development.

In the inferior colliculus (IC), a largest auditory center of the brainstem, no data is available about KCC2. We then examined the developmental expression of KCC2 using postnatal rats. Morphological properties on GABAnergic neurons were also observed in comparison between the IC and the cerebellum.

2. MATERIAL AND METHODS

Anti-KCC2 polyclonal antibodies-KCC2 fusion protein was prepared, containing a 112-amino acid segments of the carboxyl terminus of KCC2 (amino acids 932-1043). The sequence encoding the fusion protein was amplified in a polymerase chain reaction.

The fusion protein was produced in E. coli according to the manufacture's instructions (Novagen, Madison, WI). Purified KCC2 fusion protein was injected subcutaneously to rabbit with complete Freud's ajuvant. Immune antisera were used for immunohistochemistry.

Immunohistochemical analysis was done as shown previously (Nakamura *et al.*, 2001). Briefly, rats were anesthetized with sodium pentobarbital (100mg/kg b.w.) and perfused for 10 min via a left ventricle with a fixative containing 4% paraformaldehyde, 0.5% glutaraldehyde and 0.2% picric acid in 0.1M phosphate buffer. Brains were dissected and lightly postfixed. Brain slices cut at 30μm thickness with a cryostat were incubated for 5 days with rabbit antibody against KCC2 (our production) or mouse antibody against GAD65 (Chemicon International) diluted 1:500 and 1:3000 at 4°C, respectively. The sections were then incubated for 2 h with two different (red and green) fluorescence dyes conjugated anti-rabbit IgG and anti-mouse IgG (Vector, USA), respectively.

For *in situ hybridization*, Digoxigenin-labeled sense and anti-sense RNAs were generated using T7 polymerase according to the protocol (Balakrishnan *et al.*, 2003). Briefly, frozen sections were sliced in a cryostat, thaw-mounted on slides and fixed in ice-cold 4% paraformaldehyde for 30 min. Sections were transferred several minutes to 70% ethanol and stored at -80°C until used. Before hybridization, tissue sections were permeabilized in 0.1N HCl for 10min. After rinsing, sections were transferred into 70% ethanol for 5min, and then back to 100% ethanol for 5 min. Hybridization occurred at 65°C in a buffer containing 50% formamide, 4 x SCC, 10% dextran sulfate, 5 x Denhardt's solution 200 mg/ml acid-alkali cleaved salmon sperm DNA. Hybridization was performed overnight at 65°C. After RNase treatment, Digoxigenin-labeled probes were detected by using an anti-digoxigenin fragment linked to alkaline phosphatase. The nitroblue-tetrazolium-chloride/5-bromo-4-chlo-indolyl-phosphatase system was used as a substrate for alkaline phosphatase.

For RT-PCR analysis, total RNA was extracted from rat's brains as described previously (Nakamura *et al.*, 2001). We used 5μg total RNA as template to synthesize complementary DNA using RNA reverse transscriptase (Super scriptII, Life Technology, Japan). For the subsequent PCR, 1 μl of the reaction mixture was used each time. The PCR product was identified by agarose gel electrophoresis, and the band of expected size visualized with ethidium bromide was used for quantification.

3. RESULTS

Table 1. Developmental changes of KCC2 in the IC in different item

item	P0	P3	P7	P14	P21	adult
Protein expression Rate of positive cells	-	-	++	++++	+++	+++
mRNA expression *In situ hybridization*	-	+	++	+++	++++	++++
Behavioral events RT-PCR relative value	+	+	+	++	+++	++++

Using postnatal developing rats, we investigated changes of KCC2 expression in the IC. First, we stained the IC neurons by immunohistochemistry using anti-KCC2 antibody. As a control, the cerebellum was stained in the same sections. KCC2 positive neurons in the IC and the cerebellum were counted, and the ratios against Nissl stained were calculated by cell number. In the IC, KCC2 positive cells varied in size, from small to large, and the topographical difference was hardly detected along with the rostro-caudal, medio-lateral, or dorso-ventral axis. The age-related changes are summarized in Table 1. At P0-P3, KCC2 positive neurons were hardly seen. At P7, a considerable number (30-40%) of neurons became KCC2 positive. At P14, around when the physiological hearing become onset in the rat, the KCC2 positive rate became drastically increased up to about 80%. During P21-adults, the rates slightly decreased to 60-70%.

Our second experiments are *in situ hybridization* to identify KCC2 gene expression in the IC neurons. To confirm specificity of our probe made by using RT-PCR of rat brains, we processed the same sections containing the IC and cerebellum. The results obtained concerning the cerebellum are in conformity with that of Mikawa *et al.* (2002). In the IC, as shown in Table 1, even no signal was detected at P0; the signal gradually increased day by day during P3-P21. As in adults, the signal was the strongest. The labeled IC neurons have various morphological characters in size and in shape. They were distributed diffusely throughout the IC.

The third experiment is a comparison of relative abundance of RT-PCR. As a normal control, mRNA of G3PDH protein was used. Although quantification of mRNA by RT-PCR bandings may lack accuracy, it secures the tendency of the previous two experiments. The results showed that KCC2 mRNA during Po-P7 is expressed constantly at certain levels. It became stronger at P14, and it increased rapidly during P14-adults.

4. DISCUSSION

The rat auditory system is immature at birth. During the early postnatal periods, their ears are closed as well as their eyes. Physiological hearing onset is said to occur around P14. Much anatomical and physiological evidence has shown that the maturation of central auditory system does occur such as formation of neuronal circuits, synaptogenesis, and intercellular signal transmission. In the rat, seizing such late maturation of the auditory function, researchers have had favor to use these "golden period" to investigate the ontogeny and mechanisms underlying neuronal development.

It became to be known that Glycine and GABA act as excitatory neuronal transmitter in immature brain structures in various central systems (Balakrismhman *et al.*, 2003). The neuron specific co-transporter, KCC2 is now thought to play major role in rendering excitatory (depolarization) nature to inhibitory (hyperpolarization) one regulating its potassium and/or chloride pumping effects. In parallel with the previous study, we investigated developmental KCC2 expression in the rat IC, which is the key structure of the auditory brain.

A major conclusion of the present experiments is that KCC2 expression in the IC is upregulated during postnatal development. This finding contrasts with that obtained in the LSO (Balakrishnan *et al.*, 2003). While KCC2 expression is upregulated in the neocortex (Wang *et al.*, 2002), the hippocampus (Wang *et al.*, 2002; Balakrishnan *et al.*, 2003) and the cerebellum (Mikata *et al.*, 2002); KCC2 expression is downregulated in the thalamus (Wang *et al.*, 2002). It is interesting to note that the IC, an auditory midbrain situated between the medulla-pons (the LSO) and the diencephalon shows "upregulation" pattern like the neocortex, the hippocampus and cerebellum in KCC2 expression.

A closer look at the age-related change of KCC2 expression estimated by the protein revealed that there is a peak at P14. On the other hand, there is no such tendency that estimated by mRNA. We could not detect any other morphological characters between KCC2 positive cells and negative cells. In additional experiment (data not shown), IC neurons in adult rats were double labeled with KCC2 and GAD 65 using immunohistochemistry. About 30% KCC2 positive cells were double labeled, yet no more interesting data was obtained. Thus, circuit formation mechanisms remained unknown at present to speculate the functional role of KCC2 for auditory processing.

5. CONCLUSION

The present study indicates that the neuro-specific co-transporter KCC2 is expressed abundantly throughout the IC. The postnatal expression pattern of KCC2 is upregulated in the IC as well as that in the neocortex, the hippocampus and the cerebellum. Considering the importance of this co-transporter on controlling auditory function, further studies are needed in all aspects including the neuronal maturation and the switching mechanisms.

ACKNOWLEDGEMENTS

This work was supported by Grant-in-Aid for Scientific Research (C) (2) 15500234 from Japan Society for the Promotion of Science, and by AFIIS project of Doshisha University.

REFERENCES

Balakrishnan, V., Becker, M, Löhrke, S., Nothwang, H. G., Güresir, E., and Friauf, E., 2003, Expression and function of chloride transporters during development of inhibitory neurotransmission in the auditory brainstem. *J. Neurosci.* **23**: 4134-4145.

Kakazu, Y., Akaike, N., Komiyama, S., and Nabekura, J., 1999, Regulation of intracellular chloride by cotransporters in developing lateral superior olive neurons. *J. Neurosci.* **19**: 2843-2851.

Mikawa, S., Wang, C., Shu, F., Wang, T., Fukuda, A., and Sato, K., 2002, Developmental changes in KCC1, KCC2 and NKCC1 mRNAs in the rat cerebellum. *Dev. Brain Res.* **136**: 93-100.

Nakamura, T., and Kishi, A., 2001, Insulin production in a neuroectodarmal tumor that express islet factor-1, but not pancreatic-duodenal homeobox I. *J. Clin. Endo. Meta.* **84**: 1793-1800.

Rivera, C., Voipio J., Payne, J. A., Ruusvuori, E., Lahtinen, H., Lamka, K., Pirvola, U., Saarma, M., and Kalia, K., 1999, The K^+/Cl^- co-transporter KCC2 renders GABA hyperpolarizing during neuronal maturation. *Nature* **397**: 251-255.

Wang, C., Shimizu-Okabe, C, Watanabe, K., Okabe, A., Matsuzaki, H., Ogawa, T., Mori, N., Fukuda, A., and Sato, K., 2002, Developmental changes in KCC1, KCC2 and NKCC1 mRNAs in the rat brain. *Dev. Brain Res.* **139**: 59-66.

Spatial and Functional Properties of Neuronal Responses to Simulated Sound Source Motion in the Inferior Colliculus of the Cat

ELENA A. RADIONOVA
Laboratory of Hearing Physiology, I.P.Pavlov Institute of Physiology, Russian Academy of Sciences, Nab. Makarova, St.Petersburg, Russia

1. INTRODUCTION

Unlike the information on stationary sound source localization, up to now only a few publications mention some fragmentary data on spatial and functional organization of the auditory structures involved in localization of moving sound sources (Nikitin and Popelář, 1981; Altman and Kalmykova, 1986; Griffits *et al.*, 1998). In view of this fact the present study concerns spatial and functional organization of the cat's inferior colliculus (IC) under conditions of stimulation with sound signals simulating sound source motion (Radionova, 2000, 2003).

2. METHODS

Experiments were performed on nine tracheotomized anaesthetized cats (chloralose and nembutal, 30 and 12 mg/kg, respectively) with normal hearing. With binaural click series presentation, 475 series of focal evoked potentials, EP-series, from 45 points of the IC central nucleus were recorded (for 20-40 times each) with low-resistance microelectrodes.

Sound signals simulating sound source motion were binaurally presented click series of 1050 ms duration with gradually changing interaural time delay (ΔT) between the clicks of the click trains. Click repetition rate was near 20/s with click duration of 100 μs. Click series intensity did not exceed 80 dB SPL (when listened monaurally by humans). ΔT-value varied throughout the signal duration in a linear manner from a maximal value (within 2.5 ms) at the train onset to zero at the middle of the click trains and then backwards to the initial value at the end of the trains. Such signal is perceived by humans as a fused auditory image (FAI) moving from the leading ear to the head midline and backwards. The greater was the initial ΔT-value the higher was the perceived velocity of FAI motion. Since in the cat ΔT-values under natural conditions vary within about ±350 μs, in the present work calculated values of FAI motion velocity lay within 50-1200 grad/s and corresponding time of FAI motion in both directions was within 1050-150 ms. Electrostatic sound emitters with identical characteristics (±4 dB within 0.1-20 kHz, measurements with the Brüel and Kjaer 4135 microphone) were used, their tips were tightly inserted into external acoustic canals of the cat's ears.

Motion effect (ME) was defined as amplitude change in the EP-series throughout the time of FAI movement (Fig. 1). ME was estimated as slight, moderate or strong when EP amplitude changed for about 10-15%, 20 or 30% and more, respectively.

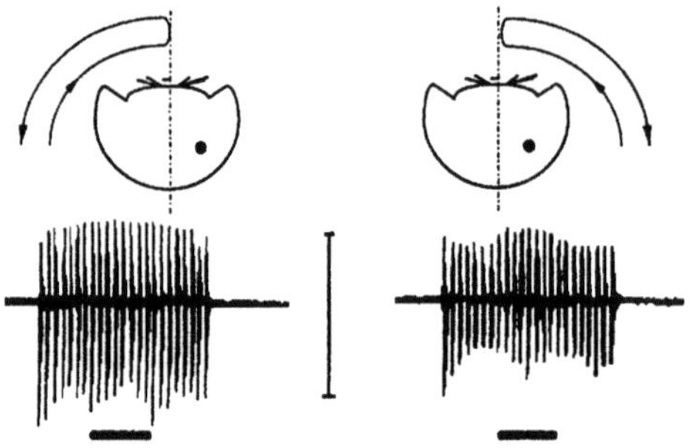

Figure 1. Motion effects (MEs) as EP-amplitude changes over the time of FAI movement (black bars, about 370 ms) on the left and on the right head sides (contr- and ipsilaterally in relation to the registration point, respectively). Amplitude calibration 0.5 mV.

3. RESULTS AND DISCUSSION

Motion effect (ME) in IC central nucleus was observable in 77% of 45 registration points, with neuronal characteristic frequencies of 0.6-11 kHz in these points. ME was most pronounced at rather high FAI movement velocities of about 360-540 grad/s.

It was found that ME was connected with prevailing effectiveness of the contralateral stimulus (when presented alone) over the ipsilateral one (Contr>Ipsi): spatial distributions of MEs and cases when Contr>Ipsi along the electrode track proved rather near (Fig. 2).

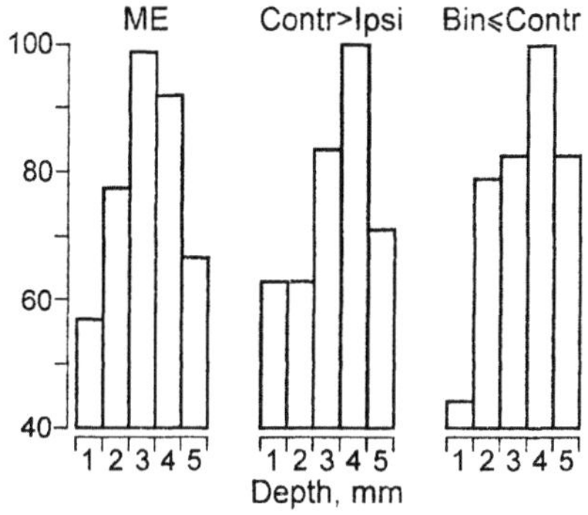

Figure 2. Motion effect (ME) over the depth of its registration in comparison with relative effectiveness of monaural and binaural stimuli.

Prevailing effectiveness of the contralateral stimulus over the ipsilateral one seems of importance for ME appearance. For instance, higher effectiveness of the contralateral stimulus (Contr>Ipsi) may be provided by higher intensity of the contralateral affererentation flow (as is well known, contralateral neuronal responses, EPs including, in about 1/3 of cases are higher in amplitude than ipsilateral responses). This would result in shorter latencies of contralateral responses in case of equal length values of both monaural pathways. However in general, contralateral pathway is longer than ipsilateral one and thus contralateral response latency would be not shorter but rather near to ipsilateral response latency ($\Delta T \rightarrow 0$). In this case, when $\Delta T \rightarrow 0$, neuron sensitivity to ΔT changes would be maximal since even a very low ΔT change (deviation from zero) will be detected.

Thus, prevalence of contralateral stimulus effectiveness may be connected with minimization of ΔT value and thereon with increase of neuronal ΔT-sensitivity which secures ME appearance.

ME distribution over the IC central nucleus (Fig. 2, ME) was also rather near to spatial distribution of cases when binaural suppression of monaural afferentations took place, i.e. when response to binaural stimulation did not exceed the response to more effective of the monaural stimuli (Fig. 2, Bin≤Contr). It may be concluded that ME is connected with binaural suppression effect.

The data presented on Figs. 2 3 show to a certain topical organization of the ME. Strong ME was observable most often at 3-4 mm depth along the electrode track (Fig. 3, ME+), whereas slight and moderate motion effects were uniformly distributed over the whole volume of the IC central nucleus (Fig. 3, Me). It should be mentioned that at the depth of 3-4 mm ME was considerably more often observed in the central part of the IC central nucleus than in its medial and lateral parts; ME amplitude was also considerably higher in the central part of the nucleus.

It seems essential that strong motion effect (ME+) was best pronounced at 3-4 mm depth of the electrode track (Fig. 3, ME+), where also great multipolar neurons were presented to the greatest extent (Fig. 3, N). As it was mentioned earlier (Radionova, 2000, 2003) just in this region great multipolar neurons receive significantly more intensive contralateral than ipsilateral influences. This property of multipolar neurons is in good correlation with the above ME property, namely with prevalence of the contralateral stimulus effectiveness over effectiveness of the ipsilateral stimulus (Fig. 2, Contr>Ipsi).

Unlike the ME+, uniform distribution of slight and moderate motion effects (Fig. 3, Me) coincides with distribution of smaller cells of elongated form (the so called "basic cells") which are uniformly distributed over the whole volume of the IC central nucleus.

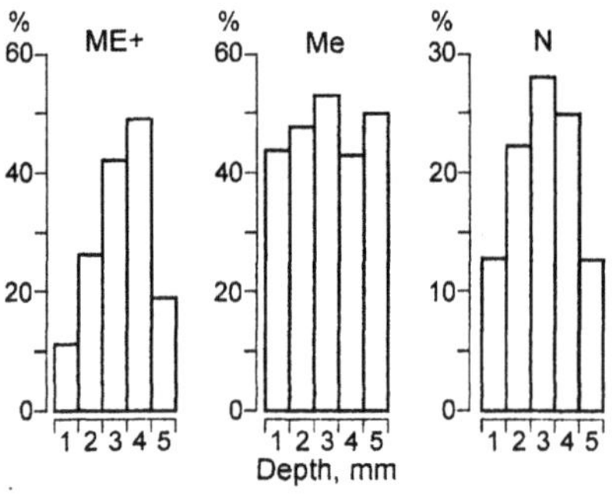

Figure 3. Strong motion effect (ME+), slight and moderate motion effects (Me) over the depth of their registration in comparison with analogous distribution of great multipolar neurons (N).

4. CONCLUSION

Thus motion effect in the IC central nucleus is topically organized (Fig. 2, ME; Fig. 3, ME+, Me). ME is connected with prevalence of contralateral over ipsilateral stimulus effectiveness (Fig. 1, Contr>Ipsi), as well as with binaural suppression of afferentation (Fig. 1, Bin≤Contr). Strong motion effect is connected with activity of great multipolar neurons (Fig.3, ME+, N) whereas slight and moderate motion effects (Fig. 3, Me) seem to be connected with activity of basic cells which are also uniformly distributed over the IC central nucleus.

ACKNOWLEDGEMENTS

This work was supported by the grant from the Russian Fond for Fundamental Investigations 03-06-80371.

REFERENCES

Altman, J. A., and Kalmykova, I. V., 1986, Role of the dog's auditory cortex in discrimination of the sound signals simulating sound source movement. *Hear. Res.* **24**: 243-253.

Griffits, T. D., Rees, J., Rees, A., and Green, T. J. R., 1998, Right parietal cortex is involved in the perception of sound movement in humans. *Nature Neurosci.* **1**: 74-79.

Nikitin, N. I., and Popelář, J., 1981, Responses of neurons from the cat's inferior colliculus to sound signals with interaural time differences. *Physiol. J. of the USSR* **67**: 672-679 (in Russian; English Summary).

Radionova, E. A., 2000, Topical organization of the cat's inferior colliculus under conditions of imitated sound source motion. *Russian Physiol. J.* **86**: 762-772 679 (in Russian; English Summary).

Radionova, E. A., 2003, *Experiments in Hearing Physiology*. Pavlov Institute of Physiology Press, Saint Petersburg (in Russian; English Contents, English Summary).

Temporal and Rate Representations of Time-Varying Signals in Auditory Cortex

XIAOQIN WANG, THOMAS LU, and LI LIANG
Department of Biomedical Engineering, Johns Hopkins University School of Medicine, 720 Rutland Avenue, Ross 424, Baltimore, MD21205, USA

1. INTRODUCTION

The neural representation of time-varying signals in the auditory cortex is of special interest to our understanding of mechanisms underlying speech processing. Time-varying signals are fundamental components of communication sounds such as human speech and animal vocalizations, as well as musical sounds. Low-frequency modulations are important for speech perception and melody recognition, while higher-frequency modulations produce other types of sensations such as pitch and roughness (Houtgast and Steeneken, 1973; Rosen, 1992). Both humans and animals are capable of perceiving the information contained in temporally modulated sounds across a wide range of time scales from millisecond to tens and hundreds of milliseconds. How the auditory cortex encodes this wide dynamic range of temporal modulations is not well understood.

Neural representations of time-varying signals begin at the auditory periphery where auditory-nerve fibers faithfully represent fine details of complex sounds in their temporal discharge patterns (Johnson, 1980; Joris and Yin, 1992; Palmer, 1982; Wang and Sachs, 1993). At subsequent nuclei along the ascending auditory pathway (CN-cochlear nucleus, IC-inferior colliculus and MGB-auditory thalamus), the precision of this temporal representation gradually degrades (e.g., CN: Blackburn and Sachs, 1989; Frisina *et al.*, 1990; Wang and Sachs, 1994, IC: Langner and Schreiner, 1988, MGB: Creutzfeldt *et al.*, 1980; de Ribaupierre *et al.*, 1980), due to biophysical properties of neurons and temporal integration of converging inputs from one station to the next.

It has long been noticed that neurons in the auditory cortex do not faithfully follow rapidly changing stimulus components (de Ribaupierre *et al.*, 1972; Goldstein *et al.*, 1959; Whitfield and Evans, 1965). A number of previous studies have showed that cortical neurons can only be synchronized to temporal modulations at a rate far less than 100 Hz (Bieser and Müller-Preuss, 1996; de Ribaupierre *et al.*, 1972; Eggermont, 1991; Eggermont, 1994; Gaese and Ostwald, 1995; Lu and Wang, 2000; Schreiner and Urbas, 1988), compared with a limit of ~1 kHz at the auditory-nerve (Joris and Yin, 1992; Palmer, 1982). The lack of synchronized cortical responses to rapid, but perceivable temporal modulation has been puzzling. Because most of the previous studies in the past three decades on this subject were conducted in anesthetized animals, with a few exceptions (Bieser and Müller-Preuss, 1996; Creutzfeldt *et al.*, 1980; de Ribaupierre *et al.*, 1972; Evans and Whitfield, 1964; Goldstein *et al.*, 1959; Whitfield and Evans, 1965), it has been speculated that the reported low temporal response rate in auditory cortex might be caused partially by anesthetics, which have been shown to alter the temporal responses properties of the auditory cortex (Goldstein *et al.*, 1959; Zurita *et al.*, 1994). Neural responses obtained under unanesthetized conditions are therefore of particular importance to our understanding of cortical representations of temporally modulated signals. This chapter summarizes recent findings from our laboratory regarding the issues related to cortical representations of time-varying signals.

2. SUMMARY OF RECENT FINDINGS

2.1 Limited Stimulus-Synchronized Discharges and Rate Responses in Auditory Cortex: A Two-Stage Mechanism in Representing Time-Varying Signals

We have re-examined cortical representations of time-varying signals in our recent experiments in awake marmoset monkeys in our recent studies. We systematically investigated responses of single neurons in the primary auditory cortex (A1) to rapid sequences of clicks and other repetitive stimuli (Lu *et al.*, 2001b, Liang *et al.*, 2002). Both wide- and narrow-band click trains with inter-click intervals (ICI) ranging from 3 ms to 100 ms were studied. Narrow-band clicks were centered at each neuron's characteristic frequency (CF). One type of response to click trains is illustrated in Fig.1A.

This neuron exhibited significant stimulus-synchronized responses to click stimuli at long ICIs (>25-30 ms). The discharges to click trains became non-synchronized at medium ICIs and diminished at short ICIs (< 20-25 ms), apparently due to inhibition (Fig.1A,C). The second type of cortical response did not exhibit stimulus-synchronized discharges (Fig.1B,D), but instead showed monotonically changing discharge rate when the ICI was shorter than ~15 ms. We have shown that the limit on stimulus-synchronized responses is on the order of 20-25 ms (median value of sampled population) in A1 of awake marmosets (Lu *et al.*, 2001b). The observation that neurons are sensitive to changes of short ICIs indicates that a discharge rate-based mechanism may be in operation when ICIs are shorter than ~20-30 ms.

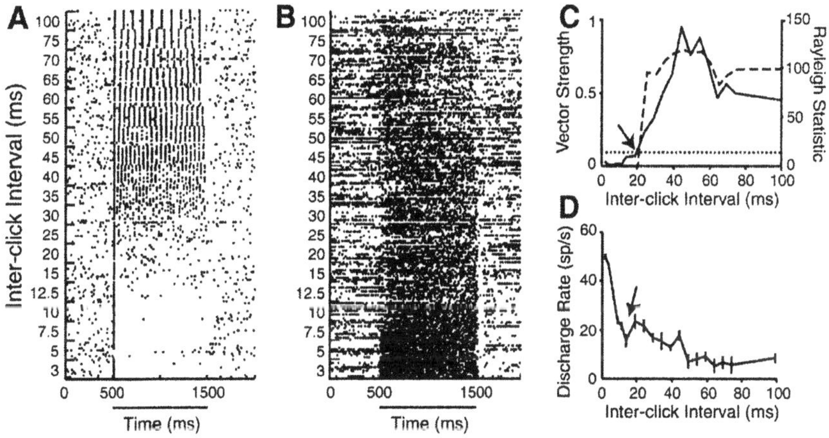

Figure 1. A, B: Examples of stimulus-synchronized and non-synchronized responses, respectively, to click trains recorded from the primary auditory cortex (A1) of awake marmosets. C: Vector strengths (dashed line) and Rayleigh statistics (solid line) analyzed for the stimulus synchronized responses shown in A. The dotted line (at the Rayleigh statistics of 13.8) indicates the threshold for statistically significant stimulus-synchronized activity ($p<0.001$). A synchronization boundary was calculated and is indicated by an arrow. D: Driven discharge rate plotted versus inter-click-interval for the non-synchronized responses shown in B. Vertical bars represent standard errors of the means (SEM). The arrow indicates calculated rate-response boundary. Results shown are based on Lu *et al.* (2001b).

Based on the above two response types (Fig.1A,B), we have identified two populations of A1 neurons, referred to as *synchronized* and *non-synchronized* populations, respectively (Fig.2). The two populations appeared to encode sequential stimuli in very different manners, by spike timing or by average discharge rate. Neurons in the synchronized population showed stimulus-synchronized discharges at long inter-click intervals, but few responses at short inter-click intervals. This population of neurons can thus represent slowly occurring temporal events *explicitly* using a temporal code. The non-synchronized population of neurons did not exhibit stimulus-synchronized discharges at either long or short inter-click intervals. This population of neurons can *implicitly* represent rapidly changing temporal intervals by their average discharge rates. Because there is an overlap between the encoding domains of these two populations of neurons, the auditory cortex can represent a wide range of time intervals in sequential, repetitive stimuli (Fig.2).

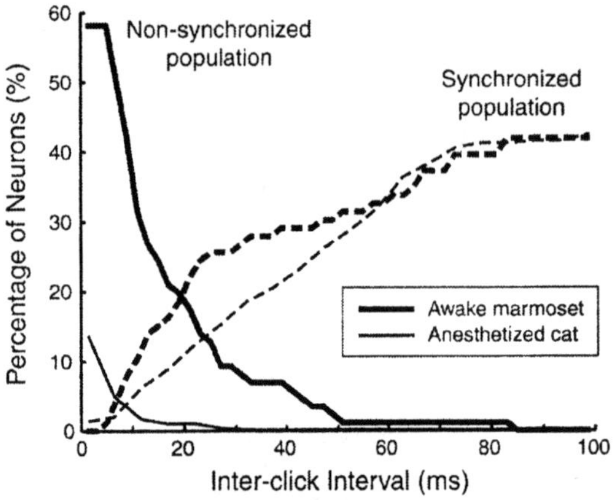

Figure 2. A combination of temporal and rate representations can encode a wide range of inter-click-intervals (ICI). Thick curves are the cumulative sum of the histograms representing response boundaries of two neural populations recorded in awake marmosets, one with stimulus-synchronized discharges (N=36) and the other with non-synchronized discharges (N=50), respectively (Lu *et al.*, 2001b). The dashed line shows the percentage of neurons with synchronization boundaries less than or equal to a given ICI. The solid line shows the percentage of neurons with rate-response boundaries greater than or equal to a given ICI. Thin curves show the data obtained from A1 of anesthetized cats using click train stimuli (Lu and Wang, 2000), analysed in the same ways as the data from marmosets.

2.2 Contrast of Response Characteristics Between Awake and Anesthetized Animals

Goldstein *et al.* (1959) showed that click-following rates of cortical evoked potentials were higher in unanesthetized cats than in anesthetized ones. We have studied responses of A1 neurons to click train stimuli in both awake marmosets (Lu *et al.,* 2001b) and anesthetized cats (Lu and Wang, 2000) in our laboratory. There were several important differences between response properties observed in these two preparations. First, in contrast to

A1 neurons in anesthetized cats, which responded strongly to both wide- and narrow-band clicks, the majority of A1 neurons in awake marmosets responded weakly or, more often, were unresponsive to wide-band (rectangular) clicks, but could be strongly driven by narrow-band clicks. Using narrow-band clicks with various bandwidths, we were able to determine that the lack of responses to rectangular clicks was due to activations of side-band inhibitions by these wide-band stimuli. It appeared that such side-band inhibitions were much stronger in A1 of awake animals than in anesthetized animals. Second, stimulus-following rates were higher in awake marmosets than in A1 of anesthetized cats (Fig.2). This finding is consistent with the earlier finding showed by Goldstein *et al.* (1959). Third, and most importantly, the large number of neurons with non-synchronized and sustained discharges at short ICIs observed in A1 of awake marmosets were not observed in A1 of anesthetized cats (Fig.2). While our comparisons were made between A1 of two species, we suspect that such response differences resulted largely from two different experimental conditions (awake versus anesthetized) rather than from two different species. A1 of both species appear to share many similar anatomical and physiological properties. It is also possible that these response differences partially resulted from laminar differences. In our recordings with awake marmosets, neurons were mostly recorded from upper cortical layers (II-III) whereas the recordings in anesthetized cats were made primarily in middle cortical layers (IV).

In summary, our observations and those of others have showed that discharge synchronization rates of auditory cortical neurons are higher in awake animals than in anesthetized animals. Non-synchronized responses are largely absent in anesthetized animals.

2.3 Differentiation of Acoustic Transients by Discharge Rate-Based Representations

The experiments discussed above suggest that A1 neurons integrate stimulus components within a time window of ~30 ms and treat components outside this window as discrete acoustic events. Humans and animals are known to discriminate changes in acoustic signals at time scales shorter than the temporal integration window of A1 neurons. Cortical neurons therefore must be able to signal such rapid changes. To demonstrate cortical neurons' sensitivity to rapid temporal changes within the putative temporal integration window, we have studied A1 neurons in awake marmosets using a class of temporally modulated signals termed *ramped and damped sinusoids.* Patterson and colleagues characterized psychophysical performance of human subjects in discriminating ramped versus damped sinusoids (Patterson, 1994a; Patterson, 1994b), therefore providing a basis for comparing cortical responses in differentiating these temporal asymmetric stimuli.

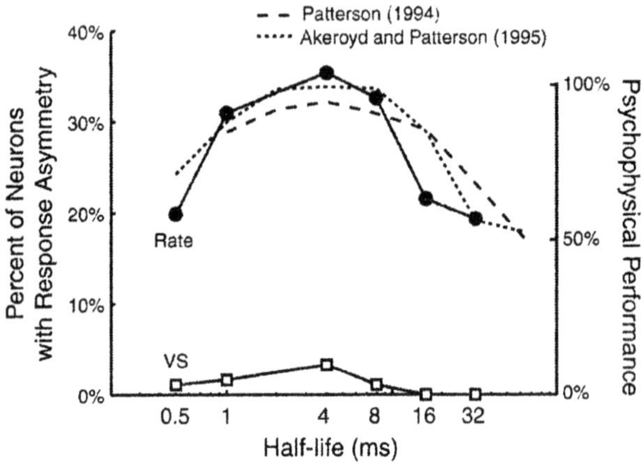

Figure 3. Comparison between asymmetry preference of cortical neurons and human psychophysical performance in discriminating ramped and damped sinusoids (Lu *et al.*, 2001a). The percentages of neurons (left ordinate) having significant asymmetry indices based on discharge rate are indicated as rate (solid line with filled circles). The percentages of neurons with significant asymmetry indices in vector strength are indicated as VS (solid line with open squares). Human psychophysical performance curves (right ordinate) using ramped and damped sinusoids is shown for tone carriers (dashed line), averaged over the different carrier frequencies (Patterson, 1994a), and noise carriers (dotted line) (Akeroyd and Patterson, 1995).

A damped sinusoid consists of a pure tone amplitude-modulated by an exponential function. It has a fast onset followed by a slow offset. The rate of amplitude decay is determined by the exponential half-life. A ramped sinusoid is a time-reversed damped sinusoid. Both types of sounds have identical long-term Fourier spectra. Our experimental stimuli consisted of ramped or damped sinusoid segments, typically with a period of 25 ms, repeated consecutively for 500 ms. For each neuron, the carrier frequency was set to a neuron's CF and the half-life was varied from 0.5 to 32 ms. Most cortical neurons that we studied showed clear preference for either ramped or damped sinusoids (i.e., they responded more vigorously to a single stimulus type), with a greater portion of neurons preferring ramped stimuli (Lu *et al.*, 2001a). Some neurons responded nearly exclusively to one stimulus type (ramped or damped). The response asymmetry was observed in average discharge rate, but not necessarily in stimulus-synchronized discharges. Generally, preference for stimulus type was consistent across half-lives. These observations demonstrated that temporal characteristics within the temporal integration window could profoundly modulate a cortical neuron's responsiveness.

In Fig.3 we compare response asymmetry of populations of A1 neurons with the psychophysical performance in discriminating ramped versus damped sinusoids by humans (Lu *et al.*, 2001a). The shape of the curve based on average discharge rate is qualitatively similar to psychophysical data with both tone carriers (Patterson, 1994a) and wide-band noise carriers (Akeroyd and Patterson, 1995). The psychophysical performance across the half-life appears to be related to the percentage of A1 neurons that showed significant response asymmetry in their average discharge rates. A population measure based on discharge synchrony, on the other hand, reveals that only a very small portion of A1 neurons (<5%) showed response asymmetry in their temporal discharge patterns for the stimulus period used (25 ms).

These results show that discharge-rate based cortical representations can serve as the basis to discriminate rapid acoustic transients, in the absence of stimulus-synchronized responses.

2.4 How Precise Is Spike Timing in Auditory Cortex Neurons?

To further explore the role of spike timing in two populations of cortical neurons, spike times over the duration of the stimulus were uniformly randomized, and the entropies of their ISI distributions were computed (Lu and Wang, 2004). The entropies of the unaltered spike times were then compared to those from the randomized version. The randomness, and therefore the entropy, of a spike train would increase if temporal structures were originally present in the spike train. The example in Fig.4A-D shows the effect of this manipulation on a synchronized unit. The discharge rate remained the same (Fig.4C) but any synchronized responses were eliminated (Fig.4D). The result of the randomization of spike times on the population of synchronized neurons was that they no longer had stimulus-synchronized spike trains (Fig. 4E).

The ratio of the entropy of the unaltered spike train over the entropy of the randomized spike train was used as a measure to indicate the amount of spike timing information contained in a stimulus-evoked spike train. The average entropy ratio, calculated from the two populations of neurons, respectively, is plotted against the click train ICI (Fig.4F, lines with symbols). For comparison, the entropy ratio was also calculated for spontaneous discharges (Fig.4F, dashed line). Note that the entropy ratios for spontaneous discharges had values less than 1, indicating that spontaneous firing was not distributed completely randomly. In the case of the synchronized population (Fig.4F, solid line with open circles), as the ICI increased, the entropy ratio began to drop near 30 ms (the mean ability of neurons to synchronize reported in Lu *et al.*, 2001b was about 21 ms ICI) and became significantly different from the entropy ratio of spontaneous discharges at 50 ms. It continued to be significantly different for all larger values of ICI, having a value nearly 0.85 at 100 ms ICI. This difference in entropy ratios demonstrated that the randomness of the spike trains for the synchronized population was less than that of their randomized versions at ICIs longer than ~30 ms. The curve for the average entropy ratio of the non-synchronized population was fairly flat over the range of ICIs tested (Fig.4F, solid line with crosses). In addition, none of the ratios were significantly different from that of the spontaneous discharges. The fact that the entropy ratio of the non-synchronized population was consistently higher than that of spontaneous discharges across ICIs indicated that stimulus driven discharges in these neurons were at least as random as those of spontaneous firing.

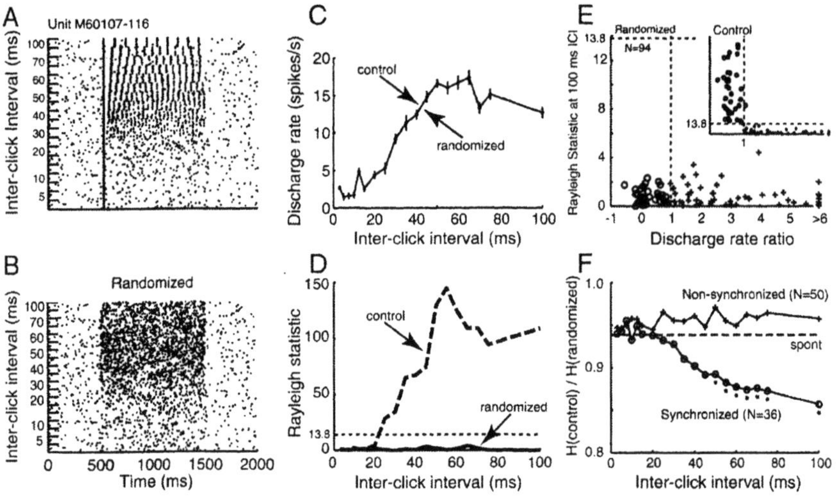

Figure 4. A: Dot raster of a cortical neuron's responses to click train stimuli. This neuron had stimulus-synchronization to regular click trains in its responses. B: Dot raster of A after spike time randomization. C: Effect of randomization on discharge rate. Curves from randomized and control responses overlap exactly. D: Effect of randomization on stimulus-synchronization measure, Rayleigh statistic. E: The effects of randomization on the two populations of neurons. Synchronized neurons are marked with open circles, and non-synchronized neurons are marked with '+'s. The abscissa is the discharge rate ratio, the ratio

of the discharge rate for a 3 ms ICI click train to that at 100 ms ICI. The ordinate is the Rayleigh statistic calculated at 100 ms ICI. Values above the dashed line at 13.8 indicate statistically significant stimulus-synchronization. The inset shows the results from the unmodified control responses. **F:** Entropy ratio of control (unaltered) to randomized spike times as a function of inter-click interval for synchronized units (line with circles) and non-synchronized units (line with crosses), respectively. The curves represent the averaged entropy of the control divided by the average entropy of the randomized spike times. The horizontal dashed line indicates the entropy ratio computed from the spontaneous activity prior to the stimuli for synchronized and non-synchronized units. Points that were significantly different from the entropy ratio of spontaneous activity are indicated with '*'s ($p<0.05$, *Wilcoxon ranksum*). Results shown are based on Lu and Wang (2004).

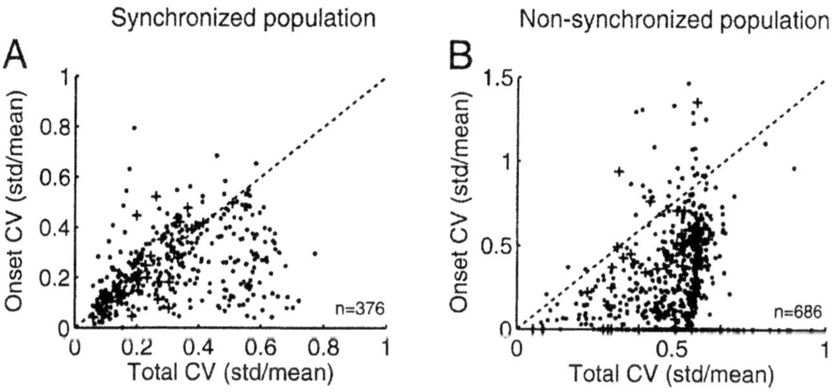

Figure 5. Spike timing precision of responses to click-trains. A: Mean CV (standard deviation of the latency divided by mean latency) of the onset response (median=0.22, [25%,75%]=[0.13, 0.35]) versus the mean CV of the total response (median=0.30, [25%,75%]=[0.15, 0.48]) to regular click trains, for the synchronized population. Each data point represents the response from a single stimulus ICI. '+'s indicate responses to the 100 ms ICI click train. B: Same analyses as in A for the non-synchronized population. *Onset*: median=0.26, [25%,75%]=[0.07, 0.50]. *Total*: median=0.53, [25%,75%]=[0.44, 0.58]. Results shown are based on Lu and Wang (2004).

We quantified the precision of spike timing of the responses to click train stimuli for each of the two populations of neurons. In general, spike timing dispersion was smaller at stimulus onset than at successive stimulus events for an ongoing stimulus. Fig.5A shows that the CVs calculated from the onset responses of regular click trains were significantly ($P<<0.001$, *Wilcoxon ranksum*) smaller than the CVs calculated from the total responses. The results shown in Fig.5A and 5B indicate that, for synchronized units, spike time precision was higher in the onset response and in responses to more sparsely occurring stimulus events. The same analyses were applied to non-synchronized units (Fig.5B). Because many non-synchronized neurons did not have a precise onset response, the data points for onset CV were widely scattered while the data points for the total CV tended to gather near 0.5. These results show that, compared to synchronized units, non-synchronized units had greater spike timing dispersion in both onset (first 50 ms window) and sustained responses.

In summary, spike timing on the occurrence of acoustic events is more precise at the first event than at successive ones, and more precise with sparsely distributed events (longer time intervals between events) than with densely packed events. These results indicate that auditory cortical neurons mark sparse acoustic events (or onsets) with precise spike timing and transform rapidly occurring acoustic events into firing rate-based representations.

3. FUNCTIONAL IMPLICATIONS OF TEMPORAL-TO-RATE TRANSFORMATION IN AUDITORY CORTEX

The reduction of the upper limit on stimulus-following discharges is an inevitable consequence of temporal integration at each processing stage along the ascending pathway. In a modelling study of the transformation of temporal discharge patterns from the auditory nerve to the cochlear nucleus, Wang and Sachs (1995) showed that the reduction of phase-locking in cochlear nucleus stellate cells (*chopper* units) can result from three mechanisms: convergence of subthreshold inputs on the soma, inhibition, and the well-known dendritic low-pass filtering (Rall and Agmon-Snir, 1998). Figure 6 shows results of simulations where reduced stimulus-synchronized activities produced by the biophysical model of stellate cells are compared with those observed experimentally in AVCN chopper units (Wang and Sachs, 1995). These basic mechanisms may also operate at successive nuclei leading to the auditory cortex, progressively reducing the temporal limit of stimulus-synchronized responses.

The significant reduction in the temporal limit on stimulus-synchronized discharges at the auditory cortex, as compared with the auditory periphery, has important functional implications.

First, it shows that considerable temporal-to-rate transformations have taken place by the time auditory signals reach the auditory cortex. The importance of "non-synchronized" neural responses is that they represent processed (instead of preserved) information.

Second, it suggests that cortical processing of sound streams operates on a "segment-by-segment" basis rather than on a "moment-by-moment" basis as found in the auditory periphery. This is necessary for complex integration to take place at this level of the auditory system, since higher-level processing tasks require temporal integration over a time window preceding and following a particular time of interest.

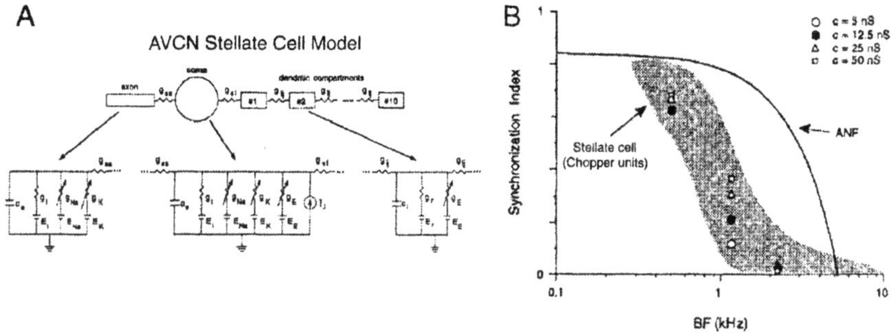

Figure 6. A: A compartmental model of anteroventral cochlear nucleus (AVCN) stellate cells (*chopper* units) with the axon compartment connected by axial conductance g_{as} to the soma, the soma connected by axial conductance g_{s1} to compartment 1, and so on. g_{ij} is the axial conductance connecting compartment i and j. Corresponding electrical circuit representations for the axon compartment, the soma, and the i-th dendritic compartment are also shown. Model parameters: C_a (membrane capacitance of axon), C_s (membrane capacitance of soma), C_i (membrane capacitance of the i-th compartment), g_r (resting conductance), g_l (leakage conductance), g_{Na} (voltage-dependent sodium conductance), g_k (voltage-dependent potassium conductance), g_E (excitatory conductance), E_r (resting potential), E_l (leakage equilibrium potential), E_{Na} (sodium equilibrium potential), E_k (potassium equilibrium potential), E_E (reversal potential for excitatory inputs), and I_J (constant current source for somatic inhibition). B: Synchronization index of the model output is plotted versus the best frequency (BF) of the input auditory-nerve fibers (ANFs) for several conductances (*c = 5, 12.5, 25 and 50 nS*). Higher conductance values simulate stronger synaptic efficacy for inputs. Ten ANFs were placed at the soma as the model inputs in these simulations. Three BFs were tested (0.5, 1.15, and 2.22 kHz, respectively). The shaded area indicates the distribution of the BF synchronization index across frequency in AVCN chopper units (based on the data from Blackburn and Sachs, 1989). The solid line is the least-squares fit to the ANF synchrony data of Johnson (1980) by the use of units with BFs ≤ 6.0 kHz. Results shown are based on Wang and Sachs (1995).

Third, the reduction in the temporal limit on stimulus-synchronized discharges at the auditory cortex is a prerequisite for multi-sensory integration in the cerebral cortex. Auditory information is encoded at the periphery at a much higher temporal modulation rate than the rates at which visual or tactile information is encoded at the periphery, but discharge synchrony rates are similar across sensory cortical areas. The slow-down of temporal response rate along the ascending auditory pathway is necessary for rapid auditory information to be integrated in the cerebral cortex with information from other sensory modalities that is intrinsically slower.

ACKNOWLEDGEMENTS

The research in our laboratory was supported by grants from NIH-NIDCD. We thank Dr. Ross Snider and Steve Eliades for their contributions to the data acquisition system used in these studies and Ashley Pistorio for assistance with animal training.

REFERENCES

Akeroyd, M. A., and Patterson, R. D., 1995, Discrimination of wideband noises modulated by a temporally asymmetric function. *J. Acoust. Soc. Am.* **98**: 2466-2474.

Bieser, A., and Muller-Preuss, P., 1996, Auditory responsive cortex in the squirrel monkey: neural responses to amplitude-modulated sounds. *Exp. Brain Res.* **108**: 273-284.

Blackburn, C. C., and Sachs, M. B., 1989, Classification of unit types in the anteroventral cochlear nucleus: post-stimulus time histograms and regularity analysis. *J. Neurophysiol.* **62**: 1303-1329.

Creutzfeldt, O., Hellweg, F.-C., and Schreiner, C., 1980, Thalamocortical transformation of responses to complex auditory stimuli. *Exp. Brain Res.* **39**: 87-104.

de Ribaupierre, F., Goldstein, M. H. Jr., and Yeni-Komshian, G., 1972, Cortical coding of repetitive acoustic pulses. *Brain Res.* **48**: 205-25.

de Ribaupierre, F., Rouiller, E., Toros, A., and de Ribaupierre, Y., 1980, Transmission delay of phase-locked cells in the medial geniculate body. *Hear. Res.* **3**: 65-77.

Eggermont, J. J., 1991, Rate and synchronization measures of periodicity coding in cat primary auditory cortex. *Hear. Res.* **56**: 153-67.

Eggermont, J. J., 1994, Temporal modulation transfer functions for AM and FM stimuli in cat auditory cortex. Effects of carrier type, modulating waveform and intensity. *Hear. Res.* **74**: 51-66.

Evans, E. F., and Whitfield, I. C., 1964, Classification of unit responses in the auditory cortex of the unanesthetized and unrestrained cat. *J. Physiol.* **171**: 476-493.

Frisina, R. D., Smith, R. L., and Chamberlain, S. C., 1990, Encoding of amplitude modulation in the gerbil cochlear nucleus: I. A hierarchy of enhancement. *Hear. Res.* **44**: 99-122.

Gaese, B. H., and Ostwald, J., 1995, Temporal coding of amplitude and frequency modulation in the rat auditory cortex. *Eur. J. Neurosci.* **7**: 438-450.

Goldstein, M. H. Jr., Kiang, N. Y.-S., and Brown, R. M., 1959, Responses of the auditory cortex to repetitive acoustic stimuli. *J. Acoust. Soc. Am.* **31**: 356-364.

Houtgast, T., and Steeneken, H. J. M., 1973, The modulation transfer function in room acoustics as a predictor of speech intelligibility. *Acustica* **28**: 66-73.

Johnson, D. H., 1980, The relationship between spike rate and synchrony in responses of auditory-nerve fibers to single tones. *J. Acoust. Soc. Am.* **68**: 1115-1122.

Joris, P. X., and Yin, T. C. T., 1992, Responses to amplitude-modulated tones in the auditory nerve of the cat. *J. Acoust. Soc. Am.* **91**: 215-232.

Langner, G., and Schreiner, C. E., 1988, Periodicity coding in the inferior colliculus of the cat. I. Neuronal mechanisms. *J. Neurophysiol.* **60**: 1799-1822.

Liang, L., Lu, T., and Wang, X., 2002, Neural representations of sinusoidal amplitude and frequency modulations in the primary auditory cortex of awake primates. *J. Neurophysiol.* **87**: 2237-2261.

Lu, T., Liang, L., and Wang, X., 2001a, Neural representation of temporally asymmetric stimuli in the auditory cortex of awake primates. *J. Neurophysiol.* **85**: 2364-2380.

Lu, T., Liang, L., and Wang, X., 2001b, Temporal and rate representations of time-varying signals in the auditory cortex of awake primates. *Nat. Neurosci.* **4**: 1131-1138.

Lu, T., and X. Wang, 2004, Information content of auditory cortical responses to time-varying acoustic stimuli. *J. Neurophysiol.* **91**: 301-313.

Lu, T., and Wang, X., 2000, Temporal discharge patterns evoked by rapid sequences of wide- and narrow-band clicks in the primary auditory cortex of cat. *J. Neurophysiol.* **84**: 236-246.

Palmer, A. R., 1982, Encoding of rapid amplitude fluctuations by cochlear-nerve fibers in the guinea-pig. *Arch. Otorhinolaryngol.* **236**: 197-202.

Patterson, R. D., 1994a, The sound of a sinusoid: Spectral models. *J. Acoust. Soc. Am.* **96**: 1409-1418.

Patterson, R. D., 1994b, The sound of a sinusoid: Time-interval models. *J. Acoust. Soc. Am.* **96**: 1419-1428.

Rall, W., and Agmon-Snir, H., 1998, Cable theory for dendritic neurons. In *Methods in Neuronal Modeling* (C. Koch and I. Segev, eds.), The MIT Press, Cambridge, MA, pp. 27-92.

Rosen, S., 1992, Temporal information in speech: acoustic, auditory and linguistic aspects. *Philos. Trans. R. Soc. Lond. B Biol. Sci.* **336**: 367-373.

Schreiner, C. E., and Urbas, J. V., 1988, Representation of amplitude modulation in the auditory cortex of the cat. II. Comparison between cortical fields. *Hear. Res.* **32**: 49-63.

Wang, X., and Sachs, M. B., 1993, Neural encoding of single-formant stimuli in the cat I. Responses of auditory-nerve fibers. *J. Neurophysiol.* **70**: 1054-1075.

Wang, X., and Sachs, M. B., 1994, Neural encoding of single-formant stimuli in the cat II. Responses of anteroventral cochlear nucleus units. *J. Neurophysiol.* **71**: 59-78.

Wang, X., and Sachs, M. B., 1995, Transformation of temporal discharge patterns in a VCN stellate cell model: Implications for physiological mechanisms. *J. Neurophysiol.* **73**: 1600-1616.

Whitfield, I., and Evans, E., 1965, Responses of auditory cortical neurons to stimuli of changing frequency. *J. Neurophysiol.* **28**: 655-672.

Zurita, P., Villa, A. E., de Ribaupierre Y., de Ribaupierre F., and Rouiller, E. M., 1994, Changes of single unit activity in the cat's auditory thalamus and cortex associated to different anesthetic conditions. *Neurosci. Res.* **19**: 303-316.

Communication-Call Representation in the Mouse Auditory Cortex: Perception vs. Recognition

GÜNTER EHRET, and DIANA B. GEISSLER
Dept. of Neurobiology, University of Ulm, D-89069 Ulm, Germany

1. INTRODUCTION

The auditory cortex (AC) of mammals consists of several fields. Often, two primary fields, the primary auditory field (AI), and the anterior auditory field (AAF) form the core of the AC. These two fields are tonotopically organized, i.e. the frequency representation in these fields reflects the cochlear tonotopy (e.g. Reale and Imig, 1980; Morel *et al.*, 1993; Thomas *et al.*, 1993; Stiebler *et al.*, 1997). Higher-order fields, such as the second auditory field (AII) with a less clear or actually lacking tonotopy surround the primary fields. Neurons in these higher-order fields seem to respond most reliably and vigorously to complex sounds of communicative and/or environmental significance and, thus, are different from neurons in primary fields that often show clear and strong responses to pure tones (Suga, 1988, 1994; Rauschecker *et al.*, 1995; Esser *et al.*, 1997; Kanwal, 1999). Hence, higher-order fields of the AC are probably closer to the representation of the biological significance of sounds than primary fields. Their role in the process of sound perception and recognition, however, is still not understood.

Modern tomographic imaging methods in humans and various other techniques such as optical recording (e.g. Fukunishi and Murai, 1995; Horikawa *et al.*, 1997) and visualization of activation-dependent gene expression (e.g. Sagar *et al.*, 1988; Sheng and Greenberg, 1990; Chaudhuri, 1997) in other mammals have opened new possibilities and perspectives for the study of functional representations in the brain. Here, we report an approach in house mice, using c-Fos immunocytochemistry, for measuring the activation in auditory cortical fields induced by complex communication sounds. This method has widely been applied for labelling neurons that have been activated by sensory, including acoustic stimuli (e.g. Ehret and Fischer, 1991; Friauf, 1992; Vischer *et al.*, 1994; Kai and Niki, 2002). Our results show that a differential activation only in higher-order fields of the AC reflects differences between perception and recognition. The sounds to be perceived and recognised in our studies are so-called wriggling calls of mouse pups. The addressees of the calls are the pups' mothers whose ACs are subject of our study (Geissler and Ehret, 2004).

2. PERCEPTION OF COMMUNICATION CALLS IN MICE

With only eight call types, the repertoire of vocalizations of house mice (*Mus musculus domesticus*) is rather small (Ehret, 1975; Haack *et al.*, 1983; Whitney and Nyby, 1983). Calls of mouse pups such as pure ultrasounds and wriggling calls are especially important, because they elicit specific responses of pup-caring behaviour in their mothers. Wriggling calls are produced by pups in a large litter pushing for the mother's nipples when she is in a nursing position on the litter. Most frequently, pups vocalize wriggling calls in series of several calls. Only such call series have a high probability to be responded by maternal behaviour. The mothers start licking their pups, change the lactation position or do some nest building (Ehret and Bernecker, 1986; Ehret and Riecke, 2002). We observed these maternal behaviours in response to wriggling calls almost immediately after birth of the pups so that the maternal responsiveness seems to be based on innate releasing mechanisms (instinctive recognition).

The acoustical properties of the wriggling calls in the frequency and time domain that are both necessary and sufficient for the release of maternal behaviour have been studied rather extensively in behavioural tests (Ehret and Riecke, 2002; Geissler and Ehret, 2002). In the frequency domain, nearly optimum calls are composed of three harmonically related frequency components (formants) near 4, 8, and 12 kHz (Fig. 1a). In the time domain, single calls should have a minimum duration of about 100 ms and a synchronous beginning of the three formants within about 30 ms. Inter-call intervals of about 200 ms have been found in series of natural calls and have also been used in call models of high releasing capability. Thus, the message or the biological significance of wriggling calls is encoded in a call series as shown in Fig. 1a. Relative to series of natural wriggling calls of the mothers' own pups, series of call model A release maternal behaviour in average 77 %

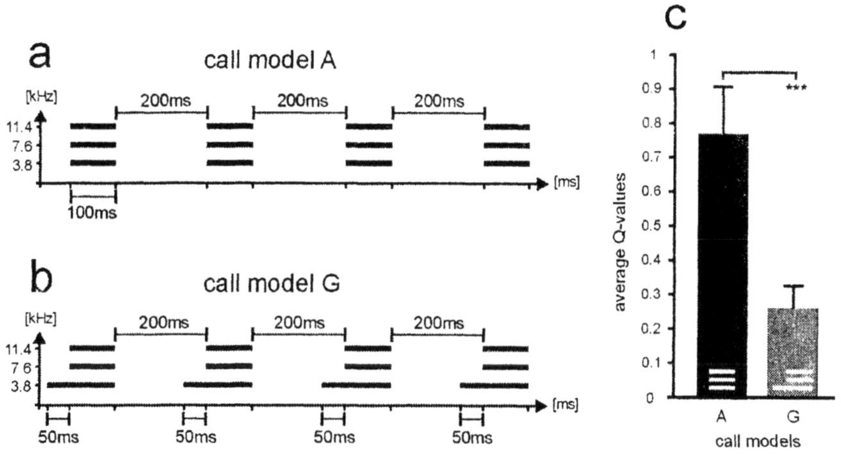

Figure 1. Frequency and time structure of call model A (a) and call model G (b) used to stimulate the mother mice in a behavioural context of nursing their pups. Call model A releases maternal behaviour at a high rate, call model G at a low rate. The rate is expressed by the Q-values (c, modified from Geissler and Ehret, 2002)), which is the relative response rate to a given call model divided by the relative response rate to natural calls of pups (Ehret and Riecke, 2002).

of the cases the call model is presented to the mother in the nursing context (Q-value of 0.77 in Fig. 1c). The biological significance of call models is strongly reduced if the three formants are not heard in synchrony. For example, series of call model G (Fig. 1b) have an average releasing capability of only about one third of that of call model A (Q-values of only 0.25 in Fig. 1c). In short, series of call model A are good releasers of maternal behaviour, the message of the call model is "recognised" by the mothers. On the other hand, series of call model G are insufficient releasers of maternal behaviours, the call model is not recognised (Geissler and Ehret, 2002).

Here, we present data on mothers stimulated either with call model A or call model G in a nursing situation in which they could show maternal behaviour to five of their own pups which did not call themselves (Geissler and Ehret, 2004). After the stimulation, the brains of the mothers were processed for c-Fos immunocytochemistry. Fos-positive cells were located in electrophysiologically mapped fields of the AC and were taken as indicators of the activation there, due to the call models used as acoustic stimuli.

3. THE AUDITORY CORTEX OF THE MOUSE

The left and right side auditory cortex of mice of the NMRI outbred strain and of hybrids of feral and NMRI mice have been studied electrophysiologically with multi-unit extracellular recordings (Stiebler *et al.,* 1997). Besides the two tonotopically organized primary auditory fields, the primary auditory field (AI) and anterior auditory field (AAF), three fields were found without clear tonotopy (Fig. 2a, b). One had neurons responding strongly to tone bursts of frequencies higher than about 45 kHz (ultrasonic field, UF), another ventral of AAF had rapidly habituating neurons and was called second auditory field (AII), and a third was called dorsoposterior field (DP) with broadly tuned neurons and rhythmic background activity. This general pattern of field geometry was common to all mice despite large variability in the sizes of the fields and the exact positions and extents on both rostrocaudal and dorsoventral coordinates of the brain. On average, the size of the right-side auditory cortex was only 76 % of the size of the left hemisphere (Stiebler *et al.*, 1997). The significant left-hemisphere advantage in size was also found in the c-Fos labelling studies of Geissler and Ehret (2004).

4. REPRESENTATION OF WRIGGLING CALLS IN THE AUDITORY CORTICAL FIELDS

Both call models A and G evoked Fos-positive cells in all five auditory cortical fields. In Fig. 2b, reconstructions of serial brain sections through the region of AI are shown for the left hemisphere of a mouse stimulated with call model A and another one stimulated with call model G. There are clearly three locations in AI of both animals with increased numbers of Fos-positive cells corresponding to the positions of 3.8, 7.6, and 11.4 kHz on the AI tonotopy (Fig. 2c). Similar increases in the number of Fos-positive cells related to the tonotopy are seen in the AAF in mothers stimulated with either call model. A quantitative evaluation of the distribution of Fos-positive cells of both hemispheres of all animals stimulated with either call model did not show significant differences between hemispheres for animals in each call model group, both for AI and AAF. In addition, no inter-group differences occurred (Geissler and Ehret, 2004). That is, the differences between the two call models in their releasing capabilities for maternal behaviour seen in the behavioural tests (Fig. 1c) are not reflected in the amount and distribution of Fos-positive cells and, thus, in the general activation pattern of the two tonotopically organised fields, AI and AAF.

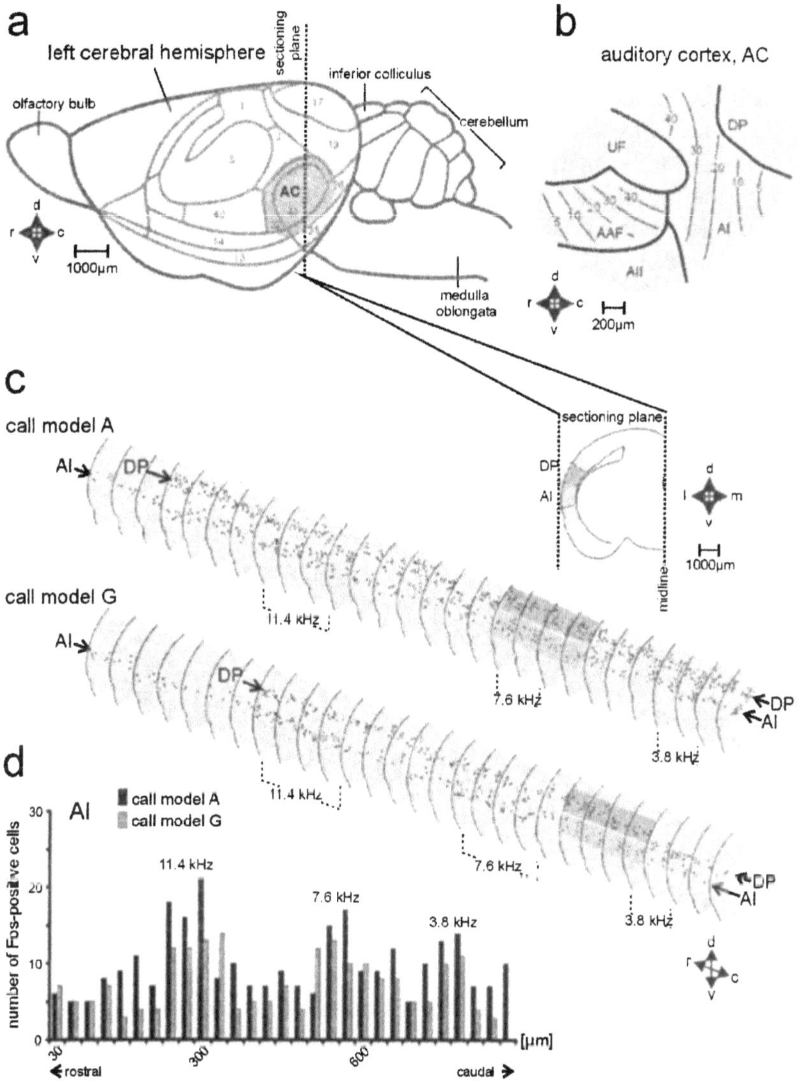

Figure 2. Lateral view on the left hemisphere of the mouse brain (a) with the position of the auditory cortex (AC) indicated and enlarged (b). Examples of serial frontal sections through the AC show (c) the region of AI of a mother stimulated with either call model A or G. Fos-positive cells are indicated by dots. The increased labelling due to the three frequency components of the calls is quantified in a histogram (d) (modified from Geissler and Ehret, 2004).

In contrast to the primary fields, both higher-order auditory fields AII and DP were differentially activated depending on the call model played to the mothers (Fig. 3). The AII shows significantly less Fos-positive cells to the recognised call model A, compared to the non-recognised call model G. In the DP, a left-hemisphere advantage in the number of Fos-positive cells is seen for call model A, but nor for call model G. That is, call recognition seems to be expressed only in the DP of the left hemisphere, while non-recognition does not produce a hemisphere difference in activation (Fig. 3). Clearly, the biological significance of wriggling call models is reflected in the activation patterns of both higher-order fields AII and DP of the mouse auditory cortex.

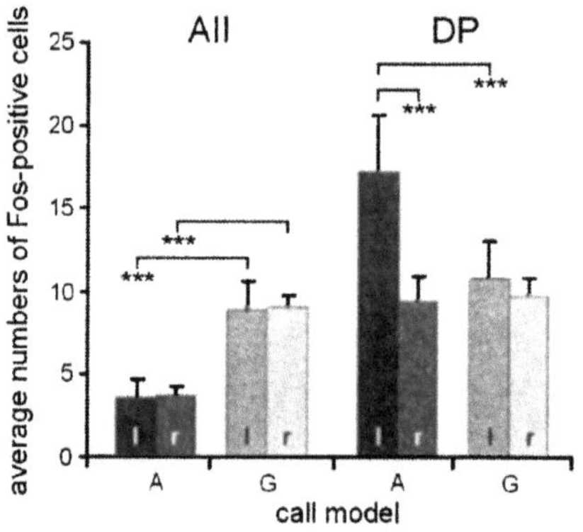

Figure 3. Quantification of Fos-positive cells in AII and DP of the left (l) and right (r) hemispheres of mothers responding either to call model A or G. Statistical differences between the means are indicated by stars ($p < 0.001$). Standard deviations are also shown (modified from Geissler and Ehret, 2004).

5. DISCUSSION

5.1 The Specific Role of AII in Call Recognition

The following approach to explain the roles of AII and DP on processing biologically significant communication calls considers the described data on Fos-labelling due to wriggling call perception (Geissler and Ehret, 2004), Fos-labelling in the mouse auditory cortex due to the perception of pup ultrasounds (Fichtel and Ehret, 1999), and role of the Fos protein in intracellular signalling in the context of long-term changes associated with learning (e.g. Abraham *et al.*, 1991; Kaczmarek and Chaudhuri, 1997; Tischmeyer and Grimm, 1999; Staiger *et al.*, 2002).

A higher-order auditory cortical field (Te3) of the rat, corresponding to AII of the mouse, demonstrates higher numbers of Fos-positive cells if the animals were stimulated with novel compared to familiar sounds (Wan *et al.*, 2001). These data correspond to the stimulation of mother mice perceiving call model G. The acoustic properties of call model G (Fig. 1b) are novel to them, because they are used to hearing wriggling calls structurally similar to call model A (Fig. 1a). It seems that familiar sounds (call model A) in a familiar behavioural context such as nursing pups, lead to only low numbers of specifically activated neurons in AII, while unfamiliar sounds in the same context lead to a significantly higher number of activated neurons (Fig. 3). We suggest that the perception of familiar sounds in the nursing context activated only those few neurons that are specifically tuned to the combination of sound properties typical for natural wriggling calls. The Fos-level in these cells that make them detectable for us in the brain sections, may indicate a response of stabilisation of the association of familiar sounds with a familiar behavioural context and thus represents a level of learning. The high number of Fos-positive neurons in the mothers perceiving call model G, may then indicate the beginning of learning the association between a familiar context and novel sounds. The amounts of Fos-labelling in the AII of mother mice responding to ultrasonic calls of their pups (Fichtel and Ehret, 1999) are similar to those of mothers perceiving wriggling call model G and, thus, are in accordance with these interpretations. In that case, the mothers heard familiar sounds in an unfamiliar behavioural context, because the ultrasonic stimuli were transmitted from above the cage, which is an unexpected, odd direction, dissociating the pups in the cage from the source of their calls. These mothers had to newly learn the association between familiar sounds and a novel context.

These data add to the proposition that the activation of neurons in AII expresses the processing and coding of the relationship between an acoustical stimulus of significance and the context of its occurrence. In summary, neural activity in AII seems to reflect auditory cognition (Geissler and Ehret, 2004). In that function, AII might be part of the ventral stream of acoustical information transfer form the AC to the frontal cortex. This stream is suggested to transfer information about "what", i.e. about the acoustical Gestalt of a perceived sound (e.g. Romanski *et al.*; 1999, Kaas and Hackett, 2000; Rauschecker and Tian, 2000).

5.2 The Specific Role of DP in Call Recognition

The dorsoposterior field may belong to the non-tonotopically organized dorsal belt region of the AC (e.g. Reale and Imig, 1980; Thomas *et al.*, 1993). Many neurons in this region code for spectrally complex sounds and are sensitive to timing parameters in the sounds (He *et al.*, 1997). This is compatible with a sensitivity to series of wriggling call models (Fig. 1a, b). The surprising fact is the left-hemisphere dominance in activation only for the recognized call model A (Fig. 3). Our interpretation of these data (Geissler and Ehret, 2004) considers the following: The surplus of activation in the left hemisphere sits on an already high activation comparable to that in AII for the non-recognized call model G (Fig. 3). Hence, DP seems to be activated in both hemispheres of mothers if they are motivated to produce an appropriate behaviour to sounds in a given situation. Mothers show an appropriate maternal behaviour at a high rate in response to call model A, because the sound properties match the behavioural context to release the maternal response. The low-rate maternal behaviour in response to call model G, however, is also an appropriate response, because the sound properties do not match the behavioural context. Therefore, we suggest that a basic level of activation in DP of both hemispheres may occur whenever animals are in a motivational/emotional state to be able to perform a specific, appropriate behaviour in response to a heard sound. The surplus of activation in the left hemisphere of mothers hearing call model A, then, may be due to the feedback of increased maternal emotion/motivation by producing maternal behaviours to the sounds at a high rate. It has been shown that positive emotions expressed by the attraction of animals to certain stimuli and by search for social contact, are based on a left-hemisphere dominant activity of the brain (e.g. Tucker, 1981; Thompson, 1985; Silberman and Weingartner, 1986; Hook-Costigan and Rogers, 1998). Since maternal emotions/motivations are certainly positive, we suggest that the high level of maternal motivations in mothers responding to call model A are the reason for the increased activation of DP of the left hemisphere.

In summary, we hypothesize that the neural activation in DP reflects a basic level of emotional/motivational state compatible with responding specifically and adequately to a perceived sound. Hemisphere differences in activation may reflect increased emotional/motivational levels by feedback from the actions in response to the sounds. Because of the laterality of emotional processing in the brain, positive emotions/motivations would selectively increase the activity of the left hemisphere DP. This hypothesis needs pathways of emotional influence from the limbic system and/or thalamus to the DP. Such pathways have now to be investigated.

5.3 Left Hemisphere Advantage in Communication Call Perception

Mother mice show a high rate of maternal behaviour in response to wriggling calls of their pups only, if the calls are heard with two ears (binaurally). This is shown in Fig. 4. It is evident, however, that, in a monaural situation, mothers are better if they can hear with the right ear (left ear is plugged) compared to the left ear (right ear is plugged) (Haase and Ehret, 1990). This right-ear advantage means a left-hemisphere brain advantage for wriggling call recognition. These behavioural data fit nicely to the data about the left-hemisphere advantage in activation of DP by wriggling calls (Fig. 3) in mothers perceiving the adequate wriggling call model A.

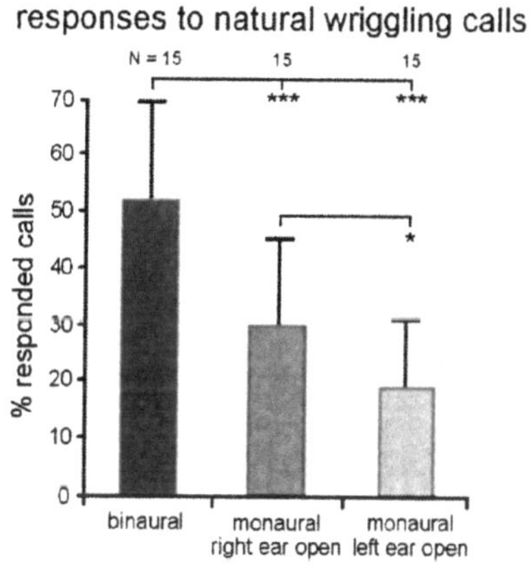

Figure 4. Mother mice with binaural hearing respond with significantly higher rates to series of wriggling calls of their own pups compared to monaural mothers ($p < 0.001$), and mice with the right ear open (the left ear is plugged) are better than animals with the left ear open (the right ear is plugged) ($p < 0.05$). Modified from Haase and Ehret (1990).

Previous studies on ultrasound perception in mice (Ehret, 1987) have demonstrated a left-hemisphere (right ear) advantage if mother mice discriminated and preferred the ultrasounds of mouse pups against irrelevant sounds in a context of maternal emotion/motivation, that is, instinctively. Ear (hemisphere) advantages in the discrimination and preference of ultrasounds did not occur in naïve females without maternal experience who were trained in an operant-reward conditioning paradigm to perform the discrimination and preference behaviour. In view of the data about auditory cortical activation in wriggling call processing (Fig. 3) we can infer that the left-hemisphere advantage in the instinctive responding to ultrasounds by the mothers may reflect a left-hemisphere advantage of activation in DP. The absence of a laterality in the naïve females may reflect the sound recognition on an cognitive basis mediated by the non-lateralised processing in AII (Fig. 3)

6. CONCLUSION

c-Fos immunocytochemistry is used for labelling activated neurons in studies of auditory cortical processing of communication sounds in mice. Activation of the primary auditory fields AI and AAF does not discriminate between sounds of high or low behavioural significance. Labelling related to call recognition occurs in the second auditory field AII. Activation in another auditory field of higher order, the dorsoposterior field DP, seems to be related to an integration of call recognition with the emotional/motivational background of the animals. A left-hemisphere advantage of activation in this field is associated with higher levels of maternal emotions. These data on auditory cortical activation fit excellently to perception data of communication calls in mice.

ACKNOWLEDGEMENTS

We thank Dr. K.-H. Esser for valuable help in preparing the call models. The study was supported by grants of the Deutsche Forschungsgemeinschaft (Eh 53/17-1,2).

REFERENCES

Abraham, W.C., Dragunow, M., and Tate, W.P., 1991, The role of immediate early genes in the stabilization of long-term potentation. *Mol. Neurobiol.* **5**: 297-314.

Chaudhuri, A., 1997, Neural activity mapping with inducible transcription factors. *NeuroReport* **8**: 3-8.

Ehret, G., 1975, Schallsignale der Hausmaus (*Mus musculus*). *Behaviour* **52**: 38-56.

Ehret, G., 1987, Left hemisphere advantage in the mouse brain for ultrasound recognition in a communicative context. *Nature* **325**: 249-251.

Ehret, G., and Bernecker, C., 1986, Low-frequency communication by mouse pups (*Mus musculus*): wriggling calls release maternal behaviour. *Anim. Behav.* **34**: 821-830.

Ehret, G., and Fischer, R., 1991, Neuronal activity in the auditory system visualized by c-fos gene expression. *Brain Res.* **567**: 350-354.

Ehret, G., and Riecke, S., 2002, Mice and humans perceive multiharmonic communication sounds in the same way. *Proc. Natl. Acad. Sci. USA* **99**: 479-482.

Esser, K. H., Condon, C. J., Suga, N., and Kanwal, J. S., 1997, Syntax processing by auditory cortical neurons in the FM-FM area of the mustached bat *Pteronotus parnellii*. *Proc. Natl. Acad. Sci. USA* **94**: 14019-14024.

Fichtel, I., and Ehret, G., 1999, Perception and recognition discriminated in the mouse auditory cortex by c-Fos labeling. *NeuroReport* **10**: 2341-2345.

Friauf, E., 1992, Tonotopic order in the adult and developing auditory system of the rat as shown by c-fos immunocytochemistry. *J. Neurosci.* **4**: 789-812.

Fukunishi, K., and Murai N., 1995, Temporal coding in the guinea pig auditory cortex as revealed by optical imaging and its pattern-time-series analysis. *Biol. Cybern.* **72**: 463-473.

Geissler, D. B., and Ehret, G., 2002, Time-critical integration of formants for perception of communications calls in mice. *Proc. Natl. Acad. Sci. USA* **99**: 9021-9025.

Geissler, D. B., and Ehret, G., 2004, Auditory perception vs. recognition: representation of complex communication sounds in the mouse auditory cortical fields. *Eur. J. Neurosci.*, in press.

Haack, B., Markl, H., and Ehret, G., 1983, Sound communications between parents and offspring. In *The Auditory Psychobiology of the Mouse* (F. J. Willott, ed.), CC Thomas, Springfield, pp. 57-97.

Haase, H., and Ehret, G., 1990, Lateralization of sound perception in the brain of the mouse (*Mus musculus*). In *Brain-Perception-Cognition.* (N. Elsner and G. Roth, eds.), Thieme, Stuttgart, p. 150.

He, J., Hashikawa, T., Ojima, H., and Kinouchi, Y., 1997, Temporal integration and duration tuning in the dorsal zone of cat auditory cortex. *J. Neurosci.* **17**: 2615-2625.

Hook-Costigan, M.A., and Rogers, L.J., 1998, Lateralized use of the mouth in production of vocalizations by marmosets. *Neuropsychologia* **36**: 1265-1273.

Horikawa, J., Hosokawa Y., Nasu, M., and Taniguchi, I., 1997, Optical study of spatiotemporal inhibition evoked by two-tone stimuli in the guinea pig auditory cortex. *J. Comp. Physiol. A.* **181**: 677-684.

Kaas, J.H., and Hackett, T.A., 2000, Subdivisions of auditory cortex and processing streams in primates. *Proc. Natl. Acad. Sci. USA* **97**: 11793-11799.

Kaczmarck, L., and Chaudhuri, A., 1997, Sensory regulation of immediate-early gene expression in mammalian visual cortex: implications for functional mapping and neural plasticity. *Brain Res.* **23**: 237-256.

Kai, N., and Niki, H., 2002, Altered tone-induced Fos expression in the mouse inferior colliculus after early exposure to intense noise. *Neurosci. Res.* **44**: 305-313.

Kanwal, J. S., 1999, Processing species-specific calls by combination-sensitive neurons in an echolocating bat. In *The Design of Animal Communication.* (M. D. Hauser, and M. Konishi, eds.), MIT Press, Cambridge MA, pp. 133-157.

Morel, A., Garraghty, P. E., and Kaas, J. H., 1993, Tonotopic organization, architectonic fields, and connections of auditory cortex in macaque monkeys. *J. Comp. Neurol.* **335**: 437-459.

Rauschecker, J. P., Tian, B., and Hauser, M., 1995, Processing of complex sounds in the macaque nonprimary auditory cortex. *Science* **268**: 111-114.

Rauschecker, J. P., and Tian, B., 2000, Mechanisms and streams for processing of "what" and "where" in auditory cortex. *Proc. Natl. Acad. Sci. USA* **97**: 11800-11806.

Reale, R. A., and Imig, T. J., 1980, Tonotopic organization in auditory cortex of the cat. *J. Comp. Neurol.* **192**: 265-291.

Romanski, L. M., Tian, B., Fritz, J., Mishkin, M., Goldman-Rakic, P. S., and Rauschecker, J. P., 1999, Dual streams of auditory afferents target multiple domains in the primate prefrontal cortex. *Nat. Neurosci.* **2**: 1131-1136.

Sagar, S. M., Sharp, F. R., and Curran, T., 1988, Expression of c-fos protein in brain: metabolic mapping at the cellular level. *Science* **240**: 1328-1331.

Sheng, M., and Greenberg, M. E., 1990, The regulation and function of c-fos and other immediate early genes in the nervous system. *Neuron* **4**: 477-485.

Silberman, E. K., and Weingartner, H., 1986, Hemispheric lateralization of functions related to emotion. *Brain Cognit.* **5**: 322-353.

Staiger, J. F., Masannek, C., Bisler S., Schleicher, A., Zuschratter, W., and Zilles, K., 2002, Excitatory and inhibitory neurons express c-Fos in barrel related columns after exploration of a novel environment. *Neurosci.* **109**: 687-699.

Stiebler I., Neulist R., Fichtel, I., and Ehret, G., 1997, The auditory cortex of the house mouse: left-right differences, tonotopic organization and quantitative analysis of frequency representation. *J. Comp. Physiol. A.* **181**: 559-571.

Suga, N., 1988, Auditory neuroethology and speech processing: complex sound processing by combination-sensitive neurons. In *Auditory Function. Neurobiological Base of Hearing* (G.M. Edelman, W.E. Gall, and W.M. Cowan, eds.), Wiley, New York, pp. 679-720.

Suga, N., 1994, Multi-function theory for cortical processing of auditory information: implications of single-unit and lesion data for future research. *J. Comp. Physiol. A* **175**: 135-144.

Thomas, H., Tillein J., Heil, P., and Scheich, H., 1993, Functional organization of auditory cortex in the Mongolian gerbil (*Meriones unguiculatus*). I. Electrophysiological mapping of frequency representation and distinction of fields. *Eur. J. Neurosci.* **5**: 882-897.

Thompson, J. K., 1985, Right brain, left brain; left face, right face: hemisphericity and the expression of facial emotion. *Cortex* **21**: 281-299.

Tucker, D. M., 1981, Lateral brain function, emotion, and conceptualization. *Psychol. Bull.* **89**: 19-46.

Tischmeyer, W., and Grimm, R., 1999, Activation of immediate-early genes and memory formation. *Cell. Mol. Life Sci.* **55**: 564-574.

Vischer, M. W., Häussler, R., and Rouiller E. M., 1994, Distribution of Fos-like immunoreactivity in the auditory pathway of the Sprague-Dawley rat elicited by cochlear stimulation. *Neurosci. Res.* **19**: 175-185.

Wan, H., Warburton, E. C., Kuśmierek, P., Aggleton, J. P., Kowalska, D. M., and Brown, M. W., 2001, Fos imaging reveals differential neuronal activation of areas of rat temporal cortex by novel and familiar sounds. *Eur. J. Neurosci.* **14**: 118-124.

Whitney, G., and Nyby, J., 1983, Sound communication among adults. In *The Auditory Psychobiology of the Mouse*. (J. F. Willott, ed.), CC Thomas, Springfield, pp. 98-129.

Molecular Mechanisms in Deafness Related Auditory Brain Stem Plasticity

RICHARD A. ALTSCHULER*, AVRIL GENENE HOLT*, MIKIYA ASAKO*, CATHERINE A. LOMAX*, MARGARET I. LOMAX*, and JOSE JUIZ#

**Kresge Hearing Research Institute, Department of Otolaryngology, University of Michigan, Ann Arbor, Michigan; and #Catedrático de Histología, Universidad de Castilla–La-Mancha, Albacete, Spain*

1. INTRODUCTION

There is increasing evidence of activity dependent plasticity in the adult central nervous system. A model has been emerging which is summarized in Figure 1. In this model neuronal changes in activity levels cause changes in calcium levels (often mediated by NMDA receptors) and calcium binding, as well as in levels of activation of transcription factors including the CREB and Fos families (Kornhauser *et al.*, 2002; Finkbeiner and Greenberg, 1998; West *et al.*, 2002). These transcription factors can induce intracellular regulatory pathways including those for neurotrophic factors such as BDNF, resulting in additional changes in gene expression as well as post-translational changes (Ghosh *et al.*, 1994a,b; Takasu *et al.*, 2002). Many of the changes are in excitatory and inhibitory amino acids and their receptors, with resulting changes in excitation and inhibition (Ghosh and Greenberg, 1995).

The auditory pathway is an excellent system to test this model, with the ability to decrease or increase activity in a "natural" fashion (deafness versus noise) and to determine functional, morphological and molecular consequences. There are also important clinical implications for central auditory plasticity, with the ability to return activity following deafness, using the highly successful cochlear prostheses. Moreover, plastic changes may contribute to a major clinical problem, central tinnitus.

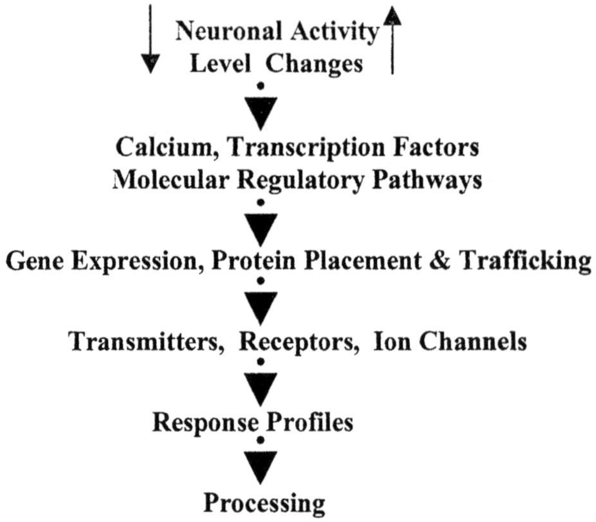

Figure 1. Model of activity related plasticity in the central nervous system.

2. AUDITORY BRAIN STEM PLASTICITY

Studies of auditory brain stem plasticity, from our lab and others, have shown that the large decrease in activity from profound deafness produces many dramatic changes in the central auditory pathways. Deafness related changes have been recently summarized in a review chapter by Syka (2002). There are deafness related morphological changes such as decreases in neuronal cell size (e.g. Lesperance *et al.*, 1995; Nishiyama *et al.*, 2000; Pasic and Rubel, 1989, 1991, 1994; Rubel *et al.*, 1990; Winskey and Jacobowitz, 1995) and changes in synapses (e.g. Benson *et al.*, 1997; Gulley *et al.*, 1978; Miller *et al.*, 1991; Moore *et al.*, 1994; Morest and Bohne, 1983, Morest *et al.*, 1996; Muly *et al.*, 2002; Potashner *et al.*, 1997; Redd *et al.*, 2000; Ryugo *et al.*, 1998). There are also neurochemical transmitter changes as well as changes in neurotransmitter receptors and binding (e.g. Altschuler *et al.*, 1995, 1997; Bledsoe *et al.*, 1995; Caspary *et al.*, 1999; Helfert *et al.*, 1999; Milbrandt *et al.*, 1997; Mossop *et al.*, 2000; Nakagawa *et al.*, 2000; Potashner *et al.*, 1997, 2000; Sato *et al.*, 2000; Suneja *et al.*, 1998; Willot and Turner, 2000). There are also changes in synapse related proteins such as protein kinases, Gap43 and calbindin (e.g. Garcia *et al.*, 2000; Idrizbegovic *et al.*, 1998; Illing *et al.*, 1997; Syka *et al.*, 2000, Zhang *et al.*, 2003a,b).

3. DECREASES IN INHIBITION

Emerging from the studies of deafness related changes in the auditory brain stem is the general observation of major decreases in inhibition accompanied by increases in spontaneous activity, excitation and changes in the tonotopic map (Syka, 2002 for review). In the inferior colliculus (IC) this appears to be mediated by changes in the inhibitory amino acid transmitter GABA. Bledsoe *et al.* (1995, 1997) used chronic dialysis to show major decreases in GABA release in the rat and guinea pig IC 2-3 weeks following deafness. This was accompanied by decreases in inhibition as well as changes in fos immunoreactive (IR) neurons evoked by cochlear electrical stimulation (Bledose *et al.*, 1995, 1997; Nagase *et al.*, 2000). The fos-IR studies not only showed increased number of neurons in a wider band evoked by cochlear electrical stimulation after 14 days of deafness, but that chronic cochlear electrical stimulation imposed its own plastic changes (Nagase *et al.*, 2000). Caspary's group (Caspary *et al.*, 1995, 1999, 2001; Milbrandt *et al.*, 1997) have shown decreases in the GABA synthesizing enzyme GAD as well as changes in GABA receptor subunit expression and GABA receptor binding in the IC with age related hearing loss. Mossop *et al.* (2000) showed deafness-related changes in GAD as well as decreases in inhibition in the IC.

In the superior olivary complex (SOC) and cochlear nucleus (CN), the deafness related changes may be more associated with the inhibitory amino acid transmitter glycine. Studies have shown that glycine is a major inhibitory transmitter in the CN and SOC (e.g. Adams and Mugnaini, 1990; Altschuler *et al.*, 1997; Bledsoe *et al.*, 1990; Caspary *et al.*, 1987, 1994; Golding and Oertel, 1996; Grothe and Sanes, 1993; Harty and Manis, 1998; Helfert *et al.*, 1992; Juiz *et al.*, 1996; Kolston *et al.*, 1992; Kuwabara and Zook, 1992, 1999; Sato *et al.*, 1995; Wickesberg and Oertel, 1990; Wickesberg *et al.*, 1994; Wenthold *et al.*, 1987; Wu and Oertel, 1986; Wu and Kelly, 1994). In studies of age related deafness, Willot *et al.* (1993, 2000) showed dramatic deafness related decreases in glycine immunoreactive neurons and puncta. Krenning *et al.*, (1998) showed changes in glycine receptor subunits in the rat CN associated with age related hearing loss.

We therefore used post-embedding immunostaining of plastic sections (as in chapter in this volume by Merchan) to examine deafness related changes in axo-somatic puncta, which Merchan has shown largely correspond to axo-somatic terminals. We examined changes in GABA in the rat IC and glycine in the rat SOC and CN, 14 days following bilateral deafening from topical intrascalar infusion of 10% neomycin. Our studies showed no significant changes in GABA-IR axosomatic puncta in the IC, while dramatic decreases in glycine IR axosomatic puncta were found in the SOC and CN (Asako *et al.*, 2003; Buras *et al.*, 2003).

3.1 Changes in Glycine in the CN and SOC

The number of immunolabeled puncta around profiles of the soma of different neuronal cell types was assessed for GABA in the IC and glycine in SOC and CN from one micron plastic sections. This was considered and compared both as the total number of immunolabeled puncta per somatic profile and also as the number of puncta per 100 μm of somata perimeter, to correct for deafness related decreases in cell size, that occurred in most (but not all) neurons assessed. We did not find any significant changes in the number of GABA IR axosomatic puncta on profiles of principal cells in the central nucleus of the IC, 14 days following deafness, compared to the number in age-matched normal hearing rats. This suggests that the GABA pool in GABAergic terminals on CIC principal cells does not have a sufficient decrease to bring it below the level of immunocytochemical detection, despite the large decrease in GABA release. Other mechanisms, such as the deafness related changes in regulators of neurotransmitter release, recently shown by Potashner's group (Suneja *et al.*, 2003; Zhang *et al.*, 2003a,b) in the auditory brain stem, may then contribute to the deafness related loss of GABA inhibition in the IC.

On the other hand, dramatic decreases in the number of glycine IR axosomatic puncta were found for major neuron types in both the SOC and CN. Decreases in glycine IR axosomatic puncta in the SOC ranged from approximately 25% for principal cells of the medial and lateral nuclei of the trapezoid body to approximately 40% for principal cells of the superior paraolivary nucleus and ventral nucleus of the trapezoid body. Decreases in the CN were even larger, 50-60% for dorsal CN fusiform cells and ventral CN spherical bushy, globular bushy and radiate cells. Thus, decreased activity from deafness has a significant effect on glycine in the CN and SOC, decreasing the pool in terminals such that it falls below the level for detection in more than half the glycinergic terminals (Asako *et al.*, 2003).

3.2 Changes in Receptor Expression

Suneja *et al.* (1998a,b) and Potashner *et al.* (2000) have reported activity related changes in glycine receptor binding in the auditory brain stem following cochlear ablation and Caspary *et al.* (1999) have reported changes in expression of GABA receptors in the IC with age related hearing loss. With the dramatic changes in Gly IR puncta 14 days following deafness in the CN, we therefore used Real Time PCR to examine changes in expression

Changes in GABA & Glycine Receptor Subunit expression in the CN 3 days and 3 weeks following bilateral cochlear ablation

	3 days	3 weeks
GlyRα1	+	
GlyRα2		
GlyRα3	++	+
GABA-ARα1	++	+++
GABA-ARα2	+	+++
GABA-ARα3		+
GABA-ARα5		
GABA-ARβ1	+	++
GABA-ARβ3	++	+++
GABA-ARγ1	+	+
GABA-ARγ2L	+	++
GABA-BR	-	

--	=	greater than 50% decrease
-	=	25-50% decrease
+	=	1.5 – 1.9 fold increase
++	=	2 – 2.9 fold increase
+++	=	3 fold and greater increase

Figure 2. Summary of changes in expression of Glycine and GABA receptor subunits in the rat cochlear nucleus at 3 days and 3 weeks following deafness from bilateral cochlear ablation based on Real Time PCR.

for several glycine and GABA receptor subunits in the rat CN, 3 days and 3 weeks following bilateral cochlear ablation. Significant changes were found at both times (Asako *et al.*, 2004). For the glycine receptor, a small increase in expression of the α1 subunit and a large increase in expression of the α3 subunit were found at 3 days following deafness. Both increases were transient, with levels of gene expression returning to baseline values at 3 weeks following deafness. On the other hand, for GABA receptor subunits, the most dramatic changes were seen at 3 weeks following deafness. Various GABA-A receptor subunits showed small increases in expression at 3 days following deafness and greater increases at 3 weeks following deafness, compared to normal hearing age matched controls (Figure 2). GABA-B receptor expression showed a small decrease in expression at 3 days following deafness. Future studies will add later time points to determine if changes in GABA-A receptor subunit expression are transient.

3.3 Gene Microarray Expression Changes

Gene Microarrays provide the ability to examine expression of thousands of genes simultaneously. Affymetrix gene chips were utilized to examine deafness related changes in gene expression in the rat IC at 3 days and 3 weeks following bilateral deafening by cochlear ablation (Holt *et al.*, 2003). More genes showed changes in expression at 3 weeks than at 3 days and more genes showed increased expression than decreased expression at both times. At three days following deafness approximately 400 genes showed increases and 40 showed decreases. Genes showing increased expression included those for GABA-A and glycine receptor subunits, consistent with our observations in CN using Real Time PCR. In addition there were also increases in glutamate receptor subunits, such as GluR2. Genes showing decreased expression included synaptic vesicle glycoprotein 2a and GAD. There were both increases and decreases in ion channel associated gene expression with, for example, the calcium channel $\alpha 1$ subunit showing increased expression and the potassium inwardly rectifying channel showing a decrease. At 21 days following deafness there were approximately 700 genes showing significant increases in expression and 300 showing decreases. Those showing increases now also included calbindin and calmodulin 1, those showing decreases included tyrosine hydroxylase and NGF1. When looking overall at groupings of genes, several stories emerged. In general "presynaptic" transmitter and vesicle related genes decreased in expression, while "post-synaptic" receptor complex related gene expression increased, perhaps in compensation. There were changes in genes related to transcriptional regulation and development. For example, BMP4 expression increased and Noggin expression decreased, whereas EGR1 expression decreased at 3 days and increased at 21 days. There were also changes in expression in stress pathway related genes, with caspase 3 expression increasing and BCL2 expression decreasing. There were also increases in the expression of heat shock proteins, SOD and cyclin. Finally, there were changes in expression of genes for proteins in dendritic spines and cytoskeletal related proteins. For example, MAP2 expression increased and MAP tau decreased, whereas laminin and ankyrin expression increased and activity related cytoskeletal protein decreased.

4. CONCLUSIONS

There is considerable capacity for plasticity in the mature auditory brain stem and plastic changes can be evoked by the change in activity resulting from profound deafness. There are deafness related changes in transmitters and receptors with resulting changes in inhibition and activity that can effect auditory processing. An understanding of the underlying mechanisms of plasticity, with gene microarray results pointing to a number of regulatory pathways, could help in development of interventions to influence plasticity. Such interventions should be helpful to provide the best circumstances for return of hearing with cochlear prostheses following deafness as well as for management of clinical problems resulting from the plasticity of partial deafness and central tinnitus.

ACKNOWLEDGEMENTS

The work from our laboratory was supported by NIDCD grant #DC00383. We would like to thank Eric Buras for important contributions to these studies.

REFERENCES

Adams, J. C., and Mugnaini, E., 1990, Immunocytochemical evidence for inhibitory and disinhibitory circuits in the superior olive. *Hear. Res.* **49**: 281-298.

Altschuler, R. A., Hoffman, D. W., and Wenthold. R.J., 1986, Neurotransmitters of the cochlea and cochlear nucleus: immunocytochemical evidence. *Am. J. Otolaryngol.* **7**: 100-106.

Altschuler, R. A., Satao, K., Dupont, J., Bonneau, J. M., and Nakagawa, H., 1997, Diversity in glycine and NMDA receptor composition in the rat cochlear nucleus and superior olivary complex and changes with Deafness, In *Acoustic Signal Processing in the Central Auditory System*, (J. Syka, ed.), Plenum Press, New York, pp. 193-20.

Asako, M., Holt, A. G., Griffith, R. D., Buras, E. D., and Altschuler, R. A., 2003, Deafness Related Changes in Glycine-Immunoreactive Staining in the Rat Cochlear Nucleus, *Assoc. Res. Otolaryngol. Abs.* 39.

Asako, M., Holt, A. G., Lomax C., Juiz J., and Altschuler, R. A., 2004, Deafness Related Changes in Glycine and GABA receptor subunit expression in the rat cochlear nucleus, *Assoc. Res. Otolaryngol. Abs.*

Benson, C. G., Gross, J. S., Suneja, S. K., and Potashner, S. J., 1997, Synaptophysin immunoreactivity in the cochlear nucleus after unilateral cochlear or ossicular removal. *Synapse* **25**: 243-57.

Bledsoe, S. C. Jr., Snead, C. R., Helfert, R. H., Prasad, V., Wenthold, R. J., and Altschuler, R. A., 1990, Immunocytochemical and lesion studies support the hypothesis that the projection from the medial nucleus of the trapezoid body to the lateral superior olive is glycinergic. *Brain. Res.* **517**:189-194.

Bledsoe, S. C., Nagase, S., Altschuler, R. A., and Miller, J. M., 1997, Changes in the central auditory system with deafness and return of activity via a cochlear prosthesis. In *Acoustic Signal Processing in the Central Auditory System*, (J. Syka, ed.), Plenum Press, New York, pp. 513-529.

Bledsoe, S. C., Nagase, S., Miller, J. M., and Altschuler, R. A., 1995, Deafness-induced plasticity in the mature central auditory system. *Neuroreport* **7**: 225-229.

Buras, E. D., Griffith, R. D., Asako, M., Holt, A. G., and Altschuler, R. A., 2003, Deafness related changes in glycine immunoreactivity in the rat superior olivary complex, *Assoc. Res. Otolaryngol. Abs.* 40.

Caspary, D. M., Pazara, K. E., Kossl, M., and Faingold, C. L., 1987, Strychnine alters the fusiform cell output from the dorsal cochlear nucleus. *Brain. Res.* **417**: 273-282.

Caspary, D. M., Backoff, P. M., Finlayson, P. G., and Palombi, P. S., 1994, Inhibitory inputs modulate discharge rate within frequency receptive fields of anteroventral cochlear nucleus neurons. *J. Neurophysiol.* **72**: 2124-2133.

Caspary, D. M., Salvi, R. J., Helfert, R. H., Nrozoski, T. J., and Bauer, C. A., 2000, Neuropharmacology of noise induced hearing loss in brainstem auditory structures. In *Abstracts from Noise Pollution Health Effects Reduction 2000 Conference (7-9 July 2000, UK), p.13.*

Caspary, D. M., Milbrandt, J. C., and Helfert, R. H., 1995, Central auditory aging: GABA changes in the inferior colliculus. *Exp. Gerontology* **30**: 349-60.

Caspary, D. M., Holder, T. M., Hughes, L. F., Milbrandt, J. C., McKernan, R. M., Naritoku, D. K., 1999, Age-related changes in GABA(A) receptor subunit composition and function in rat auditory system. *Neurosci.* **93**:307-12.

Finkbeiner, S., and Greenberg, M. E., 1998, Ca2+ channel-regulated neuronal gene expression. *J. Neurobiol.* **37**: 171-89.

Garcia, M. M., Edward, R., Brennan, G. B., and Harlan, R.E., 2000, Deafferentation-induced changes in protein kinase C expression in the rat cochlear nucleus. *Hear. Res.* **147**: 113-124.

Ghos, A., Ginty, D., Bading, H., and Greenberg, M. E., 1994a, Calcium regulation of gene expression in neuronal cells. *J. Neurobiol.* **25**: 294-303.

Ghosh, A., Carnahan, J., and Greenberg, M. E., 1994, Requirement for BDNF in activity-dependent survival of cortical neurons. *Science* **263**: 1618-23.

Ghosh, A., and Greenberg, M. E., 1995, Calcium signaling in neurons: molecular mechanisms and cellular consequences, *Science* **268**: 239-47.

Golding, N. L., and Oertel, D., 1996, Context-dependent synaptic action of glycinergic and GABAergic inputs in the dorsal cochlear nucleus. *J. Neurosci.* **16**: 2208-19.

Harty, T. P., and Manis, P. B., 1998, Kinetic analysis of glycine receptor currents in ventral cochlear nucleus, *J. Neurophysiol.* **79**: 1891-901.

Grothe, B., and Sanes, D. H., 1993, Bilateral inhibition by glycinergic afferents in the medial superior olive. *J. Neurophysiol.* **69**: 1192-1196.

Gulley, R. L., Wenthold, R. J., and Neises, G. R., 1978, Changes in the synapses of spiral ganglion cells in the rostral anteroventral cochlear nucleus of the waltzing guinea pig following hair cell loss. *Brain. Res.* **158**: 279-294.

Helfert, R. H., Sommer, T. J., Meeks, J., Hofstetter, P., and Hughes, L. F., 1999, Age-related synaptic changes in the central nucleus of the inferior colliculus of Fischer-344 rats, *J. Comp. Neurol.* **486**: 285-98.

Helfert, R. H., Juiz, J. M., Bledsoe, S. C., Bonneau, J., Wenthold, R. J., and Altschuler, R. A., 1992, Patterns of glutamate, glycine and GABA immunolabeling in four synaptic terminal classes in the lateral superior olive of the guinea pig. *J. Comp. Neurol.* **323**: 305-325.

Holt, A. G., Asako, M., Lomax, C., and Lomax, M. I., 2003, Microarray analysis of genes expressed in the inferior colliculus of the rat following cochlear ablation, *Society for Neuroscience Abs.*

Idrizbegovic, E., Bogdanovic, N., and Canlon, B., 1998. Modulating calbindin and parvalbumin immunoreactivity in the cochlear nucleus by moderate noise exposure in mice. A quantitative study on the dorsal and posteroventral cochlear nucleus. *Brain. Res.* **800**: 86-96.

Illing, R.B., Horvath, M., and Laszig, R., 1997, Plasticity of the auditory brainstem: effects of cochlear ablation on GAP-43 immunoreactivity in the rat. *J. Comp. Neurol.* **382**: 116-138.

Juiz, J. M., Helfert, R. H., Bonneau, J. M., Wenthold, R. J., and Altschuler, R. A., 1996, Three classes of inhibitory amino acid terminals in the cochlear nucleus of the guinea pig. *J. Comp. Neurol.*, **373**: 11-26.

Kolston, J., Osen, K. K., Hackney, C. M., Ottersen, O. P., and Storm-Mathissen, J., 1992, An atlas of glycine and GABA-like immunoreactivity and colocalization in the cochlear nuclear complex of the guinea pig. *Anat. Embryol.* **186**: 443-465.

Kornhauser, J. M., Cowan, C. W., Shaywitz, A. J., Dolmetsch, R. , Griffith, E., Hu, L. S., Haddad, C., Xia, Z., and Greenberg, M. E., 2002, CREB transcriptional activity in neurons is regulated by multiple, calcium-specific phosphorylation events. *Neuron* **34**: 221-33.

Krenning, J., Hughes, L. F., Caspary, D. M., and Helfert, R.H., 1998, Age related glycine receptor subunit changes in the cochlear nucleus of Fischer-344 rats, *Laryngoscope* **108**: 26-31.

Kuwabara, N., and Zook, J. M., 1992, Projections to the medial superior olive from the medial and lateral nuclei of the trapezoid body in rodents and bats. *J. Comp. Neurol.* **324**:522-538.

Kuwabara, N, and Zook, J. M., 1999, Local collateral projections from the medial superior olive to the superior paraolivary nucleus in the gerbil. *Brain Res.* **846**: 59-71.

Lesperance, M. M., Helfert, R. H., and Altschuler, R. A., 1995, Deafness induced cell size changes in rostral AVCN of the guinea pig. *Hear. Res.* **86**: 77-81.

Merchan, M., Aguilar, L., Lopez-Poveda, E. A., and Malmierca, M. S., 2003, Quantitative and qualitative analysis of inhibitory amino acids in the inferior colliculus of the rat, In *Plasticity of the Central Auditory System and Processing of Complex Acoustic Signals*, Prague 7. – 10. 7. 2003, p.5.

Milbrandt, J. C., Hunter, C., and Caspary, D. M., 1997, Alterations of GABA-A receptor subunit mRNA levels in aging Fisher 344 rat inferior colliculus, *J. Comp. Neurol.* **379**: 455-465.

Miller, J. M., Altschuler, R. A., Niparko, J. K., Hartshorn, D. O., Helfert, R. H., and Moore, J. K., 1991, Deafness-induced changes in the central auditory system and their reversibility and prevention. In *Noise Induced hearing loss* (Dancer, Henderson, Salvi, Hamernik, eds.). Mosby Year Book, St. Louis

Moore, D. R., 1994, Auditory brainstem of the ferret: long survival following cochlear removal progressively changes projections from the cochlear nucleus to the inferior colliculus. *J. Comp. Neurol.* **339**: 301-310.

Morest, D. K., and Bohne, B. A., 1983, Noise-induced degeneration in the brain and representation of inner and outer hair cells. *Hear. Res.* **9**: 145-151.

Morest, D. K., Kim, J., and Bohne, B. A., 1997, Neuronal and transneuronal degeneration of auditory axons in the brainstem after cochlear lesions in the chinchilla: cochleotopic and non-cochleotopic patterns. *Hear. Res.* **103**: 151-168.

Mossop, J. E., Wilson, M. J., Caspary, D. M., and Moore, D. R., 2000, Down-regulation of inhibition following unilateral deafening. *Hear. Res.* **147**: 183-18.

Muly, S. M., Gross, J. S., Morest, D. K., Potashner, S. J., 2002, Synaptophysin in the cochlear nucleus following acoustic trauma. *Exp. Neurol.* **177**: 202-221.

Nakagawa, H., Sato, K., Shiraishi, S., Kuriyama, H., and Altschuler, R. A., 2000, NMDAR1 isoforms in the rat superior olivary complex and changes after unilateral cochlear ablation. *Molecular Brain Res.*, **77**: 246-257.

Nagase, S., Dupont, J., Miller, J. M., Prieskorn, D., and Altschuler, R. A., 2000, Deafness related changes in fos expression from cochlear electrical stimulation in the rat inferior colliculus, *Hear. Res.* **147**: 242-250.

Nishiyama, N., Hardie, N. A., and Shepherd, R. K., 2000, Neonatal sensorineural hearing loss affects neurone size in cat auditory midbrain. *Hear. Res.* **140**: 18-22.

Pasic, T. R., Moore, D. R., and Rubel, E. W., 1994, Effect of altered neuronal activity on cell size in the medial nucleus of the trapezoid body and ventral cochlear nucleus of the gerbil. *J. Comp. Neurol.* **348**: 111-120.

Pasic, T. R., and Rubel, E. W., 1989, Rapid changes in cochlear nucleus cell size following blockade of auditory nerve electrical activity in gerbils. *J. Comp. Neurol.* **283**: 474-480.

Pasic, T. R., and Rubel, E. W., 1991, Cochlear nucleus cell size is regulated by auditory nerve electrical activity. *Otolaryngol. Head Neck Surg.* **104**: 6-1.

Potashner, S. J., Suneja, S. K., and Benson, C. G., 1997, Regulation of D-aspartate release and uptake in the adult brain stem auditory nuclei after unilateral middle ear ossicle removal and cochlear ablation. *Exp. Neurology* **148**: 222-235.

Potashner, S. J., Suneja, S. K., and Benson, C. G., 2000, Altered glycinergic synaptic activities in guinea pig brain stem auditory nuclei after cochlear ablation. *Hear. Res.* **147**: 125-136

Redd, E. E., Cahill, H. B., Pongstaporn, T., and Ryugo, D. K., 2002, The effects of congenital deafness on auditory nerve synapses: Type I and Type II multipolar cells in the anteroventral cochlear nucleus of cats. *J. Assoc. Res. Otolaryngol.* **3**: 403-417.

Redd, E. E., Pongstaporn, T., and Ryugo, D. K., 2000, The effects of congenital deafness on auditory nerve synapses and globular bushy cells in cats. *Hear. Res.* **147**: 160-174.

Rubel, E. W., Hyson, R. L., and Durham D., 1990, Afferent regulation of neurons in brain stem auditory system. *J. Neurobiol.* **21**: 169-96.

Ryugo, D. K., Rosenbaum, B. T., Kim, P. J., Niparko, J. K., Saada, A. A., 1998, Single unit recordings in the auditory nerve of congenitally deaf white cats: morphological correlates in the cochlea and cochlear nucleus. *J. Comp. Neurol.* **397**: 532-548.

Sato, K., Kuriyama, H., and Altschuler, R. A., 1995, Expression of glycine receptor subunits in the cochlear nucleus and superior olivary complex using non-radioactive in-situ hybridization. *Hear. Res.* **91**: 7-18.

Sato, K., Kuriyama, H., and Altschuler, R. A., 2000, Expression of glycine receptor subunit mRNAs in the rat cochlear nucleus, *Hear. Res.* **144**: 47-52.

Sato, K., Shiraishi, S., Nakagawa, H., Kuriyama, H., and Altschuler, R. A., 2000, Diversity and plasticity in amino acid receptor subunits in the rat auditory brain stem, *Hear. Res.* **147**: 137-144.

Suneja, S. K., Potashner, S. J., and Benson, C. G., 1998, Plastic changes in glycine and GABA release and uptake in the adult brain auditory nuclei after unilateral middle ear ossicle removal and cochlear ablation, *Exp. Neurol.* **151**: 273-288.

Suneja, S. K., Benson, C. G., and Potashner, S.J., 1998, Glycine receptors in the adult guinea pig brain auditory nuclei: regulation after cochlear ablation. *Exp. Neurol.* **154**: 473-488.

Suneja, S. K., Potashner, S. J., 2003, ERK and SAPK signaling in auditory brainstem neurons after unilateral cochlear ablation, *J. Neurosci. Res.* **73**: 235-45.

Syka, J., and Rybalko, N., 2000, Threshold shifts and enhancement of cortical evoked responses after noise exposure in rats. *Hear. Res.* **139**: 59-68.

Syka, J., 2002, Plastic changes in the central auditory system after hearing loss, restoration of function, and during learning. *Physiol. Rev.* **82**: 601-636.

Takasu, M. A., Dalva, M. B., Zigmond, R. E., and Greenberg, M. E., 2002, Modulation of NMDA receptor-dependent calcium influx and gene expression through EphB receptors. *Science* **295**: 491-5.

Wenthold, R. J., Huie, D., Altschuler, R. A., and Reeks, K. A., 1987, Glycine immunoreactivity localized in the cochlear nucleus and superior olivary complex. *Neurosci.* **22**: 897-912.

West, A. E., Chen, W. G., Dalva, M. B., Dolmetsch, R. E., Kornhauser, J. M., Shaywitz, A. J., Takasu, M. A., Tao, X., and Greenberg, M. E., 2002, Calcium regulation of neuronal gene expression. *Proc Natl Acad Sci USA* **98**: 11024-31.

Wickesberg, R. E., and Oertel, D., 1990, Delayed, frequency-specific inhibition in the cochlear nuclei of mice: a mechanism for monaural echo suppression. *J. Neurosci.* **10**: 1762-1768.

Wickesberg, R. E., Whitlon, D., and Oertel, D., 1994, In vitro modulation of somatic glycine-like immunoreactivity in presumed glycinergic neurons. *J. Comp. Neurol.* **339**: 311-327.

Willott, J. F., Aitkin, L. M., and McFadden, S. L., 1993, Plasticity of auditory cortex associated with sensorineural hearing loss in adult C57BL/6J mice. *J. Comp. Neurol.* **329**: 402-411.

Willot, J. F., Milbrandt, J. C., Bross, L. S., and Caspary, D. M., 1997, Glycine immunoreactivity and receptor binding in the cochlear nucleus of C57BL/6J and CBA/CaJ mice: Effects of cochlear impairment and aging. *J. Comp. Neurol.* **385**: 45-414.

Willott, J. F., and Turner, J. G., 2000, Neural plasticity in the mouse inferior colliculus: relationship to hearing loss, augmented acoustic stimulation, and prepulse inhibition. *Hear. Res.* **147**: 275-281.

Winsky, L., and Jacobowitz, D. M., 1995, Effects of unilateral cochlea ablation on the distribution of calretinin mRNA and immunoreactivity in the guinea pig ventral cochlear nucleus. *J. Comp. Neurol.* **354**: 564-582.

Wu, S. H., and Kelly, J. B, 1994, Physiological evidence for ipsilateral inhibition in the lateral superior olive: synaptic responses in mouse brain slice. *Hear. Res.* **73**: 57-64.

Wu, S. H., and Oertel, D., 1986, Inhibitory circuitry in the ventral cochlear nucleus is probably mediated by glycine, *J. Neurosci.* **6**: 2691-706.

Zhang, J., Suneja, S. K., and Potashner, S. J., 2003, Protein kinase A and calcium/calmodulin-dependent protein kinase II regulate D-[3H]aspartate release in auditory brain stem nuclei. *J. Neurosci. Res.* **74**: 81-90.

Zhang, J., Suneja, S. K., and Potashner, S. J., 2003, Protein kinase C regulation of glycine and gamma-aminobutyric acid release in brain stem auditory nuclei. *Exp. Neurol.* **182**: 75-86.

Challenges to a Neuroanatomical Theory of Forebrain Auditory Plasticity

JEFFERY A. WINER*, CHARLES C. LEE*, KAZUO IMAIZUMI#, and CHRISTOPH E. SCHREINER#
*Department of Molecular and Cell Biology, University of California at Berkeley, Berkeley, CA, USA; and #W.M. Keck Center for Integrative Neuroscience, University of California at San Francisco, San Francisco, CA, USA

1. INTRODUCTION

The mature brain performs paradoxical tasks. It encodes sensory experience accurately and with fidelity, and it modifies otherwise stable maps of the ears, eyes, and body to represent learning (Weinberger *et al.*, 1984), forgetting (Bakin and Weinberger, 1990), and manipulation of stimulus parameters and statistics (Zhang *et al.*, 2002). These two functional modes have elicited attention at the synaptic (Metherate and Ashe, 1995) and systems (Wall, 1988) levels of analysis, both of which have dynamic (Calford, 2002) as well as metastable (Schieber, 2001) elements. Reconciling the conflicting requirements of stability and lability in the adult map is a profound challenge for systems neuroscience that has received surprisingly little attention at the neuroanatomical level of discourse.

The finding that primary auditory cortex (AI) neurons can reorganize their tonotopic map rapidly and globally in a frequency-specific manner (Kilgard and Merzenich, 1998) suggests that the continuous impact of sensory plasticity on shaping local processing (Weinberg, 1997) may have been underestimated (Kaas, 1997). A key issue for understanding the limits of such systems-level plasticity is to propose, and ultimately to test, a theory of neural substrates that plausibly encode, or that instruct the brain to represent, experience selectively (Recanzone *et al.*, 1993), specifically

(Kaas, 1999), rapidly (Recanzone, 1998), and enduringly (Galvan and Weinberger, 2002). This chapter examines some features of forebrain auditory circuits that may be related to these processes. The aim of this exposition is frankly speculative, the intent to encourage the debate and discourse that must precede any more coherent theory.

2. THE PROBLEM OF PLASTICITY

Is there any systems-level neuroanatomical substrate, such as a specific connection or an unusual concentration of interneurons, which might support the apparently antithetical functional requirements of stability and lability in maps? Are adjustable maps of receptive field properties a general feature of auditory forebrain organization?

2.1 Maps as Conceptual Tools

In cat auditory cortex, about a dozen areas have either a tonotopic representation, or neurons that prefer acoustic stimuli, or are connected with the medial geniculate body (Winer, 1992). Only five areas (Fig. 1A: light gray) have more or less complete maps of the cochlear sensory epithelium (Imig and Reale, 1980), with local functional differences between areas (Eggermont, 1998; Linden *et al.*, 2003). Interleaved with the tonotopic maps in AI are related representations of binaural (Middlebrooks *et al.*, 1980) or amplitopic (Schreiner *et al.*, 1992) organization, and gradients for sharpness of tuning (Read *et al.*, 2001) or timing (Cheung *et al.*, 2001) (Fig. 1B). In contrast, the other fields have, at best, only a coarse gradient of characteristic frequency (Schreiner, 1995; Schreiner and Cynader, 1984), though their thalamic, corticocortical, and commissural connections are as topographic as those in AI (Lee and Winer, 2003). This raises questions as to why tonotopic maps are not more common and how functional domains in non-tonotopic cortex are organized.

Figure 1. Auditory cortex subdivisions and laminar and functional organization of layers. (A) Three functional types of auditory cortex areas in a lateral view of a cat hemisphere. (B) A module of auditory cortex ~2 × 1.5 mm (height × width) was rotated to reveal its internal structure. Multiple sensory representations are interleaved and overlap, including frequency (Cochleotopic Axis) and modules with different spectral responses (Spectral Integration). I-VI, cortical layers; AAF, anterior auditory field; AI, primary auditory cortex; AII, second auditory cortex area; BB, broadband submodule; CC, corticocollicular projection; cNB,

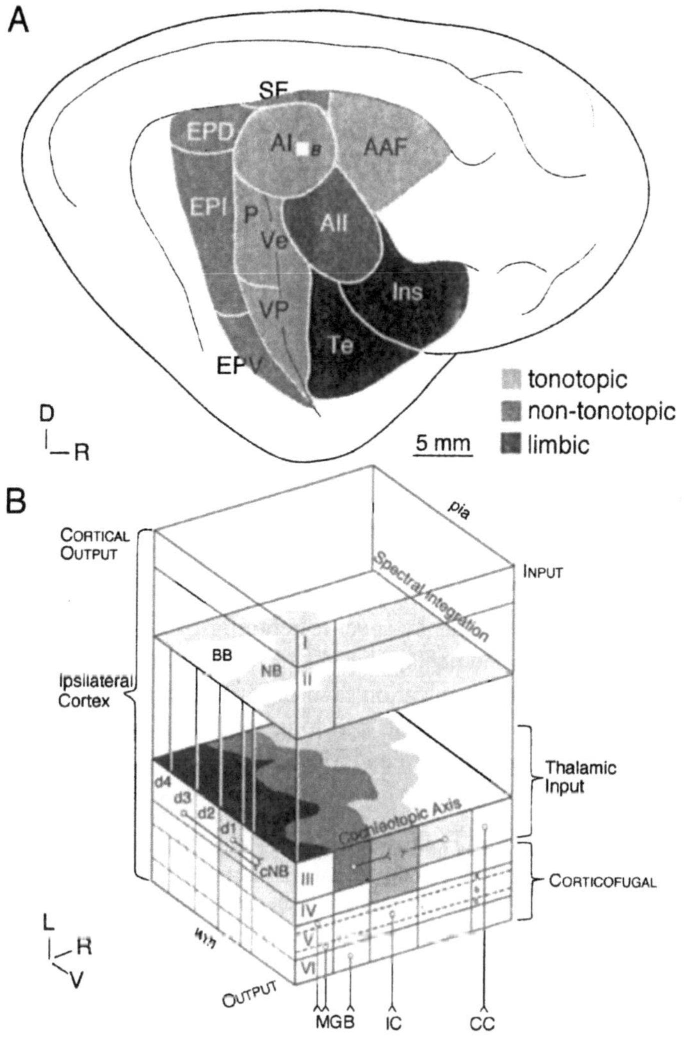

central narrowband region; D, dorsal; d1-d4, regions of sharp tuning parallel to the central narrowband region and projecting to it; EPD, EPI, EPV, dorsal, intermediate, and ventral fields of posterior ectosylvian gyrus; IC, inferior colliculus; Ins, insular cortex; L, lateral; MGB, medial geniculate body; NB, narrowband submodule; P, posterior auditory field; R, rostral; SF, suprasylvian fringe auditory area; Te, temporal cortex; Ve, ventral or ventral auditory field; VP, ventral posterior auditory field; wm, white matter. White ovals in layer III: presumptive γ-aminobutyric acid-containing interneurons mediating lateral intralaminar connectivity.

2.2 Maps as Matrices

One view of cortical (and perhaps of subcortical) maps is that they might represent an armature upon which functional subdomains are arrayed. This scaffold enables concurrent processing of different auditory tasks. It permits sequential operations, and it minimizes connectional path length in a system where spatial constraints are severe and connectivity is at a premium (Klyachko and Stevens, 2003). A corollary of this view is that such maps could provide a framework for recovery of function after peripheral damage, since the map itself may be surprisingly well preserved via its many remaining connections (Fig. 1B) even after severe trauma (Irvine *et al.*, 2000). Other than developmental plasticity, this view accentuates the modular architecture of cortex, it has little allowance for large scale reorganization, and it makes no claims as to the purpose or functional capacity of non-tonotopic areas. For brevity, we call this the classical view.

3. THE DYNAMIC MAP

One paradigm for frequency-specific retuning of AI is via activation of cholinergic neurons in the nucleus basalis of Meynert in awake rodents. This elicits a concomitant reorganization favoring the characteristic frequency of the stimulus, and evokes specific and long term changes in the frequency representation in rat AI (Kilgard and Merzenich, 1998). These experiments

Figure 2. Principal connections of auditory cortex. (A) A deposit of biotinylated dextran amines in the ventral division (V) of the medial geniculate body. Black, core of injection; gray, border of diffusion. (B) The ensuing anterograde thalamocortical connections (gray) are highly divergent. (C) A deposit of wheat germ agglutinin conjugated to horseradish peroxidase in AI. (D) Thalamocortical (black dots) and corticothalamic (gray) projections largely overlap, though there are heterotopic thalamocortical neurons (black dots) >1 mm from the center of the densest aggregate of neurons, and zones of retrograde somatic labeling without corresponding anterograde input. (E) Corticocollicular input (gray) concentrates in inferior colliculus subdivisions outside the central nucleus. Decimals, percentages from the caudal pole of the inferior colliculus or the medial geniculate body. For auditory cortex abbreviations see Fig. 1. I-IV, layers of the dorsal cortex of the inferior colliculus; CG, central gray; CIC, commissure of the inferior colliculus; CN, central nucleus of the inferior colliculus; D, dorsal nucleus of the medial geniculate body *or* dorsal; DC, dorsal cortex of the inferior colliculus; DD, deep dorsal nucleus of medial geniculate body; DS, dorsal superficial nucleus of the medial geniculate body; IcT, intercollicular tegmentum; L, lateral; LN, lateral nucleus of the inferior colliculus; M, medial division of the medial geniculate body *or* medial; Ov, *pars ovoidea* of the ventral division of the medial geniculate body; R, rostral; Sgl, suprageniculate nucleus, lateral part; Vl, ventrolateral nucleus of the medial geniculate body.

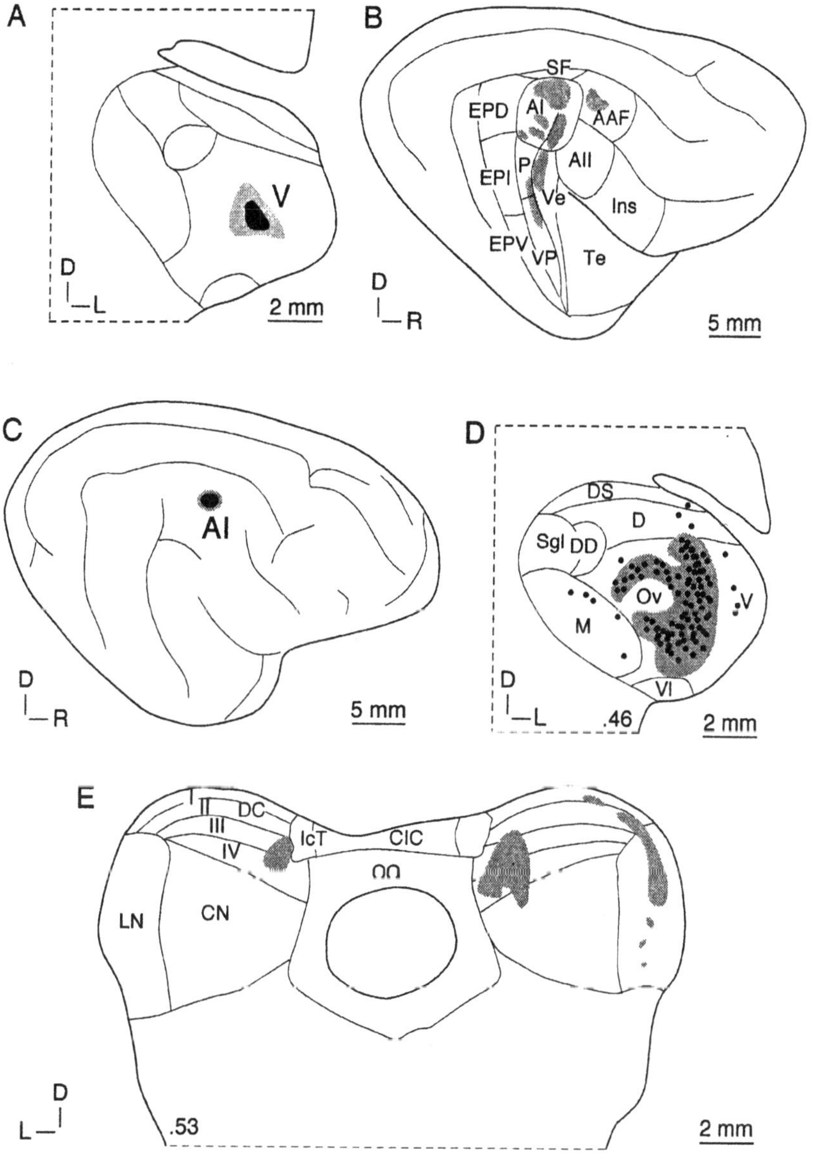
A
V
D
L
2 mm
B
SF
AI
AAF
EPD
EPI
P
AII
Ve
Ins
EPV
VP
Te
D
R
5 mm
C
AI
D
R
5 mm
D
DS
D
Sgl
DD
Ov
V
M
VI
D
L
.46
2 mm
E
I
II
III
IV
DC
IcT
CIC
LN
CN
D
L
.53
2 mm

and others in which visual input is experimentally diverted onto the auditory forebrain (Frost, 1981) and transforms it into a visual area, suggest that the implements of map formation may inhere in primary areas, irrespective of the source of peripheral input (Pallas *et al.*, 1990). These properties are not fully defined, but covariance and neighborhood relationships in the input patterns may contribute to map formation. These findings each challenge the notion of immutable representation implicit in the classical view (Weinberger, 1998).

4. NEUROANATOMICAL SUBSTRATES

Since no specific neuroanatomical substrate has been identified as essential in global plasticity, we consider several candidates. These examples are limited to the cat auditory system unless specified otherwise.

4.1 Axonal Divergence

One model of auditory thalamocortical organization is as a network of point-to-point relations between thalamic neurons and their axon terminal domains in AI (Brandner and Redies, 1990). This model is based on retrograde tract tracing and is at odds with studies in which small deposits of anterograde tracer in the thalamus label large cortical zones (Fig. 2A,B) (Huang and Winer, 2000). The latter result is congruent with the widespread divergence of thalamocortical axons within one area in the visual (Ferster and LeVay, 1978) and somatic sensory (Landry and Deschênes, 1981) systems. A single fiber can span several millimeters of cortex, well beyond the topographic limits predicted by the point-to-point model. This suggests that a native and mechanistic, rather than a computational and flexible, substrate enables topographically organized thalamic neurons to activate cortical regions beyond their expected topographic domain.

Analogous patterns of divergence are present in the corticothalamic (Winer *et al.*, 2001) and corticocollicular (Fig. 2E) (Winer *et al.*, 1998) pathways, in which small cortical deposits (Fig. 2C) of anterograde tracer reveal divergent projections beyond the isofrequency domain.

4.2 Heterotopic Projections

This family of convergent connectional relations is documented by retrograde tract tracing in the cortex and thalamus. It suggests that there is an ample basis for topographically unexpected (heterotopic) projections, much like those documented in divergence experiments.

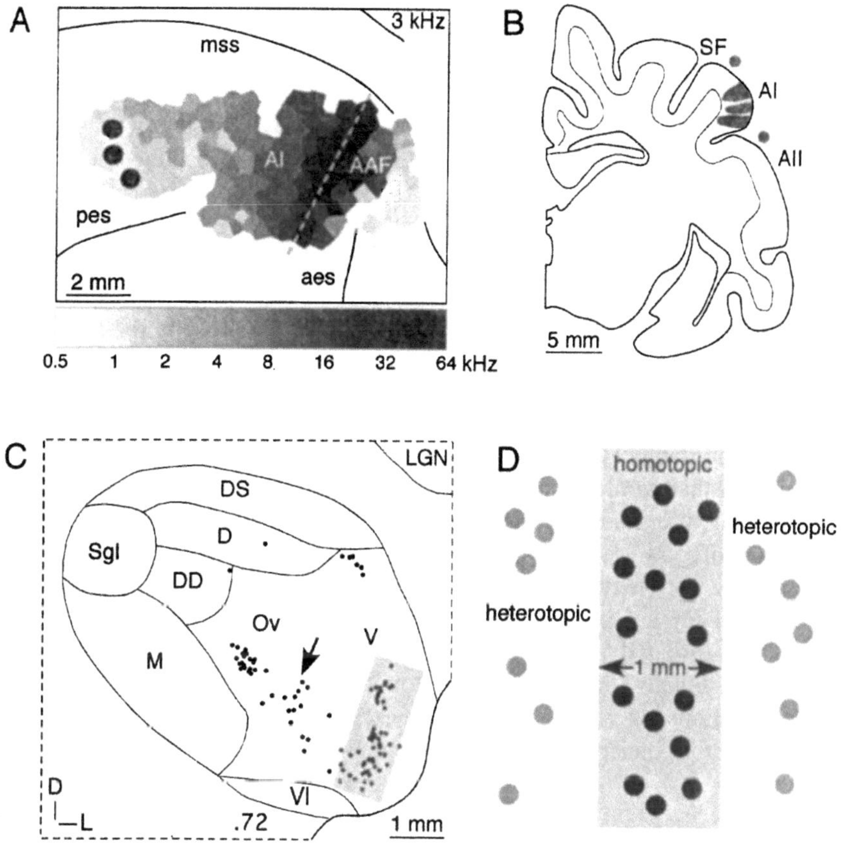

Figure 3. Defining heterotopic projections. (A) Map of characteristic frequency (gray squares) in area AI and the three deposit sites (black circles) of cholera toxin β fragment (CTβ). The Voronoi-Dirichlet tessellation was used to analyze the spread of tracer in an isofrequency region in a deposit centered at 3 kHz. (B) A transverse view of the deposit sites (black, including diffusion). (C) Retrogradely labeled medial geniculate body neurons (dots); gray square, geometric center of retrograde labeling, with medial dispersion (outside gray box). (D) Identification of homo- and heterotopically labeled neurons; the gray box spans ~1 octave, ~1 mm of neural space, a conservative estimate of auditory thalamic (Morel and Imig, 1987) and cortical (Merzenich *et al.*, 1975) octave width (see panel A); neurons outside this region are in frequency-inappropriate, heterotopic regions. For medial geniculate body abbreviations, see Fig. 1. AAF, anterior auditory field; aes, anterior ectosylvian sulcus; AI, primary auditory cortex; AII, second auditory cortical area; D, dorsal; L, lateral; LGN, lateral geniculate nucleus; mss, middle suprasylvian sulcus; pes, posterior ectosylvian sulcus; SF, suprasyvian fringe auditory area.

If the point-to-point model is valid, then cortical tracer deposits restricted to frequency-specific loci (Fig. 3A,B) should label thalamic neurons exclusively in a matching tonotopic region. This is not the case (Fig. 3C): up to 30% of the neurons lay beyond the geometric center of a 1 mm-wide zone (Fig. 3D) that is far larger than the appropriate thalamic frequency region (Imig and Morel, 1985).

Studies of corticocortical interareal connectivity also label a heterotopic population (Fig. 4A) far beyond their expected locus, in tonotopic (Fig. 4A:AI, arrow) and non-tonotopic (Fig. 4A:AII, arrows) areas (Lee *et al.*, 2004).

The same heterotopic principle of input is found in the commissural pathway (Fig. 4B:AI, arrow), suggesting that such projections might be a feature of all extrinsic connections of auditory cortex and perhaps in other cortical areas. While overall labeling is much reduced relative to homotopic corticocortical projections, heterotopic commissural input up to 5 mm from the center of homotopic projections is present.

Topographically inappropriate (heterotopic) connections satisfy several criteria that would seem essential to a theory of global plasticity. They can furnish information that is not limited to the immediate parametric neighborhood of the target region, thus providing the substrate for large-scale receptive field adjustments. They are sufficiently robust for such a role. They are present in all types of thalamic and cortical connectivity, implying that dynamic changes may be engendered through any of these routes. Finally, they occur in non-tonotopic areas which are either incapable of plasticity or whose capacity for experience-dependent reorganization is outside the conceptual span of any contemporary theory of plasticity.

5. A MODEL OF PLASTICITY

Even though the processes underlying map formation and reorganization are known imperfectly, several functional aspects must be involved. These impose pragmatic and conceptual challenges for theory.

Figure 4. Cortical retrograde connections after CTβ deposits in AI. (A) Ipsilateral corticocortical projections are topographic. Arrows, heterotopic neurons (cf. Fig. 3C) are identified in all labeled areas. (B) Commissural heterotopic labeling is sparse, but clustered. Abbreviations as in Figs. 1 and 3.

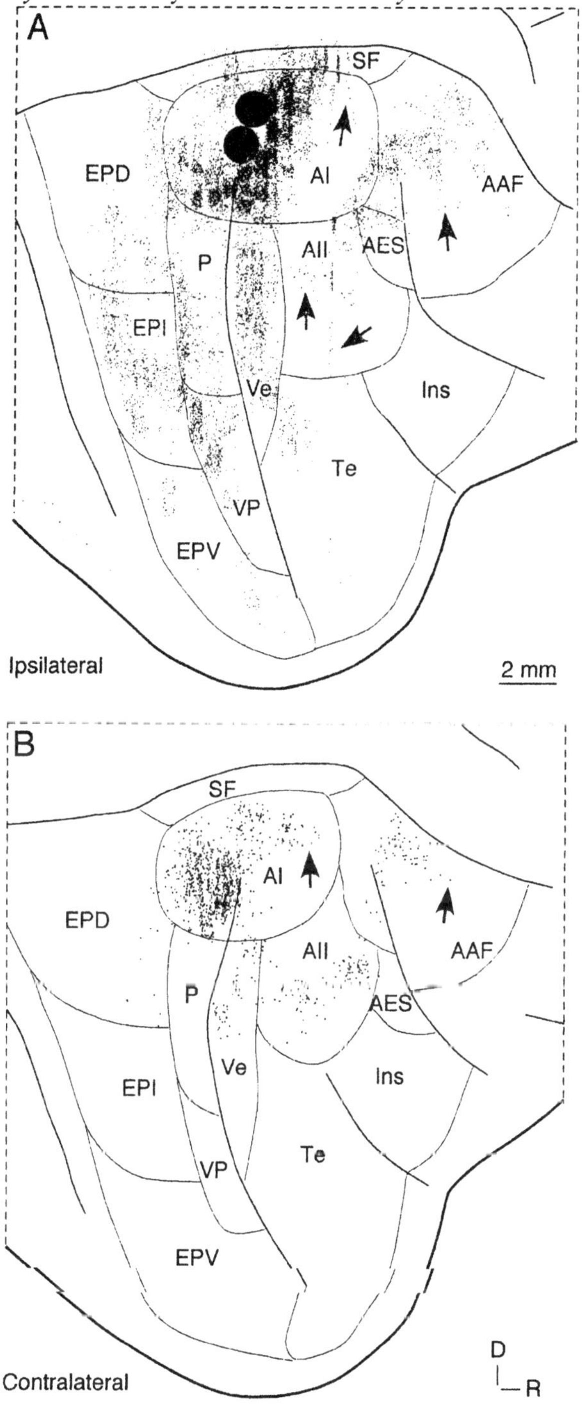
A
SF
EPD
AI
AAF
AII
AES
P
EPI
Ve
Ins
Te
VP
EPV
Ipsilateral
2 mm
B
SF
AI
EPD
AII
AAF
P
AES
Ve
EPI
Ins
Te
VP
EPV
Contralateral
D
R

5.1 Specificity

Does AI reorganization entail a parallel tonotopic realignment in other tonotopic fields, in associated subcortical centers, and in other cortical fields connected to AI (Rouiller *et al.*, 1991)? If subcortical centers are not involved (Rajan and Irvine, 1998), then their cholinergic input from sources other than the nucleus basalis (Fitzpatrick *et al.*, 1989) may be unrelated to such reorganization. Likewise, the plasticity of other tonotopic fields and of non-tonotopic cortex, each of which is likely the target of nucleus basalis input, remains to be demonstrated. If AI alone shows these changes, it must be explained how areas and nuclei robustly connected to it are insulated from transneuronal and polysynaptic plastic effects.

5.2 Initiation

How does nucleus basalis stimulation trigger or enable widespread topographic experimental changes (Kilgard and Merzenich, 1998)? Brainstem afferents to nucleus basalis are limited and arise from secondary auditory sources such as the peripeduncular nucleus and from the hypothalamus (Jones *et al.*, 1976). Since nucleus basalis has no known role in hearing, the mechanism which synchronizes its activity in behaving animals with that of auditory cortex is unknown (Fig. 5B).

5.3 Recruitment

How does the topographic map expand? Heterotopic projections could be recruited via existing connections (Fig. 5C: dark gray boxes) by unmasking excitatory (Winer and Prieto, 2001) or inhibitory (Winer *et al.*, 1999) circuits through unspecified lateral processes. Such substrates likely exist in the thalamocortical (Huang *et al.*, 1999), corticothalamic (Villa *et al.*, 1991) and corticocollicular (Mitani *et al.*, 1983) systems.

Figure 5. A multistage model of auditory cortex plasticity. (A) A hypothetical isofrequency module (see Fig. 1B) reduced to two dimensions; only one neuron is shown for illustrative purposes, the layer V pyramidal cell (black triangle) (Winer and Prieto, 2001). I, III, V, cortical layers; ACh, cholinergic; AI, primary auditory cortex; B7-B9, aminergic cell groups; GABA, γ-aminobutyric acid; Glu, glutamate; MGB, medial geniculate body; NA, noradrenergic; 5-HT, 5-hydroxytryptamine; arrows, putative anterograde connections (cf. Fig. 1B). (B) Nucleus basalis activation triggers cortical plasticity. (C) Corticothalamic cells activate heterotopic medial geniculate body neurons. (D) The activation of heterotopic neurons expands the frequency specific cortical module (gray) from its static position (clear vertical rectangle in panels A-C). (E) Activation of the thalamic reticular nucleus confers attentional impact to the reorganizing cortex. (F) Consolidation of corticocortical and commissural subsystems follows to coordinate the global response. See text for details.

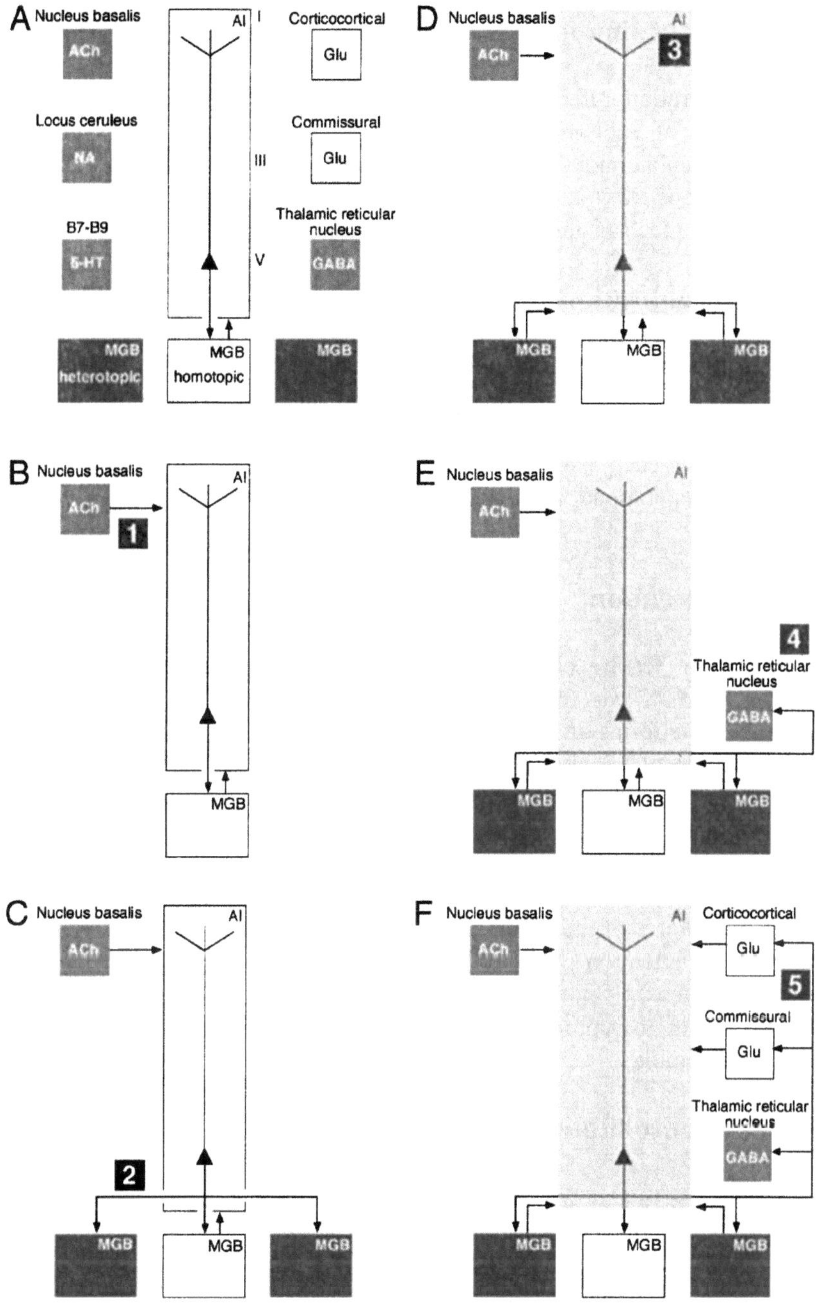

A
Nucleus basalis
ACh
Locus ceruleus
NA
B7-B9
5-HT
AI
I
III
V
Corticocortical
Glu
Commissural
Glu
Thalamic reticular nucleus
GABA
MGB heterotopic
MGB homotopic
MGB
B
Nucleus basalis
ACh
1
AI
MGB
C
Nucleus basalis
ACh
AI
2
MGB
MGB
MGB
D
Nucleus basalis
ACh
AI
3
MGB
MGB
MGB
E
Nucleus basalis
ACh
AI
4
Thalamic reticular nucleus
GABA
MGB
MGB
MGB
F
Nucleus basalis
ACh
AI
Corticocortical
Glu
5
Commissural
Glu
Thalamic reticular nucleus
GABA
MGB
MGB
MGB

5.4 Coordination

This operation refers to the necessity to synchronize activity in the subdivisions of auditory cortex, or between the auditory cortex and the medial geniculate body (Fig. 5D: arrows). Given the several cortical tonotopic maps, it would seem essential that they maintain continuity and use common mechanisms of statistics and scaling.

5.5 Calibration

This is the counterpart to coordination, except that it occurs at the level of perception and across modalities. It would require, for example, that the spatial map for sound localization and for head movements in the superior colliculus are each calibrated in their internal registration and outputs. Analogous operations are implicit in the premotor and proprioceptive domains.

5.6 Attention

A striking feature of one demonstration of map lability (Kilgard and Merzenich, 1998) was that attentional mechanisms were not implicated, and certainly some forms of instrumental learning do not require attention per se as a condition of learning (Paller *et al.*, 2003). In adult monkeys, changes induced by perceptual learning are expressed only in the attended modality (Recanzone *et al.*, 1992; Recanzone *et al.*, 1993). By contrast, exposure of juveniles prior to a critical period or in adults undergoing nucleus basalis stimulation may not require attention to induce plasticity. In these cases, statistical features of the dominant stimulus environment shape the changes. Nonetheless, attention would seem intuitively necessary (Fig. 5E: thalamic reticular nucleus), if not to initiate change, then to imbue it with cognitive significance or to suppress responses to otherwise irrelevant or negatively reinforced stimuli.

5.7 Consolidation

Some as-yet-unknown process must either maintain the dynamic map (Fig. 5F) in a new steady state or supervise its return to the original state.

5.8 Caveats

The model outlined above is preliminary and provisional. Its several elements can only be validated by experimental test. It makes no claim to inclusiveness. In its extended form it would include synaptic events and Hebbian processes. Several assumptions implicit in it require comment.

5.8.1 Locus of Plasticity

No conclusion that the cortex is the singular source of plasticity is warranted. Subcortical sites (Kaas *et al.*, 1999) may be required to enable, implement, or express it, and it would be difficult to see how they would be isolated from its effects since they are part of the ascending pathway as well as a target of descending control (Kaas, 1999).

5.8.2 Temporal Sequence of Expression

The scheme presented here (Fig. 5) is illustrative only, and its internal order is arbitrary and contingent on further investigation. A presumptive eliciting role is proposed for subcortical cholinergic (and perhaps for brainstem aminergic) projections.

5.8.3 Tempo of Reorganization

Experiments with digit inactivation (Calford *et al.*, 1993) or sound exposure (Diamond and Weinberger, 1986) suggest a rapid phase for certain facets of cortical reorganization and plasticity. It remains to be seen how this might relate to changes evoked by stimulation, and what mechanism governs the pace of plasticity.

5.8.4 Extending the Theory

A valid theory of forebrain plasticity should include conscious performance. If heterotopic connections are active in awake, behaving mammals, how might they be manifested? Perhaps their impact would degrade the otherwise precise tonotopic map via extensive axonal divergence in the thalamocortical system, contributing to a cortical representation of characteristic frequency whose features include broadened tuning of individual neurons, and a gradient rather than a strict map of frequency. It may not be coincidental that these features are functional hallmarks of the tonotopic arrangement of the awake, dynamic auditory thalamus (Morel *et al.*, 1987) and cortex (Evans *et al.*, 1965).

The study of plasticity must enter a new phase which incorporates map interactions within and between cortical and subcortical structures. A more systematic theory would include sensory, motor, and integrative frames of reference in the behaving organism (Calford, 2002). A further issue is the plastic potential of other parametric representations (Fig. 1B).

6. CONCLUDING REMARKS

Few auditory cortex areas have a tonotopic organization (Schreiner *et al.*, 2000). Nonetheless, those areas devoid of such organization have thalamic, commissural, and corticocortical input that is as precise and topographic as that of their tonotopic neighbors, and they have comparable heterotopic projections (Lee and Winer, 2003). How these non-tonotopic areas might likewise reorganize is a question we cannot yet contemplate from our present, tonotopocentric viewpoint.

ACKNOWLEDGEMENTS

These studies were supported by National Institutes of Health grants to J.A.W. (R01 DC02319-24) and C.E.S. (R01 DC2260-07 and POI NS34835-07).

REFERENCES

Bakin, J. S., and Weinberger, N. M., 1990, Classical conditioning induces CS-specific receptive field plasticity in the auditory cortex of the guinea pig. *Brain Res.* **536**: 271-286.

Brandner, S., and Redies, H., 1990, The projection of the medial geniculate body to field AI: organization in the isofrequency dimension. *J. Neurosci.* **10**: 50-61.

Calford, M. B., 2002, Dynamic representational plasticity in sensory cortex. *Neuroscience* **111**: 709-738.

Calford, M. B., Rajan, R., and Irvine, D. R. F., 1993, Rapid changes in the frequency tuning of neurons in cat auditory cortex resulting from pure-tone-induced temporary threshold shift. *Neuroscience* **55**: 953-964.

Cheung, S. W., Bedenbaugh, P. H., Nagarajan, S. S., and Schreiner, C. E., 2001, Functional organization of squirrel monkey primary auditory cortex: responses to pure tones. *J. Neurophysiol.* **85**: 1732-1749.

Diamond, D. M., and Weinberger, N. M., 1986, Classical conditioning rapidly induces specific changes in frequency receptive fields of single neurons in secondary and ventral ectosylvian auditory cortical fields. *Brain Res.* **372**: 357-360.

Eggermont, J. J., 1998, Representation of spectral and temporal sound features in three cortical fields of the cat. Similarities outweigh differences. *J. Neurophysiol.* **80**: 2743-2764.

Evans, E. F., Ross, H. F., and Whitfield, I. C., 1965, The spatial distribution of unit characteristic frequency in the primary auditory cortex of the cat. *J. Physiol. (London)* **179**: 238-247.

Ferster, D., and LeVay, S., 1978, The axonal arborizations of lateral geniculate neurons in the striate cortex of the cat. *J. Comp. Neurol.* **182**: 923-944.

Fitzpatrick, D., Diamond, I. T., and Raczkowski, D., 1989, Cholinergic and monoaminergic innervation of the cat's thalamus: comparison of the lateral geniculate nucleus with other principal sensory nuclei. *J. Comp. Neurol.* **288**: 647-675.

Frost, D. O., 1981, Orderly anomalous retinal projections to the medial geniculate, ventrobasal, and lateral posterior nuclei of the hamster. *J. Comp. Neurol.* **203**: 227-256.

Galvan, V. V., and Weinberger, N. M., 2002, Long-term consolidation and retention of learning-induced plasticity in the auditory cortex of the guinea pig. *Neurobiol. Learn. Mem.* **77**: 78-108.

Huang, C. L., Larue, D. T., and Winer, J. A., 1999, GABAergic organization of the cat medial geniculate body. *J. Comp. Neurol.* **415**: 368-392.

Huang, C. L., and Winer, J. A., 2000, Auditory thalamocortical projections in the cat: laminar and areal patterns of input. *J. Comp. Neurol.* **427**: 302-331.

Imig, T. J., and Morel, A., 1985, Tonotopic organization in ventral nucleus of medial geniculate body in the cat. *J. Neurophysiol.* **53**: 309-340.

Imig, T. J., and Reale, R. A., 1980, Patterns of cortico-cortical connections related to tonotopic maps in cat auditory cortex. *J. Comp. Neurol.* **192**: 293-332.

Irvine, D. R. F., Rajan, R., and McDermott, H. J., 2000, Injury-induced reorganization in adult auditory cortex and its perceptual consequences. *Hear. Res.* **147**: 188-199.

Jones, E. G., Burton, H., Saper, C. B., and Swanson, L. W., 1976, Midbrain, diencephalic and cortical relationships of the basal nucleus of Meynert and associated structures in primates. *J. Comp. Neurol.* **167**: 385-420.

Kaas, J. H., 1997, Topographic maps are fundamental to sensory processing. *Brain Res. Bull.* **44**: 107-112.

Kaas, J. H., 1999, Is most of neural plasticity in the thalamus cortical? *Proc. Natl. Acad. Sci. U.S.A.* **96**: 7622-7623.

Kaas, J. H., Florence, S. L., and Jain, N., 1999, Subcortical contributions to massive cortical reorganizations. *Neuron* **22**: 657-660.

Kilgard, M. P., and Merzenich, M. M., 1998, Cortical map reorganization enabled by nucleus basalis activity. *Science* **279**: 1714-1718.

Klyachko, V., and Stevens, C. F., 2003, Connectivity optimization and the positioning of cortical areas. *Proc. Natl. Acad. Sci. U.S.A.* **100**: 7937-7941.

Landry, P., and Deschênes, M., 1981, Intracortical arborizations and receptive fields of identified ventrobasal thalamocortical afferents to the primary somatic sensory cortex in the cat. *J. Comp. Neurol.* **199**: 345-372.

Lee, C. C., Imaizumi, K., Schreiner, C. E., and Winer, J. A., 2004, Concurrent tonotopic processing streams in auditory cortex. *Cereb. Cortex* **14**: in press.

Lee, C. C., and Winer, J. A., 2003, Topographic projections in cat auditory cortex. *Proc. Soc. Neurosci.* **28**: 592.15.

Linden, J. S., Liu, R. C., Sahani, M., Schreiner, C. E., and Merzenich, M. M., 2003, Spectrotemporal structure of receptive fields in areas AI and AAF of mouse auditory cortex. *J. Neurophysiol.* **90**: 2660-2675.

Merzenich, M. M., Knight, P. L., and Roth, G. L., 1975, Representation of cochlea within primary auditory cortex in the cat. *J. Neurophysiol.* **38**: 231-249.

Metherate, R., and Ashe, J. H., 1995, Synaptic interactions involving acetylcholine, glutamate, and GABA in rat auditory cortex. *Exp. Brain Res.* **107**: 59-72.

Middlebrooks, J. C., Dykes, R. W., and Merzenich, M. M., 1980, Binaural response-specific bands in primary auditory cortex (AI) of the cat: topographic organization orthogonal to isofrequency contours. *Brain Res.* **181**: 31-48.

Mitani, A., Shimokouchi, M., and Nomura, S., 1983, Effects of stimulation of the primary auditory cortex upon colliculogeniculate neurons in the inferior colliculus of the cat. *Neurosci. Lett.* **42**: 185-189.

Morel, A., and Imig, T. J., 1987, Thalamic projections to fields A, AI, P, and VP in the cat auditory cortex. *J. Comp. Neurol.* **265**: 119-144.

Morel, A., Rouiller, E., de Ribaupierre, Y., and de Ribaupierre, F., 1987, Tonotopic organization in the medial geniculate body (MGB) of lightly anesthetized cats. *Exp. Brain Res.* **69**: 24-42.

Pallas, S. L., Roe, A. W., and Sur, M., 1990, Visual projections induced into the auditory pathway of ferrets. I. Novel inputs to primary auditory cortex (AI) from the LP/pulvinar complex and the topography of the MGN-AI projection. *J. Comp. Neurol.* **298**: 50-68.

Paller, K. A., Hutson, C. A., Miller, B. B., and Boehm, S. G., 2003, Neural manifestations of memory with and without awareness. *Neuron* **38**: 507-516.

Rajan, R., and Irvine, D. R. F., 1998, Absence of plasticity of the frequency map in dorsal cochlear nucleus of adult cats after unilateral partial cochlear lesions. *J. Comp. Neurol.* **399**: 35-46.

Read, H. L., Winer, J. A., and Schreiner, C. E., 2001, Modular organization of intrinsic connections associated with spectral tuning in cat auditory cortex. *Proc. Natl. Acad. Sci. U.S.A.* **98**: 8042-8047.

Recanzone, G. H., 1998, Rapidly induced auditory plasticity: the ventriloquism aftereffect. *Proc. Natl. Acad. Sci. U.S.A.* **95**: 869-875.

Recanzone, G. H., Merzenich, M. M., and Jenkins, W. M., 1992, Frequency discrimination training engaging a restricted skin surface results in the emergence of a cutaneous response zone in cortical area 3a. *J. Neurophysiol.* **67**: 1057-1070.

Recanzone, G. H., Schreiner, C. E., and Merzenich, M. M., 1993, Plasticity in the frequency representation of primary auditory cortex following discrimination training in adult owl monkeys. *J. Neurosci.* **13**: 87-103.

Rouiller, E. M., Simm, G. M., Villa, A. E. P., de Ribaupierre, Y., and de Ribaupierre, F., 1991, Auditory corticocortical interconnections in the cat: evidence for parallel and hierarchical arrangement of the auditory cortical areas. *Exp. Brain Res.* **86**: 483-505.

Schieber, M. H., 2001, Constraints on somatotopic organization in the primary motor cortex. *J. Neurophysiol.* **86**: 2125-2143.

Schreiner, C. E., 1995, Order and disorder in auditory cortical maps. *Curr. Opin. Neurobiol.* **5**: 489-496.

Schreiner, C. E., and Cynader, M. S., 1984, Basic functional organization of second auditory cortical field (AII) of the cat. *J. Neurophysiol.* **51**: 1284-1305.

Schreiner, C. E., Mendelson, J. R., and Sutter, M. L., 1992, Functional topography of cat primary auditory cortex: representation of tone intensity. *Exp. Brain Res.* **92**: 105-122.

Schreiner, C. E., Read, H. L., and Sutter, M. L., 2000, Modular organization of frequency integration in primary auditory cortex. *Ann. Rev. Neurosci.* **23**: 501-529.

Villa, A. E. P., Rouiller, E. M., Simm, G. M., Zurita, P., de Ribaupierre, Y., and de Ribaupierre, F., 1991, Corticofugal modulation of the information processing in the auditory thalamus of the cat. *Exp. Brain Res.* **86**: 506-517.

Wall, J. T., 1988, Variable organization in cortical maps of the skin as an indication of the lifelong adaptive capacities of circuits in the mammalian brain. *Trends Neurosci.* **11**: 549-557.

Weinberg, R. J., 1997, Are topographic maps fundamental to sensory processing? *Brain Res. Bull.* **44**: 113-116.

Weinberger, N. M., 1998, Tuning the brain by learning and by stimulation of the nucleus basalis. *Trends Cog. Sci.* **2**: 271-273.

Weinberger, N. M., Hopkins, W., and Diamond, D. M., 1984, Physiological plasticity of single neurons in auditory cortex of the cat during acquisition of the pupillary conditioned response: I. Primary field (AI). *Behav. Neurosci.* **98**: 171-188.

Winer, J. A., 1992, The functional architecture of the medial geniculate body and the primary auditory cortex, In *Springer Handbook of Auditory Research, volume 1, The Mammalian Auditory Pathway: Neuroanatomy* (D. B. Webster, A. N. Popper, R. R. Fay, eds.), Springer-Verlag, New York, New York, pp.222-409.

Winer, J. A., Diehl, J. J., and Larue, D. T., 2001, Projections of auditory cortex to the medial geniculate body of the cat. *J. Comp. Neurol.* **430**: 27-55.

Winer, J. A., Larue, D. T., Diehl, J. J., and Hefti, B. J., 1998, Auditory cortical projections to the cat inferior colliculus. *J. Comp. Neurol.* **400**: 147-174.

Winer, J. A., Larue, D. T., and Huang, C. L., 1999, Two systems of giant axon terminals in the cat medial geniculate body: convergence of cortical and GABAergic inputs. *J. Comp. Neurol.* **413**: 181-197.

Winer, J. A., and Prieto, J. J., 2001, Layer V in cat primary auditory cortex (AI): cellular architecture and identification of projection neurons. *J. Comp. Neurol.* **434**: 379-412.

Zhang, L. I., Bao, S., and Merzenich, M. M., 2002, Disruption of primary auditory cortex by synchronous auditory inputs during a critical period. *Proc. Natl. Acad. Sci. U.S.A.* **99**: 2309-2314.

Rewiring Cortex: Functional Plasticity of the Auditory Cortex during Development

JESSICA R. NEWTON, and MRIGANKA SUR
Department of Brain & Cognitive Sciences, Picower Center for Learning & Memory, Massachusetts Institute of Technology, Cambridge, Massachusetts 02139 USA

1. INTRODUCTION

How does the primary auditory cortex (A1) derive its function? Is it primarily determined by intrinsic factors expressed within cortex (i.e. molecular gradients, patterns of gene expression) or extrinsic factors that originate outside cortex (i.e. sensory experience)? This is a pivotal issue for understanding the development and function of A1, yet one that is difficult to address using traditional experimental methods because these techniques are unable to separate out the relative contributions of intrinsic and extrinsic factors. However, cross-modal experiments where the inputs of one sensory modality are re-directed to a different modality, in effect "rewiring" the brain, are able to distinguish the two and provide insight. Rewiring experiments also reveal how input activity modifies cortical organization as well as the limitations imposed on this plasticity by the underlying cortical substrate. In this chapter we review rewiring experiments in which visual inputs are induced to innervate the auditory pathway, and describe its physiological and behavioral consequences as well as the implications for normal cortical processing and human cross-modal plasticity.

2. INTRINSIC AND EXTRINSIC FACTORS IN CORTICAL DEVELOPMENT

Much work on the influence of intrinsic and extrinsic factors on cortical development has focused on the developing visual cortex. The formation of cortical layers and the arealization of cortex seem to be primarily influenced by intrinsic factors such as differential expression of gene families and molecular gradients in the developing cortical plate, before the arrival of axons from the thalamus (Sur and Leamey, 2001). However extrinsic factors, such as the amount and pattern of electrical activity in input pathways, also contribute to cortical development. Electrical activity generated within the developing brain may be sufficient for the establishment of thalamo-cortical connections, as suggested by the existence of retinal waves of spontaneous activity (Meister *et al.*, 1991; Wong *et al.*, 1993) and the presence of ocular dominance columns before eye opening (Rakic, 1976; Crowley and Katz, 2000; Crair *et al.*, 2001).

Visual experience, which influences the amount and pattern of electrical activity in pathways to the cortex, appears to be critical for the maintenance and refinement of cortical connections. Classic experiments by Hubel and Wiesel demonstrated that visual cortex displays remarkable plasticity during development, and is profoundly influenced by visual experience. In particular, depriving one eye of vision during a critical period of development, when visual experience has a maximal effect on cortical structure, induces robust changes in the anatomy and physiology of visual cortex (Wiesel and Hubel, 1965; Hubel *et al.*, 1977; LeVay *et al.*, 1980). Similarly artificially induced strabismus, which alters spatial correlations between the two eyes but not the level of activity, causes ocular dominance columns to become exclusively monocular and modifies intracortical connections (Lowel and Singer, 1992). Depriving both eyes produces remarkably little change in ocular dominance columns, indicating that the balance of activity rather than the absolute level of activity is critical for the formation of intracortical connections during the critical period (Crowley and Katz, 1999). Limiting overall light exposure also impacts cortical plasticity, such that rearing animals in the dark prolongs the critical period. After the critical period, visual experience has a minimal effect on cortical organization and function. Although these experiments demonstrate the importance of input activity in developmental plasticity, they cannot separate the relative contributions of intrinsic and extrinsic factors in determining the organization and function of a cortical area.

3. REWIRING THE BRAIN: THE PHYSIOLOGICAL CONSEQUENCES

In contrast to visual deprivation, "rewiring" experiments allow the role of patterned visual activity in specifying the function and organization of a cortical area to be distinguished from the influence of intrinsic factors. The limitations imposed on this cross-modal plasticity by the underlying cortical substrate, in this case A1, are also exposed. In these experiments visual input is re-directed to the auditory pathway by inducing retinal ganglion cell axons to innervate the medial geniculate nucleus (MGN) through surgical removal of its normal inputs at birth (see Fig. 1). This creates an alternative target for retinal axons and allows functional connections to form in the auditory thalamus, conveying visual information through existing thalamo-cortical connections to A1. Such re-routing has been done in *mice* (Lyckman *et al.*, 2001), *ferrets* (Sur *et al.*, 1988; Roe *et al.*, 1990; Roe *et al.*, 1992; Roe *et al.*, 1993; Sharma *et al.*, 2000), and *hamsters* (Schneider, 1973; Kalil and Schneider, 1975; Frost, 1982; Frost and Metin, 1985).

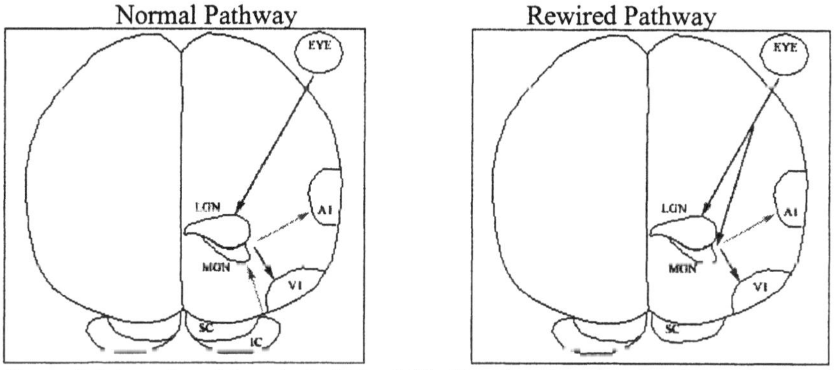

Figure 1. Schematics of the principal visual (black) and auditory (gray) pathways in normal animals *(left)* and rewired animals *(right)*. A1= primary auditory cortex, IC = inferior colliculus, LGN = lateral geniculate nucleus, MGN= medial geniculate nucleus, SC= superior colliculus, V1= primary visual cortex.

Visual activity, which has a very different spatial and temporal pattern than auditory activity, leads to visual responses in "rewired" A1 that resemble responses in primary visual cortex (V1) (Sur *et al.*, 1988; Roe *et al.*, 1990; Roe *et al.*, 1992; Sharma *et al.*, 2000). For instance, neurons in rewired A1 develop visual response features such as orientation selectivity (Roe *et al.*, 1992; Sharma *et al.*, 2000). Neurons in rewired A1 also develop direction-selectivity (Roe *et al.*, 1992; Sharma *et al.*, 2000) and an orderly retinotopic map (Roe *et al.*, 1990). Thus individual neurons within A1 are selective for different attributes of a visual stimulus such as a direction of stimulus motion, a particular line orientation or a retinotopic location (size and receptive field location in visual space) of the stimulus. Each neuron has a slightly different preference for these features compared to its neighbors, such that a coherent stimulus feature map should develop in the cortex. Optical imaging of intrinsic signals reveals that a systematic map of orientation information does develop in rewired A1, and that it is similar to the orientation map found in V1 (Sharma *et al.*, 2000). Rewired A1 also contains iso-orientation domains similar to V1, where the neurons all respond to the same preferred orientation, which are organized around pinwheel centers (see Fig. 2).

Orientation selectivity in V1 is believed to be generated in the cortex by a weak orientation selectivity bias conveyed by thalamic afferents, which is enhanced within the cortex by recurrent cortical connections (Somers *et al.*, 1995). Although the development of orientation tuning does not require visual experience (Hubel and Wiesel, 1963; Crair *et al.*, 1998), orientation selectivity can be altered by patterned visual experience during the critical period (Sengpiel *et al.*, 1999). Thus it is interesting that visual inputs directed to rewired A1 also shape its local and long-range connections such that they resemble connections in V1 (Sharma *et al.*, 2000). This suggests that patterned input activity can have a profound influence on the structure and organization of the A1 cortical substrate.

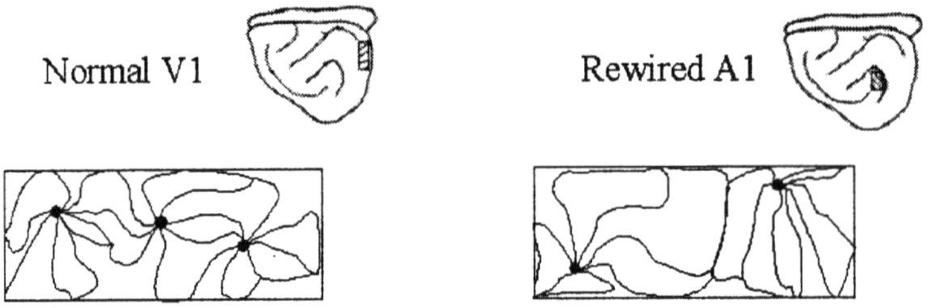

Figure 2. Schematics of the orientation maps in normal V1 (left) and rewired A1 (right). The black circles indicate the pinwheel centers, which are surrounded by iso-orientation domains where neurons all respond to the same preferred orientation.

Rewired A1 neurons form connections between domains with the same orientation preference, just like V1 neurons. In addition, the patchy connections seen in V1, which are often elongated along the orientation axis of the injection site, are also observed in rewired A1 (Gao and Pallas 1999; Sharma *et al.*, 2000). In contrast, normal A1 connections tend to be band-like and extend along the iso-frequency axis of the cortical sound frequency map. Although the organization of visual information and connectivity of rewired A1 is similar to V1, there are several notable differences. For instance, the orientation domains in rewired A1 are larger and less orderly than in V1. In addition, horizontal connections in rewired A1 are less orderly than in V1, and the spatial acuity of the rewired auditory pathway is lower than the normal visual pathway (von Melchner *et al.*, 2000). This is probably a result of the fact that retinal W cells form the primary source of inputs to the MGN (Roe *et al.*, 1993). These differences may also reflect underlying structural constraints imposed by A1 that cannot be modified by experience (i.e. the structure of A1 cortical layers). Even though receptive fields and orientation modules in rewired A1 are larger than in visual cortex, these rewiring experiments provide powerful evidence that patterned visual activity influences the functional role and organization of a cortical area. That is, input activity plays an instructive role in the establishment of cortical connections.

4. REWIRING THE BRAIN: THE BEHAVIORAL CONSEQUENCES

The impact of rewiring on behavioral function has also been explored. A study of rewired ferrets (von Melchner *et al.*, 2000) suggests that in addition to specifying cortical function, patterned inputs can influence behavior. Unilateral rewired ferrets were trained to discriminate between light and sound. Sound stimuli were presented at several spatial locations, and the ferrets received a juice reward at an "auditory" reward spout for correctly identifying the stimulus as auditory. Similarly, light stimuli were presented in the normal visual field, and ferrets received a juice reward for correctly identifying the stimulus as visual at a different visual reward spout. After training the ferrets were tested with light stimuli presented in the rewired visual field. The ferrets responded overwhelmingly at the visual reward spout, which is not surprising given that the rewired hemisphere receives both the normal visual projections to visual cortex and the rewired projection from the retina to the MGN to auditory cortex. The normal visual projection from LGN/LP to the rewired hemisphere was then ablated, and after a period of recovery the ferrets were re-tested with visual stimuli presented to the rewired visual field. The ferrets still responded overwhelmingly at the visual reward spout, indicating that the intact projection from the retina to the MGN to auditory cortex is capable of mediating the response to the visual stimulus. Finally the auditory cortex was ablated, and the ferrets were again re-tested after a period of recovery. The ferrets now responded at chance levels at the visual reward spout, indicating that the animals were no longer able to identify the visual stimulus, presumably because they were blind in the rewired visual field. Thus, the rewired projection from the retina through the MGN to auditory cortex is able to mediate visual behavior and this visual input influences the behavioral function of the auditory cortex.

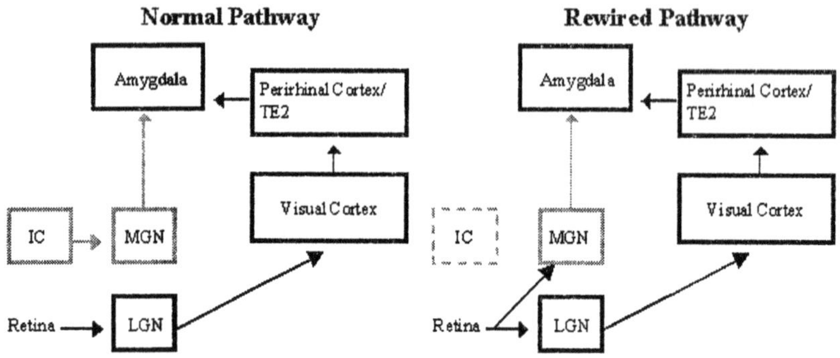

Figure 3. Simplified schematic of the principal visual (black) and auditory (gray) cued fear conditioning pathways in normal (*left*) and rewired mice (*right*). The IC (shown as a dotted box) was lesioned bilaterally in neonatal mice to induce retinal projections to the MGN. IC = inferior colliculus, LGN = lateral geniculate nucleus, MGN= medial geniculate nucleus.

Rewired visual projections in mice also influence affective behavior mediated by sub-cortical pathways, such as conditioned fear (Newton *et al.*, 2002). In fear conditioning experiments a discrete auditory cue is paired with a mild foot shock which quickly induces conditioned fear after as few as one tone-shock pairing (Fendt and Fanselow, 1999; LeDoux, 2000). In contrast, a discrete visual cue is less effective, requiring many more light-shock pairings to elicit a defensive response to the light alone (Heldt *et al.*, 2000). Dense direct connections from the MGN to the lateral nucleus of the amygdala (Fig. 3) are thought to be crucial for auditory cued conditioned fear responses (Rogan and LeDoux, 1995; Doron and LeDoux, 1999). An indirect thalamo-cortical-amygdala pathway from the MGN via auditory cortex to perirhinal cortex also conveys information to the amygdala (LeDoux, 2000; Namura *et al.*, 1997). However, lesions of the auditory cortex do not affect the magnitude or duration of freezing responses after fear conditioning (LeDoux *et al.*, 1984). In addition, single unit recordings suggest that this cortical pathway shows slower learning-induced changes than the direct thalamo-amygdala pathway, and hence is unlikely to be the principal auditory cued conditioned fear pathway (Quirk *et al.*, 1995; Quirk *et al.*, 1997). In contrast to the direct auditory pathway from the MGN to the amygdala, visual inputs primarily reach the amygdala through indirect pathways (Doron and LeDoux, 1999; Shi and Davis, 2001). Visually cued conditioned fear is thought to be mediated by projections from the LGN to

V1/V2 to visual association area TE2/perirhinal cortex (Pr) to the amygdala, or by projections from LP to V2/TE2/Pr to the amygdala (see Fig.3; Shi and Davis, 2001).

Adult sham lesion and rewired mice underwent three sessions of fear conditioning with either a visual or an auditory cue (3 cue-shock pairings per session), and behavioral testing after each session. The cued testing behavior of the different groups after one fear conditioning session are depicted in Figure 4. Consistent with previous studies, after one session of fear conditioning, light conditioned sham lesion mice did not freeze significantly more during the cue presentation compared to the habituation period (Fig. 4). Light conditioned rewired mice, however, froze significantly more during the cue presentation after only one session of fear conditioning (*$p<0.05$, paired t-test), as did tone conditioned sham lesion mice (***$p<0.01$, paired t-test). After three sessions of fear conditioning, sham lesion mice also showed greater freezing during the cue presentation.

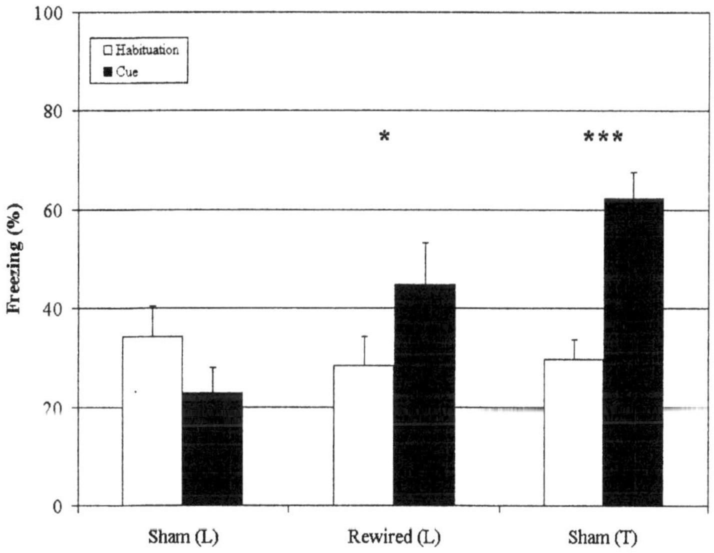

Figure 4. The mean freezing per group during the habituation (white bar) and cue presentation (black bar) periods of the cued testing session after one session of fear conditioning, with error bars denoting the standard error of the mean.

These findings indicate that the behavioral function of a target (in this case, the amygdala) is influenced by its inputs, and that it can draw upon intrinsic properties of the target. Existing pathways can convey novel information to central structures, and this information is capable of mediating behavior. The use of a natural behavior such as conditioned fear, without the infliction of additional lesions, demonstrates unequivocally that central pathways derive function from their inputs.

5. HUMAN CROSS-MODAL PLASTICITY AND NORMAL CORTICAL PROCESSING

Human cross-modal plasticity experiments also provide evidence for cortical areas deriving function from their inputs. Studies of congenitally blind humans demonstrate that the visual cortex is activated during non-visual somatosensory tasks (Sadato *et al.*, 1996; Kujala *et al.*, 1997) as well as some auditory tasks (Weeks *et al.*, 2000). Similarly, the auditory cortex in congenitally deaf individuals is involved in visual tasks (Neville *et al.*, 1983; Finney *et al.*, 2001; Bavelier and Neville, 2002; Finney *et al.*, 2003).

There is some evidence that cortical areas have an inherent propensity for the processing of subtypes of information. For instance visual and auditory cortex tend to be the most precise at processing spatial and temporal information respectively (Kitagawa and Ichihara, 2002; Welch *et al.*, 1986). This is apparent during normal cortical processing when sensory information from two modalities are in conflict, producing sensory illusions. For instance, the "ventriloquism effect" involves a discrepancy between the spatial location of an auditory and a visual stimulus, resulting in the perceived location of the event originating from the spatial location of the visual stimulus (Howard and Templeton, 1966). Similarly a spatial localization task where the visual stimulus is in conflict with proprioceptive information, known as "visual capture" results in the perceived location being determined by the visual information (Hay *et al.*, 1965). Although visual signals dominate these spatial tasks, the perceived temporal characteristics of visual signals can be modulated by conflicting auditory information. For instance, when a single flash is presented coincident with an auditory beep, a second auditory beep produces an illusory second flash (Shams *et al.*, 2002). The perceived duration and flicker rate (Gebhard and Mowbray, 1959; Shipley, 1964) of a visual stimulus can also be influenced by conflicting auditory signals.

Thus, inherent processing biases could impact functional plasticity in the cortex. For instance, visual function elicited in rewired A1 may be influenced by the temporal processing properties intrinsic to neurons or networks of the auditory cortex. Such properties may be evident as more detailed experiments are carried out on structural, functional and behavioral dynamics of rewired auditory cortex. Regardless of limits, however, rewiring experiments demonstrate the extraordinary capacity of the auditory cortex for adapting to its inputs as it develops.

REFERENCES

Bavelier, D., and Neville, H. J., 2002, Cross-modal plasticity: where and how? *Nat. Rev. Neurosci.* **3**: 443-452.

Crair, M. C., Gillespie, D. C., and Stryker, M. P., 1998, The role of visual experience in the development of columns in cat visual cortex. *Science* **279**: 566-570.

Crair, M. C., Horton, J. C., Antonini, A., and Stryker, M. P., 2001, Emergence of ocular dominance columns in cat visual cortex by 2 weeks of age. *J. Comp. Neurol.* **430**: 235-249.

Crowley, J. C., and Katz, L. C., 1999, Development of ocular dominance columns in the absence of retinal input. *Nat. Neurosci.* **2**: 1125-1130.

Crowley, J. C., and Katz, L. C., 2000, Early development of ocular dominance columns. *Science* **290**: 1321-1324.

Doron, N. N., and LeDoux, J. E., 1999, Organization of projections to the lateral amygdala from auditory and visual areas of the thalamus in the rat. *J. Comp. Neurol.* **412**: 383-409.

Fendt, M., and Fanselow, M. S., 1999, The neuroanatomical and neurochemical basis of conditioned fear. *Neurosci. Biobehav. Rev.* **23**: 743-760.

Finney, E. M., Fine, I., and Dobkins, K. R., 2001, Visual stimuli activate auditory cortex in the deaf. *Nat. Neurosci.* **4**: 1171–1173.

Finney, E. M., Clementz, B. A., Hickok, G., and Dobkins, K. R., 2003, Visual stimuli activate auditory cortex in deaf subjects: evidence from MEG. *Neuroreport* **14**: 1425-1427.

Frost, D. O., 1982, Anomalous visual connections to somatosensory and auditory systems following brain lesions in early life. *Brain Res.* **255**: 627-635.

Frost, D. O., and Metin, C., 1985, Induction of functional retinal projections to the somatosensory system. *Nature* **317**: 162-164.

Gao, W.-J., and Pallas, S. L., 1999, Cross-modal reorganization of horizontal connectivity in auditory cortex without altering thalamocortical projections. *J. Neurosci.* **19**: 7940-7950.

Gebhard, J. W., and Mowbray, G. H., 1959, On discriminating the rate of visual flicker and auditory flutter. *Am. J. Psychol.* **72**: 521-529.

Hay, J. C., Pick, H. L., and Ikeda, K., 1965, Visual capture produced by prism spectacles. *Psychonomid. Sci.* **2**: 215-216.

Heldt, S., Sundin, V., Willott, J. F., and Falls, W. A., 2000, Posttraining lesions of the amygdala interfere with fear-potentiated startle to both visual and auditory conditioned stimuli in C57BL/6J mice. *Behav. Neurosci.* **114**: 749-759.

Howard, I. P., and Templeton, W. B., 1966, *Human spatial orientation*. Wiley, London.

Hubel, D. H., and Wiesel, T. N., 1963, Receptive fields of cells in striate cortex of very young, visually inexperienced kittens. *J. Neurophysiol.* **26**: 994-1002.

Hubel, D. H., Wiesel, T. N., and LeVay, S., 1977, Plasticity of ocular dominance columns in monkey striate cortex. *Philos. Trans. R. Soc. Lond. B Biol. Sci.* **278**: 377-409.

Kalil, R. E., and Schneider, G. E., 1975, Abnormal synaptic connections of the optic tract in the thalamus after midbrain lesions in newborn hamsters. *Brain Res.* **100**: 690-698.

Kitagawa, N., Ichihara, S., 2002, Hearing visual motion in depth. *Nature* **416**: 172-174.

Kujala, T. *et al.*, 1997, Electrophysiological evidence for cross-modal plasticity in humans with early- and late-onset blindness. *Psychophysiology* **34**: 213-216.

LeDoux, J. E., 2000, Emotion circuits in the brain. *Annu. Rev. Neurosci.* **23**: 155-184.

LeDoux, J. E., Sakaguchi, A., and Reis, D.J., 1984, Subcortical efferent projections of the medial geniculate nucleus mediate emotional responses conditioned to acoustic stimuli. *J. Neurosci.* **4**: 683-698.

LeVay, S., Wiesel, T. N., and Hubel, D. H., 1980, The development of ocular dominance columns in normal and visually deprived monkeys. *J. Comp. Neurol.* **191**: 1-51.

Lowel, S., and Singer, W., 1992, Selection of intrinsic horizontal connections in the visual cortex by correlated neuronal activity. *Science* **255**: 209-212.

Lyckman, A.W. *et al.*, 2001, Enhanced plasticity of retinothalamic projections in an ephrin-A2/A5 double mutant. *J. Neurosci.* **21**: 7684-7690.

Meister, M., Wong, R. O., Baylor, D. A., and Shatz, C. J., 1991, Synchronous bursts of action potentials in ganglion cells of the developing mammalian retina. *Science* **252**: 939-943.

Namura, S., Takada, M., Kikuchi, H., and Mizuno, N., 1997, Collateral projections of single neurons in the posterior thalamic region to both the temporal cortex and the amygdala: a fluorescent retrograde double-labeling study in the rat. *J. Comp. Neurol.* **384**: 59-70.

Neville, H. J., Schmidt, A., and Kutas, M., 1983, Altered visual-evoked potentials in congenitally deaf adults. *Brain Res.* **266**: 127-132.

Newton, J. R., Ellsworth, C., Miyakawa, T., Tonegawa, S., and Sur, M., 2002, Retinal axons directed to the auditory pathway accelerate visual cued fear conditioning in mice. *Program No. 820.2. 2002 Abstract Viewer/ItineraryPlanner. Washington, DC: Society for Neuroscience, Online.*

Quirk, G. J., Repa, C., and LeDoux, J. E., 1995, Fear conditioning enhances short-latency auditory responses of lateral amygdala neurons: parallel recordings in the freely behaving rat. *Neuron* **15**: 1029-1039.

Quirk, G. J., Armony, J. L., and LeDoux, J. E., 1997, Fear conditioning enhances different temporal components of tone-evoked spike trains in auditory cortex and lateral amygdala. *Neuron* **19**: 613-624.

Rakic, P., 1976, Prenatal genesis of connections subserving ocular dominance in the rhesus monkey. *Nature* **261**: 467-471.

Roe, A. W., Pallas, S. L., Hahm, J. O., and Sur, M., 1990, A map of visual space induced in primary auditory cortex. *Science* **250**: 818-820.

Roe, A. W., Pallas, S. L., Kwon, Y. H., and Sur, M., 1992, Visual projections routed to the auditory pathway in ferrets: receptive fields of visual neurons in primary auditory cortex. *J. Neurosci.* **12**: 3651-3664.

Roe, A. W., Garraghty, P. E., Esguerra, M., and Sur, M., 1993, Experimentally induced visual projections to the auditory thalamus in ferrets: evidence for a W cell pathway. *J. Comp. Neurol.* **334**: 263-280.

Rogan, M. T., and LeDoux, J. E., 1995, LTP is accompanied by commensurate enhancement of auditory-evoked responses in a fear conditioning circuit. *Neuron* **15**: 127-136.

Sadato, N. *et al.*, 1996, Activation of the primary visual cortex by Braille reading in blind subjects. *Nature* **380**: 526-528.

Schneider, G. E., 1973, Early lesions of superior colliculus: factors affecting the formation of abnormal retinal projections. *Brain Behav. Evol.* **8**: 73-109.

Sengpiel, F., Stawinski, P., and Bonhoeffer, T., 1999, Influence of experience on orientation maps in cat visual cortex. *Nat. Neurosci.* **2**: 727-732.

Shams, L., Kamitani, Y., and Shimojo, S., 2002, Visual illusion induced by sound. *Cogn. Brain Res.* **14**: 147-152.

Sharma, J., Angelucci, A., and Sur, M., 2000, Induction of visual orientation modules in auditory cortex. *Nature* **404**: 841-847.

Shi, C., and Davis, M., 2001, Visual pathways involved in fear conditioning measured with fear-potentiated startle: behavioral and anatomic studies. *J. Neurosci.* **21**: 9844-9855.

Shipley, T., 1964, Auditory flutter-driving of visual flicker. *Science* **145**: 1328-1330.

Somers, D. C., Nelson, S. B., and Sur, M., 1995, An emergent model of orientation selectivity in cat visual cortical simple cells. *J. Neurosci.* **15**: 5448-5465.

Sur, M., Garraghty, P. E., and Roe, A. W., 1988, Experimentally induced visual projections into auditory thalamus and cortex. *Science* **242**: 1437-1441.

Sur, M., and Leamey, C. A., 2001, Development and plasticity of cortical areas and networks. *Nature Rev. Neurosci.* **2**: 251-262.

von Melchner, L., Pallas, S. L., and Sur, M., 2000, Visual behaviour mediated by retinal projections directed to the auditory pathway. *Nature* **404**: 871-876.

Weeks, R. *et al.*, 2000, A positron emission tomographic study of auditory localization in the congenitally blind. *J. Neurosci.* **20**: 2664-2672.

Wiesel, T. N., and Hubel, D. H., 1965, Comparison of the effects of unilateral and bilateral eye closure on cortical unit responses in kittens. *J. Neurophysiol.* **28**: 1029-1040.

Welch, R. B., Duttonhurt, L. D., and Warren, D. H., 1986, Contributions of audition and vision to temporal rate perception. *Percept. Psychophys.* **39**: 294-300.

Wong, R. O., Meister, M., and Shatz, C. J., 1993, Transient period of correlated bursting activity during development of the mammalian retina. *Neuron* **11**: 923-938.

Plasticity of Tonotopic and Correlation Maps in Cat Primary Auditory Cortex

JOS J. EGGERMONT
Departments of Physiology and Biophysics, and Psychology, University of Calgary, Calgary, Alberta, Canada, T2N 1N4

1. INTRODUCTION

Cat primary auditory cortex (AI) exhibits a tonotopic organization in which there is an orderly caudal-rostral (low-high frequency) progression of iso-frequency bands (Merzenich *et al.*, 1975; Reale and Imig, 1980). Peripheral and central insults have been shown to affect topographic maps in sensory cortex (Merzenich and Kaas, 1982; Arckens *et al.*, 2000). In the present study, I will compare the effects of three experimental procedures that produce changes in neural response properties in AI of the cat. These procedures are intra-cortical microstimulation (ICMS, Maldonado and Gerstein, 1996a,b), kindling through repeated electrical stimulation of auditory cortex, and exposure to loud sounds that lead to a restricted cochlear damage. The experiments were conducted in adult cats.

The effects of ICMS and the immediate effects of noise trauma were studied in ketamine-anesthetized animals (Valentine and Eggermont, 2003). The acute effects of noise trauma were also induced and studied in ketamine-anesthetized animals (Noreña *et al.*, 2003). Thus, in these experiments the effects were produced under anesthesia. Because ketamine is an NMDA blocker, the plastic changes observed in these experiments are likely not involving changes in NMDA receptors. The kindling stimulations were conducted in awake animals (Valentine *et al.*, in press) and so were the exposures to traumatizing sound in the chronic noise experiments (Eggermont and Komiya, 2000; Seki and Eggermont, 2002). So, in these cases NMDA receptor alterations could be involved in the plastic changes observed. The recordings were always done under ketamine anesthesia.

The response properties that were affected by the experimental procedures were: 1) The tonotopic map, i.e., the gradient of characteristic frequency (CF) of neurons primary auditory cortex (AI) in the caudal-rostral direction. 2) The bandwidth of the frequency-tuning curves at 20 dB above threshold (BW_{20dB}). 3) The strength of the cross-correlation calculated for simultaneously recorded spike trains from multiple electrodes.

2. INTRA-CORTICAL MICROSTIMULATION

Here, 40 ms-long trains of 13 bipolar pulses (pulse rate 300 Hz) were presented once per second for 1 hour and a current of 10 μA (Nudo *et al.*, 1990). The current stimulation was paired with a tone burst that had a frequency equal to the CF of the stimulation site in AI. The tone bursts onset occurred 20 ms prior to delivery of each ICMS pulse train, to ensure that the neural activity produced by the tone burst and the electrical stimulation affected the neurons at the same time. Recordings were carried out with either multiple independent electrodes, or with electrode-arrays with a fixed inter-electrode distance of 0.5 mm. The electrode that delivered the ICMS was also used to record activity before and after ICMS. The repetition rate of the pulse trains was in the rate regimen of long-term depression (LTD). This LTD has been demonstrated in an ICMS study in S1 brain slice (Heusler *et al.*, 2000) where sustained reduction of synaptic efficacy could be induced when bicuculine was present.

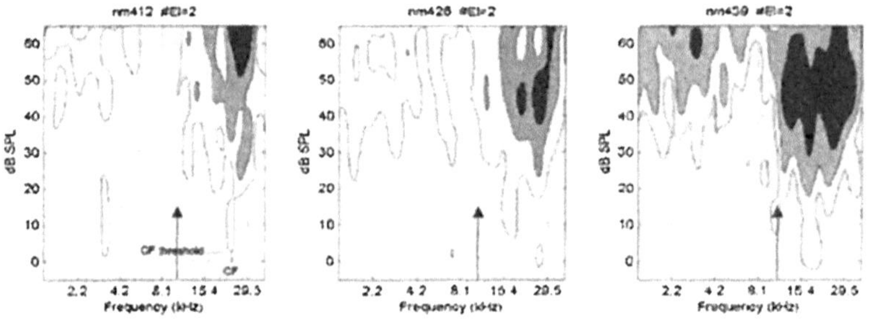

Figure 1. Changes in frequency-tuning curves after ICMS. Arrows indicate the frequency of the ICMS site. One notices an increased sensitivity and expansion of the region with frequencies below the ICMS-site CF, combined with a shift in CF towards the ICMS-site CF. This may reflect an overall increase in responsiveness as a result of reduced inhibition around the ICMS site resulting from localized LTD.

A typical observation from a recording before and after ICMS from a nearby site is shown in Figure 1. The CF of the ICMS site is indicated with an arrow. The 25% and the 50 and 75% iso-response contours (of maximum response in the post-stimulus time histogram, PSTH) are shown prior and following one hour, respectively two hours of ICMS. The 25% contour was considered as the frequency-tuning curve. One observes that the CF of the frequency-tuning curve decreases from 24 kHz to 16 kHz, and that the response becomes more robust especially for lower frequencies. The BW_{20dB} increased.

Across all animals, the CF changed in proportion to the CF-difference between the ICMS site and the recording site. When the recording site CF was higher than the ICMS site CF (positive values on the abscissa in Figure 2a), the recording site CF values decreased (ratio less than 1). When the recording site CF was lower than the stimulus site CF (negative values on the abscissa), the effect was less clear because all but one of these sites had CFs differing less than 5 kHz from the ICMS site CF (Figure 2a). In the region within 5 kHz from the ICMS-site CF on average no change in CF was found. This region then may reflect the spatial extension of the LTD caused by the ICMS. Since the ICMS site CF was generally around 10 kHz, the effect thus extends for at most ± 1 mm around the ICMS site (cf. Figure 4).

Concurrently with these changes in CF, the BW_{20dB} increased for units with CFs more than 5 kHz higher than the stimulation site CF (Figure 2b).

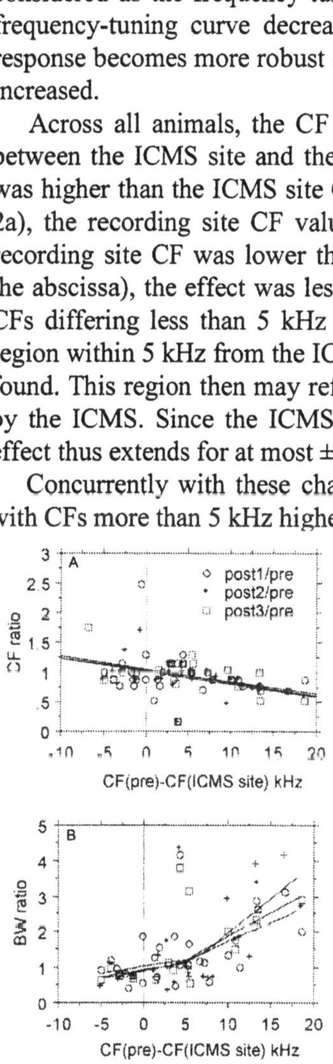

Figure 2. Effect of ICMS on CF and BW of the frequency-tuning curves. The CF change is in the direction of the ICMS-site CF and proportional to the original difference in CF with the ICMS site. The BW concurrently increases also in proportion to the distance from the ICMS site.

The centripetal CF shift (toward the ICMS site CF) and the broadening of frequency-tuning curves can also be interpreted as an expansion of the

ICMS-site CF resulting from diminished inhibition surrounding that site, and potentially the result of LTD (Heusler *et al.*, 2000).

ICMS enhanced the strength of the neural cross-correlation under spontaneous firing conditions but only for activity recorded from electrodes other than the ICMS electrode (details of quantifying the strength of the cross correlation can be found in Eggermont, 1992). Across all animals, the peak cross-correlation coefficient, R, was significantly ($p < 0.0001$) increased after the second (post2) ICMS period (Figure 3a) with respect to both the pre-ICMS condition and the post1 ICMS condition, which were nearly identical. R increased both for low and high spontaneously active neurons (Figure 3b). In this graph, a random selection of 50% of the data points was plotted to avoid crowding of the scattergram. Since ICMS also caused an increase in spontaneous firing rates (Valentine and Eggermont, 2003), and increased firing rates generally result in increased neural correlation strength, some of the increased correlation may be attributed to reduced inhibition.

Thus ICMS is accompanied by a local reduction in inhibition, resulting in an expansion of the ICMS site CF, a significant broadening of frequency-tuning curves for recording sites with CF's more than 5 kHz above that of the ICMS site, and an overall increase in neural synchrony.

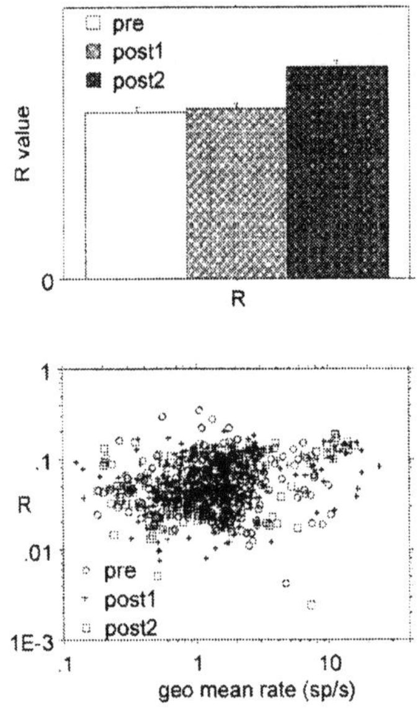

Figure 3. Cross-correlation strength increases after ICMS. The changes did not take effect after the first session, either required more time or extra stimulation was needed. The changes occurred independently of the spontaneous firing rate of the neurons involved.

3. KINDLING

Kindling refers to a highly persistent modification of brain functioning in response to repeated application of electrical stimulation that results in the development and spread of seizure activity (Goddard *et al.*, 1969). The cat AI provides an excellent model system for examining how hyper-synchronous (seizure) activity alters neocortical sensory organization and unit firing characteristics. Animals were chronically implanted with bipolar electrodes under electrophysiological control. The electrodes were prepared from twisted Teflon-coated stainless steel wire, 178 μm in diameter (A-M Systems). The two poles of each electrode were separated by 1.0 to 1.5 mm to ensure that the poles would span the superficial (layer I-II) to the deep (layers V-VI) cortical layers, providing differential recordings. The electrodes were targeted at the center of primary auditory cortex (recording). The vertical placement of the recording and stimulating electrodes were adjusted during surgery to provide optimal evoked response amplitudes. Following baseline recordings, the after-discharge thresholds (ADT) were determined as the weakest current to evoke an after-discharge (AD) of 4 seconds or longer. The kindling stimulation was applied daily and consisted of one-second trains of biphasic square wave pulses, 1.0 ms in duration, at a frequency of 60 Hz. The intensity of the stimulation was applied at 100 μA above the ADT on all subsequent kindling sessions. The animals received approximately 40 sessions of kindling stimulation prior to the acute recording procedure. The EEG activity was recorded during each kindling session and the behavioral manifestation was scored. Five cats were fully kindled (reaching stage 5) and 5 cats were implanted but not stimulated and served as sham controls.

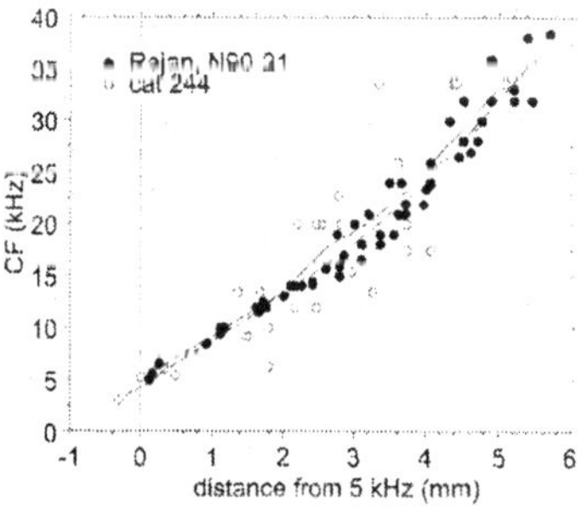

Figure 4. CF-distance plot in a sham control animal (cat 244) in comparison with a similar plot from Rajan *et al.* (1993). Distance was measured parallel to the midline from the site tuned to 5 kHz.

Tonotopic maps were obtained from the right AI, contralateral to the stimulated or implanted side. For sham control animals, the dependence of CF on the distance perpendicular to the isofrequency contours was very similar to CF-distance functions in normal cat (data adapted from Rajan *et al.*, 1993) as shown in Figure 4 for one particular example. In kindled animals, tonotopic maps were altered with an over representation of certain CF regions. These regions corresponded with the "estimated" CF on the stimulation side. In general, tonotopic order was destroyed in AI of kindled animals (Valentine *et al.*, in press).

Frequency-tuning curve bandwidths were lower in both sham and kindled animals compared to normal control cats. The mean BW_{20dB} was about 10% lower in kindled animals compared to sham controls, but this difference was not significant. Increased lateral inhibition is likely in the kindled cats because neighboring units have very similar CFs, but does not explain the findings in the sham controls.

Peak cross-correlation coefficients under spontaneous firing conditions were about 1.5 times higher in kindled animals compared to normal controls and 1.3 times higher as in sham controls (Figure 5 top). The difference between kindled animals and sham controls (and normal controls) was significant ($p < 0.005$) A scatterplot of peak cross-correlation coefficients against the geometric mean of the CFs of the units involved suggested that the increase occurred throughout the entire CF range (Figure 5 bottom). The increase in peak cross-correlation strength cannot be attributed here solely to an increase in spontaneous activity since the spontaneous firing rates were significantly lower in the kindled cats compared to the sham controls, albeit that both were significantly higher than in the normal controls.

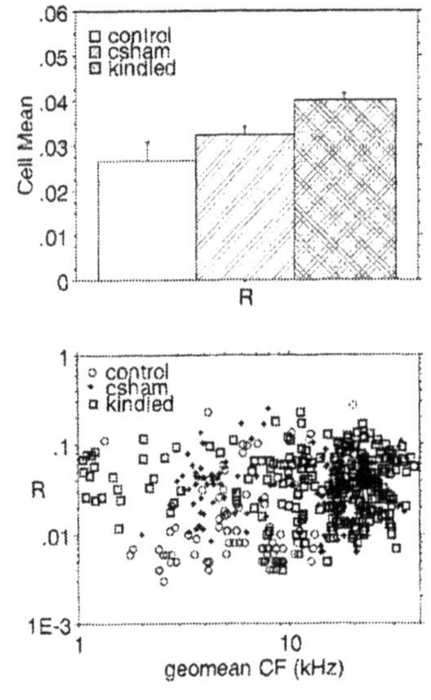

Figure 5. Kindling increases the strength of cross-correlation, compared to control and sham control animals. Because of the over representation of high CFs in kindled animals, in this region the control and sham data points are not well distinguishable.

Kindling likely resulted in an increase in the synchrony of firing for neurons located around the stimulation site. This resulted in a strengthening of synapses of radiating horizontal fibers, which resulted in an expansion of the CF range surrounding the stimulation site. This expansion is envisioned to encompass most of the kindled auditory cortex. This was accompanied by a mirroring of that CF expansion into the contralateral auditory cortex. The synaptic strengthening likely involved thalamo-cortical synapses initiated through the centrifugal action of cortico-thalamic fibers as well as horizontal fibers.

4. NOISE TRAUMA

4.1 Acute Effects

We recorded neural activity before and at various times after exposure to a 1 hr, 5 kHz, 120 dB SPL tone from the same 16 electrode positions, using two 4x2 micro-electrode arrays. Typically, we recorded neural activity to a variety of stimuli for about 4 hours prior to the exposure to the trauma tone. Immediately after we recorded frequency-tuning curves with a fast method that took 9 minutes to complete (Norena *et al.*, 2003). We presented tone pips at a rate of 2/s with frequency and intensity randomly selected. We used 27 frequencies to cover 5 octaves in logarithmic fashion (about 0.2 octave difference between frequencies) and 6 stimulus levels between 15 and 65 dB in 10 dB steps. Spontaneous firing rates were initially unchanged but started to increase over the first few hours after the trauma, especially for neurons with CFs more than one octave above the trauma-tone frequency. Figure 6 shows frequency-tuning curves for multi-unit activity recorded at the same electrode site immediately before exposure, and at 15 minutes, 1 hr 40 min and 3hr 10 min after the exposure.

The effect of the exposure to the trauma tone is two fold. Neural thresholds in the frequency range above 8 kHz are increased dramatically. As a result of decreasing inhibition from these neurons to neurons with CFs in the unaffected frequency range, it unmasks new excitatory inputs at frequencies below 8 kHz. Initially the thresholds for the unmasked regions are above 50 dB SPL, but after a few hours they decrease toward 25 dB SPL. The thresholds in the frequency range above 8 kHz remain at least elevated by 60 dB. Frequency-tuning curve bandwidths are increased after the trauma, likely as a result of diminished inhibition.

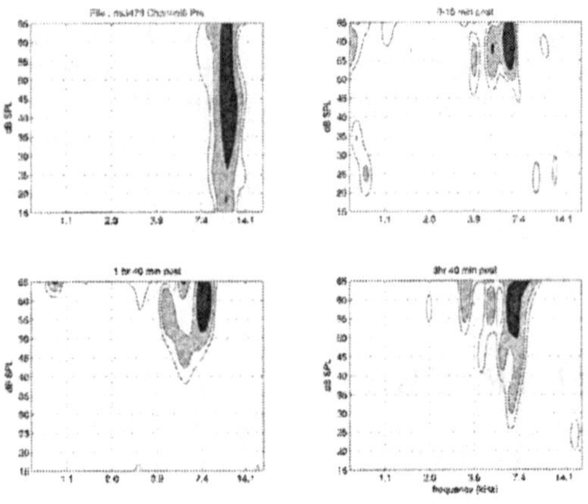

Figure 6. Frequency tuning curves at one recording site shifted as a result of the 5 kHz pure-tone exposure and show increased thresholds that recover partially over the next three and a half hours.

A comparison across all recordings (Figure 7 top) suggests that the CF's generally shifted toward lower values in proportion with the original difference between the neuron's pre-exposure CF and the trauma tone frequency. The effect was variable for units with initial CFs that were within 1 octave of the trauma-tone frequency, but quite pronounced for units with initial CFs > 1 octave higher. The shift was larger a few hours after the trauma than in the first 30 minutes. Concurrently, the BW_{20dB} increased for units that showed a shift in CF (Figure 7 bottom) but the effect was much more variable.

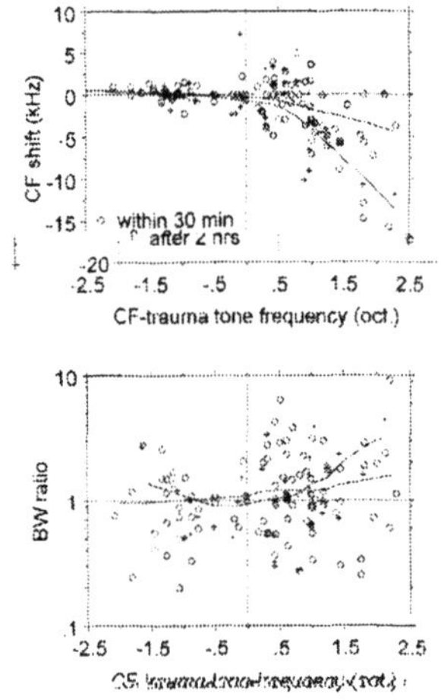

Figure 7. Changes in frequency-tuning curves immediately after noise trauma. CF values that were originally above the trauma tone frequency shift downward, those below the trauma tone frequency were not affected. The bandwidth increases significantly for CFs originally above the trauma tone frequency.

Neural synchrony increased significantly immediately after the end of the exposure and further increased for recordings made at least 2 hours after the exposure (Figure 8 top). Increased R-values were found for units with CFs above and with CFs below the trauma tone frequency (Figure 8 bottom), but on average the increased neural correlation was only significant above the trauma tone frequency.

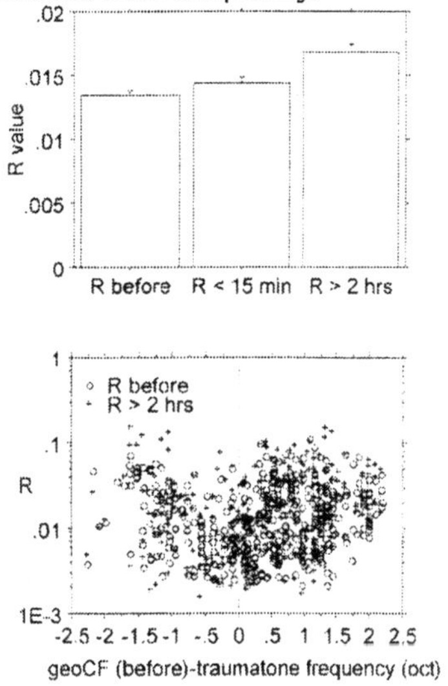

Figure 8. Cross-correlation strength increased immediately after noise trauma, and continued to increase with time after exposure. The changes were found throughout the frequency range, but significant only for neuron pairs with CFs above the trauma-tone frequency.

Thus, immediately after noise-exposure the trauma-tone CF site expands as a result of unmasking of its inputs to neurons with CFs in the frequency range of the damage, the frequency-tuning curve bandwidth increases and neural synchrony increases.

4.2 Long-standing Effects

Chronic effects of pure tone exposure (duration 2 hrs, 6 kHz tone, stimulus level 120 dB SPL) were evaluated at least 3 weeks after the exposure. If the exposure produced a high-frequency hearing loss that extended to the highest frequencies tested (32 kHz), the tonotopic map was reorganized (Eggermont and Komiya, 2000). This is illustrated here in the altered CF-distance plot (Figure 9 top). We show the CF-distance plot for one control animal (+ symbols) and for a typical pure-tone trauma animal (audiogram shown in Figure 9 bottom shows threshold increases above 12 kHz) indicated by the open circles. Both CF-distance plots were referenced to the 5 kHz site and one observes that clear map deviations start to occur for distances 3 mm rostral from the 5 kHz point. The scatterplot clearly indicates

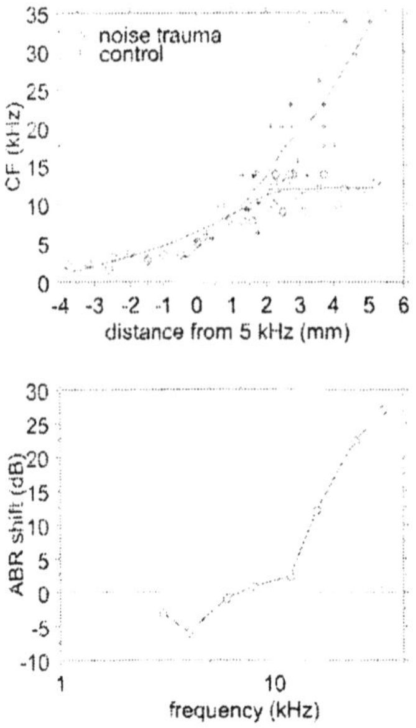

Figure 9. CF-distance plot for a control cat and one exposed to pure tone trauma. The CF does not increase beyond 14 kHz in the trauma cat (top). The high-frequency hearing loss is shown in the bottom part. As measured by the shift in auditory brainstem response threshold differences with a normal hearing control group.

the lack of CFs>10 kHz in reorganized cortex, coinciding with the edge frequency of the hearing loss (Figure 9 bottom), compared to non-reorganized cortex. For the chronic pure-tone trauma cats, the BW_{20dB} was not changed in the reorganized parts of auditory cortex compared to the normal part of the cortex. However, both parts showed elevated frequency-tuning curve bandwidths compared to normal controls. Neural synchrony increased significantly in reorganized parts of cortex (Seki and Eggermont, 2002).

5. COMMON FINDINGS

We summarize our findings with respect to our three main findings in Table I.

Tonotopic maps were changed as a result of an expansion of the CF range around the ICMS site with CF generally around 10 kHz. This change was likely caused by reduced neural activity at the ICMS site resulting from long-term depression. This reduced activity, in turn, caused the reduction in lateral inhibition for neurons with pre-ICMS CFs in the range of 15-30 kHz. As a result, the ICMS site CF expanded and, for neurons with pre-trauma CF > 15 kHz, the frequency-tuning curve's low-frequency border shifted to lower frequencies thereby increasing the frequency-tuning curve bandwidth

Tonotopic maps were also changed around the purported kindling site CF, likely by increased synaptic strengths in the horizontal connections. The dominant finding was an increased area with very similar CF that could in some cases encompass nearly the entire AI. Most of these "new" CFs were in the frequency range above 15 kHz.

Table I. Common effects of ICMS, kindling and noise trauma.

	Tonotopic maps	FTC bandwidth	Neural interaction
ICMS	ICMS-site CF expands	BW_{20dB} increases	R increases
Kindling	Kindling-site CF expands	BW_{20dB} decreases	R increases
Noise trauma acute effects	Trauma-tone-CF expands	BW_{20dB} first increases	R increases
Noise trauma long-standing effects	Edge-CF expands	BW_{20dB} decreases	R increases

Tonotopic maps were also changed after pure-tone trauma as indicated by an expansion from the edge CF. A relatively large region of AI showed the same CF and thresholds at these new CFs were in the normal range.

BW_{20dB} was increased in acute experiments (ICMS and pure-tone trauma) that were conducted under ketamine anesthesia, and also for chronic small and localized hearing losses following pure-tone exposure. In the latter case the tuning curves often became double-tipped and expanded both toward the normal hearing ranges below and above the hearing loss region.

In contrast, BW_{20dB} did not change after kindling and in cases with clear tonotopic reorganization following pure-tone trauma. One can presume that here the lateral inhibition was restored to normal values. The distinction between unmasking of new excitatory areas and use-dependent plasticity may well be reflected in the increased versus unchanged BW_{20dB} values.

In all experimental conditions, neural synchrony increased, most dramatically after kindling. These changes were in effect immediately after a pure-tone trauma suggesting that synchrony changes may well form the basis for subsequent changes in tonotopic maps.

Tonotopic map changes and increases in neural interaction are thus intimately linked, regardless of whether they are the result of unmasking of new excitatory connections as after ICMS and acute noise trauma, or of adaptive plasticity as after kindling or long-standing noise trauma. The common link between tonotopic map reorganization and increases in neural synchrony is likely found in changes in the effectiveness of central neural inhibition, either resulting from a down regulation of GABAergic mechanisms or from a relative enhancement of gutaminergic mechanisms. In any case, changes in the balance between excitatory and inhibitory processes are at the root of these changes. After adaptive plastic changes, the balance between excitation and inhibition might be partially restored as evidenced by the return of normal frequency-tuning curve bandwidths.

Both acute (Norena and Eggermont, 2003; Valentine and Eggermont, 2003) and long-standing changes (Seki and Eggermont, 2003) appeared to be accompanied also by localized increases in spontaneous activity, which also suggests a localized decrease in the effectiveness of central inhibitory mechanisms.

ACKNOWLEDGEMENTS

This report was based on studies conducted in collaboration with Pamela Valentine, Masahiko Tomita, Satoshi Seki, and Arnaud Norena. These investigations were supported by grants from the Alberta Heritage Foundation for Medical Research, the Neuroscience Canada Foundation, the Natural Sciences and Engineering Research Council of Canada, the Canadian Institutes of Health, the Canadian Language and Literacy Network (CLLRnet), and by the Campbell McLaurin Chair for Hearing Deficiencies.

REFERENCES

Arckens, L., Schweigart, G., Qu, Y., Wouters, G., Pow, D. V., Vandesande, F., Eysel, U. T., and Orban. G. A, 2000, Cooperative changes in GABA, glutamate and activity levels: the missing link in cortical plasticity. *Eur. J. Neurosci.* **12**: 4222-4232.

Eggermont, J. J., 1992, Neural interaction in cat primary auditory cortex. Dependence on recording depth, electrode separation and age. *J. Neurophysiol.* 68: 1216-1228.

Eggermont, J. J., and Komiya, H., 2000, Moderate noise trauma in juvenile cats results in profound cortical topographic map changes in adulthood. *Hear. Res.* **142**: 89-101.

Goddard, G. V., McIntyre, D. C., and Leech, C. K., 1969, A permanent change in brain function resulting from daily electrical stimulation. *Exp. Neurol.* **25**: 295-330.

Maldonado, P. E., and Gerstein, G. L., 1996a, Reorganization in the auditory cortex of the rat indiced by intracortical microstimulation: a multiple single-unit study. *Exp. Brain Res.* **112**: 420-430.

Maldonado, P. E., and Gerstein, G. L., 1996b, Neuronal assembly dynamics in the rat auditory cortex during reorganization induced by intracortical stimulation. *Exp. Brain Res.* **112**: 431-441.

Merzenich, M. M., and Kaas, J. H., 1992, Reorganization of mammalian somatosensory cortex following peripheral nerve injury. *TINS* **5**: 434-436.

Merzenich, M. M., Knight, P. L., and Roth, G. L., 1975, Representation of cochlea within primary auditory cortex in the cat. *J. Neurophysiol.* **38**: 231-49.

Noreña, A. J., and Eggermont, J. J., 2003, Changes in spontaneous neural activity immediately after an acoustic trauma: implications for neural correlates of tinnitus. *Hear. Res.* **183**: 137-153.

Noreña, A. J., Tomita, M., and Eggermont, J. J., 2003, Neural changes in cat auditory cortex after a transient pure-tone trauma. *J. Neurophysiol.* **90**: 2387-401.

Nudo, R. J., Jenkins, W. M., Merzenich, M. M., 1990, Repetitive microstimulation alters the cortical representation of movements in adult rats. *Somatosens. Mot Res.* **7**: 463-83.

Rajan, R., Irvine, D. R., Wise, L. Z., Heil, P., 1993, Effect of unilateral partial cochlear lesions in adult cats on the representation of lesioned and unlesioned cochleas in primary auditory cortex. *J. Comp. Neurol.* **338**: 17-49.

Reale, R. A., and Imig, T. J., 1990, Tonotopic organization in auditory cortex of the cat. *J. Comp. Neurol.* **192**: 265-291.

Seki, S., and Eggermont, J. J., 2002, Changes in cat primary auditory cortex after minor-to-moderate pure-tone induced hearing loss. *Hear. Res.* **173**: 172-186.

Seki, S., and Eggermont, J. J., 2003, Changes in spontaneous firing rate and neural synchrony in cat primary auditory cortex after localized tone-induced hearing loss. *Hear. Res.* **180**: 28-38.

Valentine, P. A., and Eggermont, J. J., 2003, Intracortical microstimulation induced changes in spectral and temporal response properties in cat auditory cortex. *Hear. Res.* **183**: 109-125.

Valentine, P. A., Teskey, G. C., Eggermont, J. J. Kindling changes burst firing, neural synchrony and tonotopic organization of cat primary auditory cortex. *Cerebral Cortex* (in press).

Small Cochlear Damage Causes Unmasking and Plasticity in Supra-Threshold Cortical Responses

Effects of small receptor organ damage on A1

RAMISH RAJAN
Department of Physiology, Monash University, Monash, VIC 3800, Australia

1. INTRODUCTION

There have been many studies in adult animals of the effects of cochlear damage on neuronal responses at various levels of the central auditory system, including at the level of auditory cortex (generally always in primary auditory cortex, A1). In general, these studies created large or very large hearing losses over at least some part of the cochlea, and the studies focused on the neuronal changes in regions of central auditory structures receiving input from the cochlear regions suffering the large hearing losses.

We have been interested in the question of what happens at central auditory sites when hearing damage is moderate or mild. Our interest stems from studies in A1 on plasticity of topographic maps of neuronal CF (characteristic frequency; frequency of greatest sensitivity) (Rajan and Irvine, 1998; Rajan *et al.*, 1993; Robertson and Irvine, 1989; Schwaber *et al.*, 1993). In our studies we demonstrated that plasticity of topographic maps in A1 occurred only when there was a cochlear region in which there were very large hearing losses (Rajan and Irvine, 1993), though in such cases plasticity of the CF map in A1 occurred not just in the A1 regions that used to receive input from the highly- or totally-damaged cochlear regions but also in adjacent A1 regions which were getting input from cochlear regions suffering only small hearing losses. Plasticity of the topographic CF map in A1 did not occur when hearing losses at the cochlea were small across all affected regions of the cochlea (Rajan and Irvine, 1998).

This did not mean that small cochlear damage produced no changes in A1 neuronal responses or only responses that simply reflect the small/moderate desensitizing changes that are seen in auditory nerve fibers. We had previously found that temporary hearing losses (temporary threshold shifts; TTSs) caused at the cochlea by acoustic trauma produced no changes or auditory nerve-like effects on only a very small minority of A1 neurons immediately after induction of the TTSs (Calford *et al.*, 1993). In the great majority of neurons the excitatory response area changed, either increasing or decreasing in size, and in another relatively large proportion the neurons simply stopped responding immediately after the acoustic trauma to the cochlea. These effects were found when the peak hearing losses at the cochlea were always < 40 dB, and in most cases < 25 dB. These changes in A1 neuronal response areas appeared to us (Calford *et al.*, 1993) to be interpretable by considering the effects on A1 neurons to be exercised through changes in surround inhibition in those neurons. Thus, it appeared likely that while small hearing losses caused only a small desensitization in auditory nerve fibers, it could cause profound changes in the central auditory system by affecting neuronal properties emergent in the CNS.

I have examined this hypothesis in studies in which chronic small hearing losses were created at the cochlea. These results of these studies show that small cochlear damage can cause simple threshold and near-threshold effects but, by abolishing surround inhibition, can result in dramatic changes in supra-threshold neural responses. These results and their implications are summarized below; details of the changes can be found elsewhere (Rajan, 1998, 2001, 2003).

2. EFFECT OF CHRONIC SMALL HEARING LOSSES ON A1 NEURONS

2.1 Pattern of Cochlear Damage

The studies reported here were carried out in adult cats in which there were small sensori-neural hearing losses over some part of the audiogram. Exemplars of these hearing losses, either idiopathic losses (likely due to loud sound) or induced experimentally by loud sound some months previously, are shown in Figure 1. The figure plots the VIIIth Nerve Compound Action Potential (CAP) thresholds in 3 test animals, expressed as the difference between the frequency-specific CAP thresholds from the normal mean CAP threshold at the corresponding frequency. Thus, increasing positive numbers indicate that there was a threshold desensitization (or loss) compared to normal hearing sensitivity.

The figure illustrates the two essential features of these hearing losses: (a) the losses were small (always ≤ 25 dB across all test animals), and (b) occurred over a region of the audiogram which included the most sensitive range of hearing in cats. At frequencies on either side of this affected range of frequencies, hearing sensitivity was normal, and could even be very sensitive compared to normal hearing sensitivity.

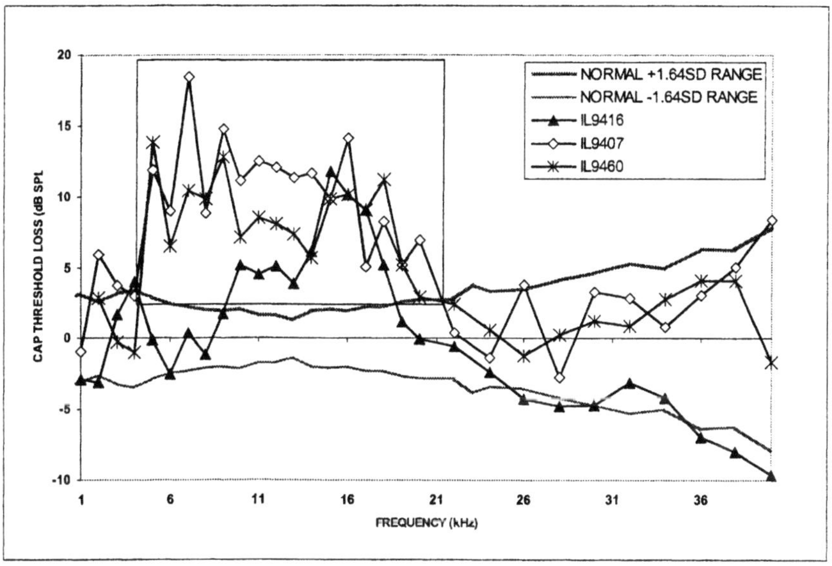

Figure 1. Compound Action Potential (CAP) threshold losses in 3 test animals. The boxed area shows the range of frequencies in which CAP thresholds were consistently elevated beyond the statistically normal range (+1.64 SD, at α=0.05) in all three animals. The light horizontally-oriented dashed lines on either side of 0 dB loss indicate the range of difference from mean normal CAP thresholds which would lie within the normal range of variations in CAP thresholds in normal animals.

This pattern of hearing losses has a great advantage since it means that in the test animals, there is a region of A1 that receives CF input from the cochlear region where there is the small damage, and as well in the same animals, there are other regions of A1 that receive input from the normal regions of the cochlea. These latter regions provide a within-animal control to confirm that any changes seen in the A1 neurons getting CF input from the damaged regions of the cochlea are specific to these neurons, and therefore likely to be directly related to the chronic small cochlear hearing loss, rather than being some generalized effect of the acoustic trauma. In the following discussion of A1 neurons in the test animals, we shall term the A1 neurons with CF at frequencies corresponding to the frequencies at which there were CAP threshold losses as "within-SNHL" A1 neurons (to indicate that these A1 neurons had CF at a frequency within the cochlear frequency range suffering from sensori-neural hearing losses), and A1 neurons with CF at other frequencies at which there were no CAP threshold losses as "outside-SNHL" A1 neurons (to indicate that these A1 neurons had CF at frequencies outside the cochlear frequency range with sensori-neural hearing losses). Data from A1 neurons from normal animals, with no peripheral hearing losses, will be termed as being from normal A1 neurons. Later in this report these normal A1 neurons will be differentiated according to whether they did or did not possess surround inhibition.

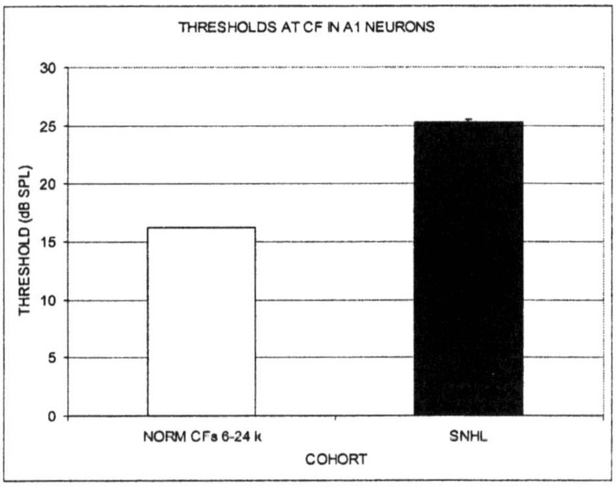

Figure 2. Mean threshold at CF in A1 neurons in normal animals (open bar) and in test animals with small CAP threshold losses (black bar). For the normal animals data shown are CF threshold data for neurons with CF from 6-24 kHz, as these were the frequencies at which there were CAP threshold losses in the test animals. For the test animals the data plotted are for CF thresholds for A1 neurons with CF at a frequency at which there was a statistically-significant elevation in CAP thresholds. The difference in mean thresholds, though small, was highly statistically significant ($p_{1\text{-tailed}} < 0.0003$)

2.2 Small Cochlear Damage Results in a Small Desensitization in A1 Neurons But No Change in Maximum Response Strength, at CF

In keeping with the fact that the hearing losses at the cochlea were small, there was only a small (but statistically-significant) elevation in threshold at CF in A1 neurons with CF at frequencies corresponding to the frequencies at which there were CAP threshold losses. This is illustrated in Figure 2, which shows that thresholds at CF in within-SNHL A1 neurons were about 10 dB higher than thresholds at the same CFs in A1 neurons in normal animals.

Such a small desensitization would not be expected to cause any significant change in maximum firing rates in peripheral auditory neurons. Similarly, in the test animals, the within-SNHL A1 neurons had maximum firing rates (at CF) that were not significantly different from maximum firing rates in normal A1 neurons with the same CFs ($p_{1\text{-tailed}} > 0.1$).

2.3 Small Cochlear Damage Results in Loss of Surround Inhibition in the Majority of A1 Neurons

A striking change in neural properties in the within-SNHL A1 neurons was a loss of surround inhibition, that form of afferent inhibition that lies outside the excitatory response area of sensory neurons (Laskin and Spencer, 1979). In our study surround inhibition was defined using a conditioning or forward/competitive-masking paradigm (Calford and Semple, 1995; Brosch and Schreiner, 1997; Rajan, 1998; Sutter *et al.*, 1999) involving the presentation of two successive tones. The second of these tones (the probe tone) was kept fixed at the A1 neuron's CF while the first was varied in frequency and intensity. Suppression of the responses to the probe tone was defined as a significant reduction (at $\alpha = 0.05$) in probe responses in the two-tone condition compared to the probe-alone condition. The suppression frequency–intensity area was overlaid on the excitatory frequency–intensity area and surround inhibition was defined to be present when the suppression area extended beyond the excitatory area boundaries (i.e., outside the range of frequencies and intensities which excite the cell) by at least two test frequencies and over an intensity range of at least 20 dB. To minimize subcortical contributions (unpublished results) to inhibitory effects in A1, only suppression with a delay of 50 ms between end of the masker and beginning of the probe tone were used to define suppression boundaries.

We found that there was a marked reduction in the proportion of within-SNHL neurons that possessed surround inhibition in comparison to either normal A1 neurons or outside-SNHL A1 neurons, as shown in Table 1. There were no significant differences between the proportions of the normal A1 neurons and outside-SNHL A1 neurons, confirming that the loss of surround inhibition was specific to the A1 regions where neurons had CF at a frequency corresponding to a frequency with small cochlear damage. Thus, the loss of surround inhibition could specifically be associated with the cochlear hearing loss.

Table 1. Percentage of A1 neurons with surround inhibition

Normal A1 neurons	Within-SNHL A1 neurons in test animals with small cochlear damage	Outside-SNHL A1 neurons in test animals with small cochlear damage
~47%	~3%	50%

In the visual, auditory and somatosensory systems, surround inhibition is believed to constrain the size of excitatory areas of neurons (Janig *et al.*, 1977, 1979; Martin and Dickson, 1983; Dykes *et al.*, 1984; Müller and Scheich, 1988; Ramoa *et al.*, 1988; Alloway *et al.*, 1989; Alloway and Burton, 1986, 1991; Kaneko and Hicks, 1990; Oka and Hicks, 1990; Caspary *et al.*, 1991; Vater *et al.*, 1992; Yang *et al.*, 1992; Suga, 1995; Palombi and Caspary, 1996), to constrain and shape neuronal responses to broad band stimuli spreading across the receptor surface (Martin and Dickson, 1983; Müller and Scheich, 1987, 1988; Yang *et al.*, 1992; Calford and Semple, 1995; Brosch and Schreiner, 1997), and to constrain the area of cortex activated by a stimulus (Horikawa *et al.*, 1996). We therefore examined a number of other response metrics in the test within-SNHL A1 neurons and compared them to normal A1 neurons. Here we differentiate between normal A1 neurons according to whether or not they possessed surround inhibition, since we would expect that within-SNHL cells, which appear generally to lack surround inhibition, should now resemble normal A1 neurons that do not possess surround inhibition.

2.4 Loss of Surround Inhibition Broadens Excitatory Bandwidth Through Unmasking and Plasticity

Consistent with the postulated role for surround inhibition in constraining responses of auditory neurons in the frequency and intensity dimensions, the loss of surround inhibition in within-SNHL A1 neurons in the test animals resulted in greater spread of responses across the frequency and intensity domains in individual neurons, with less selectivity of neuronal coding in these domains. One way to demonstrate this is to measure the bandwidth of the excitatory response areas of A1 neurons. This was done at each of a number of different levels above threshold and is illustrated in Figure 3, for normal A1 neurons that possess surround inhibition, normal A1 neurons lacking surround inhibition, and for within-SNHL A1 neurons from test animals with small cochlear damage. (Note that in the figure data are presented as bandwidths, rather than as Q values as in a previous report - Rajan, 2001.)

The figure demonstrates that with increasing intensity (i.e., level > threshold) within-SNHL cells show the same systematic increase in bandwidth of their excitatory response area as seen in normal A1 cells lacking surround inhibition. The pattern of effects in these two classes of A1 neurons is in direct contrast to that seen in normal A1 cells that possess surround inhibition in which bandwidths remain constant against increases in level. Thus, the data in within-SNHL cells are consistent with the absence of surround inhibition in these cells, and the general pattern of effects in these cells is similar to that in normal A1 cells that lack surround inhibition.

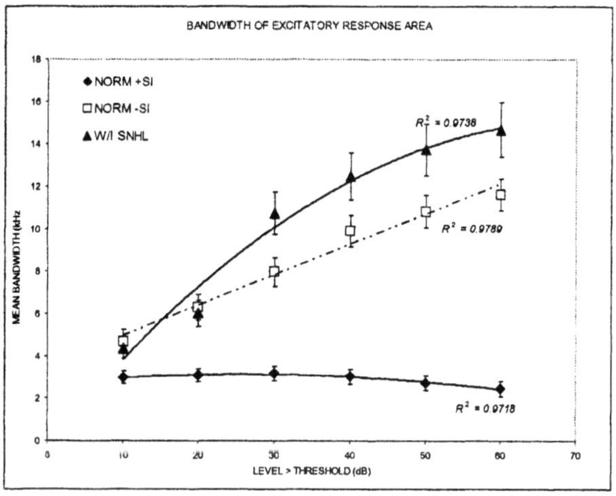

Figure 3. Bandwidth of the excitatory response area in A1 neurons. Data plotted are mean bandwidths, at levels from 10-60 dB above threshold at the CF. Error bars indicate standard error of the mean. The best lines of best fit, and correlation coefficients, are indicated in the figure. In the legend "W/I SNHL" indicates the test within-SNHL cells from the animals with small cochlear damage, "NORM -SI" indicates normal A1 cells lacking surround inhibition, and "NORM +SI" indicates normal A1 cells possessing surround inhibition.

However, careful examination of the data shows that while within-SNHL cells show the same general pattern as seen in normal A1 cells lacking surround inhibition, there are significant differences between the two groups of cells at levels commencing from 30 dB > CF threshold. At all levels from 30-60 dB > CF threshold, the excitatory response areas in within-SNHL cells were always broader than in the A1 cells lacking surround inhibition. Thus, loss of surround inhibition in within-SNHL A1 cells does not simply cause unmasking of excitatory inputs that had been suppressed by the surround inhibition, and reversion to the native state of A1 cells without surround inhibition. The greater bandwidth of excitatory response areas suggests that as well as unmasking of previously suppressed inputs there has been addition of some more excitation bandwidth.

The accretion of new excitation to the bandwidth in within-SNHL cells can also be demonstrated on examination of the profiles of the excitatory response area in A1 neurons. For this purpose, excitatory response area profiles will be divided into three categories as depicted in Figure 4.

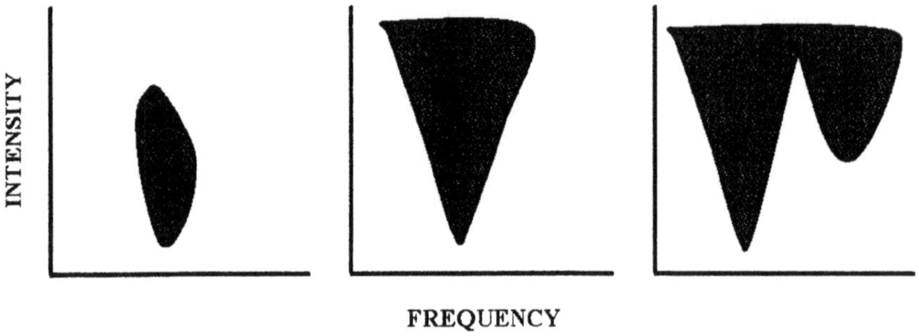

Figure 4. Three prototypical response area profiles of A1 neurons: Circumscribed (left), Open (middle), and Dual-lobed (right).

As indicated in Table 2, there is a systematic difference between the proportion of these three types of response areas in normal A1 neurons according to the presence or absence of surround inhibition. In normal A1 the great majority of neurons that possess surround inhibition have circumscribed response areas that are constrained in both the frequency and intensity domains. This is consistent with the role of surround inhibition in constraining A1 neuronal responses to a sub-set of the convergent inputs A1 neurons are likely to receive. In contrast the great majority of A1 cells lacking surround inhibition have response area profiles that are very much more like auditory nerve response areas, with no constraints in the intensity dimension at frequencies within the excitatory response area. Further a significant proportion of the A1 cells lacking surround inhibition can even possess the dual response lobes indicating convergence of across-frequency inputs.

Table 2. Percentage of the different types of excitatory response area profiles in different cohorts of A1 neurons. The first two columns are data from normal A1 neurons, either possessing surround inhibition (first column) or lacking surround inhibition (second column), while the final column is data for the within-SNHL cells from the test animals.

	Normal A1 (+SI)	Normal A1 (-SI)	Within-SNHL
Circumscribed	75.8%	1.5%	6.9%
Open	13.6%	79.7%	51.7%
Dual-lobed	10.6%	18.8%	41.4%

The within-SNHL A1 neurons, which lack surround inhibition, also show little restriction of responses in the frequency and intensity domains and only a small proportion of these neurons have circumscribed response areas. The overall distribution of response area types did not differ between within-SNHL A1 neurons and normal A1 neurons lacking surround inhibition but the distribution in both groups significantly differed (p<0.0001) from the distribution of response area types in normal A1 neurons with surround inhibition. However, note that the proportion of dual-lobed response areas in within-SNHL cells is about 2.5 times that found in normal A1 cells that do not possess surround inhibition. This reinforces the argument that loss of surround inhibition consequent to a chronic hearing loss sets in train unmasking processes that reveal previously suppressed excitatory inputs as well new plasticity processes that add some more excitation bandwidth through across-frequency convergent inputs.

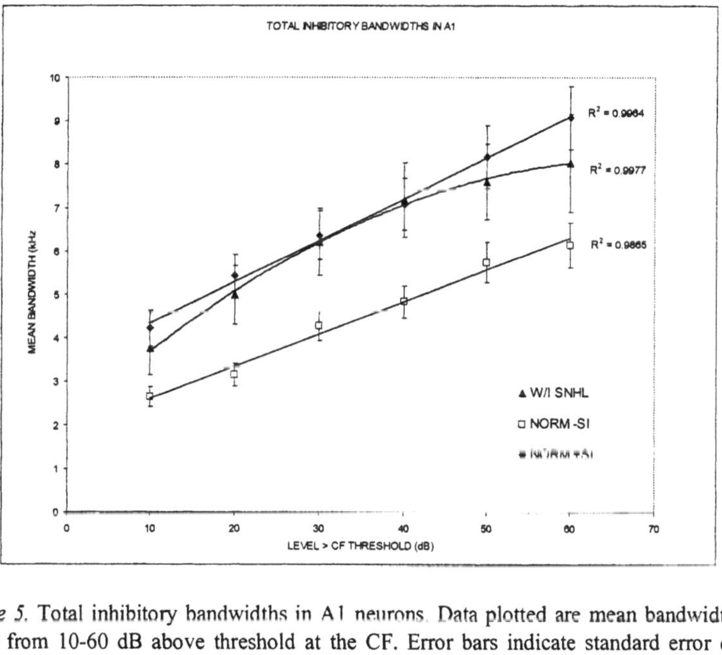

Figure 5. Total inhibitory bandwidths in A1 neurons. Data plotted are mean bandwidths, at levels from 10-60 dB above threshold at the CF. Error bars indicate standard error of the mean. The best lines of best fit, and correlation coefficients, are indicated in the figure. Legend conventions as for Figure 3.

2.5 Loss of Surround Inhibition Broadens Inhibitory Bandwidth

The small cochlear damage also resulted in changes in inhibitory bandwidth. However, the loss of surround inhibition in within-SNHL cells did *not* result in a decrease in inhibitory bandwidth, as might be expected. This can be seen in Figure 5 that plots the total bandwidth of the inhibitory areas in the three groups of A1 neurons considered to date: within-SNHL A1 neurons from the test animals with the small cochlear damage, normal A1 neurons possessing surround inhibition, and normal A1 neurons lacking surround inhibition. The figure demonstrates the surprising result that within-SNHL cells actually had inhibitory bandwidths that were greater than bandwidths in normal A1 neurons lacking surround inhibition but that were equal to the larger inhibitory bandwidths seen in A1 neurons possessing surround inhibition.

Thus, although the within-SNHL cells lack one component of afferent inhibition this does not appear to be due to a decrease in the gain of inhibition to these cells. More generally, this result indicates that the increase in excitation bandwidths and accretion of new response lobes in these cells cannot be explained by invoking a decrease in the gain of a unitary inhibition.

2.6 Consequences of the Loss of Surround Inhibition and Increased Excitation Bandwidth on Response Selectivity for Narrow and Broad Band Stimuli

The consequence of these changes in neural responses in within-SNHL cells is a loss of selectivity for narrow band and broadband stimuli. This effect can be easily demonstrated by considering the patterns of responses to white noise stimuli, i.e., the intensity functions to white noise stimuli. These data are presented in Table 3.

The great majority of normal A1 cells that possess surround inhibition are not responsive to the white noise stimulus. This absence of response to the broadband stimulus is not because these cells are generally unresponsive to sound but rather reflects the selectivity of these cells to narrow band stimuli. These cells were all responsive to tonal stimuli, with maximum responses at CF that were only slightly lower than maximum responses in A1 neurons that did not possess surround inhibition, and excitatory response areas and surround inhibitory areas could be easily defined for these A1 neurons with tonal stimuli. Thus the absence of responses to the broadband stimulus is specific to a stimulus extending beyond the excitatory response area of the A1 neuron, and therefore able to simultaneously activate the surround inhibitory response areas. In other studies not reported here due to space constraints, we have been able to directly relate the presence of surround inhibition and responses to narrow band stimuli extending beyond the excitatory response area. As narrow band stimuli are progressively broadened in frequency content from initially containing frequencies only within the excitatory response area of this category of A1 neurons, to progressively containing frequencies approaching the boundary of the excitatory response area, and then to containing frequencies extending outside the excitatory response area, there is a systematic decrease in responses. Thus, as a narrow band stimulus is broadened to include frequencies from outside the excitatory response area (and thereby engage the surround inhibitory area) of A1 neurons that possess surround inhibition, these cells decrease their responses and even stop responding.

Table 3. Percentage of the different types of intensity functions (also known as input-output functions) to white noise stimuli in different cohorts of A1 neurons. The first two columns are data from normal A1 neurons, either possessing surround inhibition (first column) or lacking surround inhibition (second column), while the final column is data for the within-SNHL cells from the test animals

Type of intensity function	Normal A1 (+SI)	Normal A1 (-SI)	Within-SNHL
Monotonic	3.2	82.5	88.9
Non-monotonic	32.2	14.3	3.7
Non-responsive	64.5	3.2	7.4

In normal A1 cells lacking surround inhibition, responses generally (but not always) tend to increase with increasing frequency bandwidth within the excitatory response area and, as frequency bandwidth extends beyond the excitatory response area, simply saturate. This is reflected in the fact, indicated in Table 3, that the vast majority of these normal A1 cells that lack surround inhibition have monotonic intensity functions to the broadband white noise stimulus. Exactly the same profile of response patterns is found for the within-SNHL cells. Thus, both these categories of cells lack selectivity for narrow band stimuli, as is found in A1 cells that possess surround inhibition.

3. CONCLUSION

These results indicate that chronic small cochlear damage results in simple effects at the characteristic frequency of neurons getting input from the cochlear region suffering this small damage. At the CF, there is a small desensitization but no change in maximum response rates in such neurons. However across the response area (in frequency and intensity domains) of such neurons, there is an increase in excitatory bandwidth. This appears to reflect both unmasking of excitatory inputs that were previously suppressed by the surround inhibition as well as addition of new excitatory inputs, most likely through long-term plasticity processes. Consequently, the excitatory bandwidth of the within-SNHL cells did not simply increase to equal that in normal A1 cells that also lack surround inhibition but, at supra-threshold levels from 30 dB > CF threshold onwards, exceeded the bandwidth of these normal A1 cells. Further, a much greater proportion of the within-SNHL cells also had two, well-differentiated, lobes of excitation compared to the normal A1 cells that lack surround inhibition, again consistent with the idea that there was addition of new excitation compared to the latter A1 cells.

There was also the addition of new inhibitory inputs to the response area of the affected cells. The bandwidth of the inhibitory areas in the within-SNHL cells was much greater than that in normal A1 cells lacking surround inhibition and was as great as that in normal A1 cells possessing surround inhibition. Again, this argues that the within-SNHL A1 cells did not simply revert to a "native" state of normal A1 cells that lack surround inhibition.

It is likely that the changes described here reflect both plasticity of neuronal responses as well as "unmasking" of pre-existing inputs. The distinction between these two terms has been well made in the context of changes in the somatosensory CNS after damage or other manipulations to the receptor surface, where it has been pointed out that plasticity in the CNS cannot simply mean the expression of pre-existing inputs once these have been released from inhibition (i.e., "unmasked") after removal of a dominant input (Calford, 2002). Instead, plasticity must involve the conversion of implicit but unexpressed inputs into explicit responses expressed as part of the normal supra-threshold repertoire of the cell, whereas unmasking involves the expression of explicit inputs that are otherwise blocked by other dominant (generally, but theoretically not exclusively, inhibitory) inputs. As such, plasticity must involve changes in synaptic weights (Buonomano and Merzenich, 1998; Calford, 2002) whereas this would not occur in unmasking. We have argued previously for this distinction between plasticity and unmasking in A1 (e.g., Rajan, 2001) and it has also lately been made in the case of changes observed in the inferior colliculus (Snyder *et al.*, 2000), and we have taken the conservative view that in cases where both unmasking and plasticity could account for any observed changes, it may be more prudent to assume that unmasking has occurred unless there is compelling evidence for plasticity. In the present study, no overt attempt was made to study whether the A1 changes produced by the small cochlear damage were due to plasticity or to unmasking. However, as noted above, the loss of surround inhibition in within-SNHL cells in A1 did not simply result in these cells "reverting" to the native status of normal A1 cells that do not possess surround inhibition. Instead the increase in excitatory and inhibitory bandwidth in the within-SNHL cells provides evidence that there was plasticity of neuronal responses in A1 even after mild hearing losses at the cochlea. Thus, while our previous studies show that changes at the CF, and particularly changes *in* the CF (i.e., plasticity of the topographic CF map) requires large hearing losses at the cochlea (see Introduction), the preset study indicates that more subtle changes across the neuronal response area can occur even with mild hearing losses.

What could be the perceptual consequences of the changes in A1 neuronal coding, especially the loss of surround inhibition? Consistent with the postulated role of surround inhibition, in other sensory systems, in shaping neuronal selectivity for narrow and broadband stimuli, we have been able to directly correlate the presence of surround inhibition with selectivity for such stimuli in A1. Here we have shown that loss of surround inhibition results in loss of selectivity for such stimuli. One consequence of the loss of surround inhibition could be a reduced ability to discriminate signals in background noise, i.e., in segregating foreground from background. We (Cainer, Dineen, James and Rajan, unpublished) have commenced studies to examine the psychophysics of auditory processing in humans with chronic mild hearing losses. Consistent with our predictions, mild hearing losses (of no more than about 10 dB worse than the normal audiometric range) did not affect factors such as loudness minimum comfort levels or speech discrimination per se, but did significantly reduce discrimination of speech in noise. Further hearing loss (moderate or severe) only further exacerbated the problem, but the effect was already established with mild hearing loss. These results have profound implications in a society where hearing loss is an increasing phenomenon (especially mild hearing losses which are a relatively common phenomenon, in our experience) and relatively high noise levels are part of normal external milieu in which humans operate.

REFERENCES

Alloway, K. D., and Burton, H., 1986, Bicuculline-induced alterations in neuronal responses to controlled tactile stimuli in the second somatosensory cortex of the cat: a microiontophoretic study. *Somatosens. Res.* **3**: 195-211

Alloway, K. D., and Burton, H., 1991, Differential effects of GABA and bicuculline on rapidly- and slowly-adapting neurons in primary somatosensory cortex of primates. *Exp. Brain Res.* **85**: 598-610

Alloway, K. D., Rosenthal, P., and Burton, H., 1989, Quantitative measurements of receptive field changes during antagonism of GABAergic transmission in primary somatosensory cortex of cats. *Exp. Brain Res.* **78**: 514-532

Brosch, M., and Schreiner, C. E., 1997, Time course of forward masking tuning curves in cat primary auditory cortex. *J. Neurophysiol.* **77**: 923-943.

Buonomano, D. V., and Merzenich, M. M., 1998, Cortical plasticity: from synapses to maps. *Ann. Rev. Neurosci.* **21**:149-86.

Caspary, D. M., Backoff, P. M., Finlayson, P. G., and Palombi, P. S., 1991, Inhibitory inputs modulate discharge rate within frequency receptive fields of anteroventral cochlear nucleus neurons. *J. Neurophysiol.* **72**: 2124-2133 .

Calford, M. B., 2002, Dynamic representational plasticity in sensory cortex. *Neuroscience* **111**: 709-738.

Calford, M. B., Rajan, R., and Irvine, D. R. F., 1993, Rapid changes in the frequency tuning of neurons in cat auditory cortex resulting from pure-tone-induced temporary threshold shift. *Neuroscience* **55**, 953-964.

Calford, M. B., and Semple, M. N., 1995, Monaural inhibition in cat auditory cortex. *J. Neurophysiol.* **73:** 1876–1891.

Dykes, R. W., Landry, P., Metherate, R., and Hicks, T. P., 1984, Functional role of GABA in cat primary somatosensory cortex: shaping receptive fields of cortical neurons. *J. Neurophysiol.* **52:** 1066-1093.

Horikawa, J., Hosokawa, Y., Kubota, M., Nasu, M., and Taniguchi, I., 1996, Optical imaging of spatiotemporal patterns of glutamatergic excitation and GABAergic inhibition in the guinea pig auditory cortex in vivo. *J. Physiol.* **497**: 629-638.

Janig, W., Schoultz, T., and Spencer, W. A., 1977, Temporal and spatial parameters of excitation and afferent inhibition in cuneothalamic relay neurons. *J. Neurophysiol.* **40**: 822-835.

Janig, W., Spencer, W. A., and Younkin, S. G., 1979, Spatial and temporal features of afferent inhibition in thalamocortical relay cell. *J. Neurophysiol.* **42**: 1450-1460.

Kaneko, T. P., and Hicks, T. P., 1990, GABAB-related activity involved in synaptic processing of somatosensory information in SI cortex of the anaesthetized cat. *Brit. J. Pharmacol.* **100**: 689-698.

Laskin, S. E., and Spencer, W. A., 1979, Cutaneous masking II. Geometry of excitatory and inhibitory receptive fields of single units in somatosensory cortex of the cat. *J. Neurophysiol.* 42: 1061-1082.

Martin, M. R., and Dickson, J. W., 1983, Lateral inhibition in the anteroventral cochlear nucleus of the cat : a microiontophoretic study. *Hear. Res.* **9**: 35-41.

Müller, C. M., and Scheich, H., 1987, GABAergic inhibition increases the neuronal selectivity to natural sounds in the avian auditory forebrain. *Brain Res.* **41**: 376-380.

Müller, C. M., and Scheich, H., 1988, Contribution of GABAergic inhibition to the response characteristics of auditory units in the avian forebrain. *J. Neurophysiol.* **59**: 1673-1689.

Oka, J.-I., and Hicks, T. P., 1990, Benzodiazepines and synaptic processing in the spatial domain within the cat's primary somatosensory cortex. *Canadian J. Physiol. Pharmacol.* **68**: 1025-1040.

Palombi, P. S., and Caspary, D. M., 1996, GABA inputs control discharge rate primarily within frequency receptive fields of inferior colliculus neurons. *J. Neurophysiol.* **75**: 2211-2219.

Rajan, R., 1998, Receptor organ damage causes loss of cortical surround inhibition without topographic map plasticity. *Nature Neurosci.* **1**: 138-143.

Rajan, R., 2001, Plasticity of excitation and inhibition in the receptive field of primary auditory cortical neurons after limited receptor organ damage. *Cerebral Cortex* **11**: 171-182.

Rajan, R., and Irvine, D. R. F., 1998, Neuronal responses across cortical field A1 in plasticity induced by peripheral auditory damage. *Audiol. & Neuro-Otol.* **3**: 123-144.

Rajan, R., Irvine, D. R. F., Wise, L. Z., and Heil, P., 1993, Effect of unilateral partial cochlear lesions in adult cats on the representation of lesioned and unlesioned cochleas in primary auditory cortex. *J. Comp. Neurol.* **338**: 17-49.

Ramoa, A. S., Paradiso, M. A., and Freeman, R. D., 1988, Blockade of intracortical inhibition in kitten striate cortex: effects on receptive field properties and associated loss of ocular dominance plasticity. *Exp. Brain Res.* **73**: 285-296.

Robertson, D., and Irvine, D. R. F., 1989, Plasticity of frequency organization in auditory cortex of guinea pigs with partial unilateral deafness. *J. Comp. Neurol.* **282**: 456-471.

Schwaber, M. K., Garraghty, P. E., and Kaas, J. H., 1993, Neuroplasticity of the adult primate auditory cortex following cochlear hearing loss. *Am. J. Otology* **14**: 252-258.

Snyder, R. L., and Sinex, D. G., 2002, Immediate changes in tuning of inferior colliculus neurons following acute lesions of cat spiral ganglion. *J. Neurophysiol.* **87**: 434-452.

Suga, N., 1995, Sharpening of frequency tuning by inhibition in the central auditory system: tribute to Yasuji Katsuki. *Neurosci. Res.* **21**: 287-299.

Sutter, M. L., Schreiner, C. E., McLean, M., O'Connor, K. N., and Loftus, W. C., 1999, Organization of inhibitory frequency receptive fields in cat primary auditory cortex. *J. Neurophysiol.* **82**: 2358-2371.

Vater, M., Habbicht, H., Kössl, M., and Grothe, B., 1992, The functional role of GABA and glycine in monaural and binaural processing in the inferior colliculus of horseshoe bats. *J. Comp. Physiol. A* **171**: 541-553.

Yang, L., Pollak, G. D., and Resler, C., 1992, GABAergic circuits sharpen tuning curves and modify response properties in the mustache bat inferior colliculus. *J. Neurophysiol.* **68**: 1760-1774.

Changes in Auditory Function Following Auditory Cortex Inactivation

JOSEF SYKA, JIŘÍ POPELÁŘ, NATALIA RYBALKO, FIDEL C. NWABUEZE-OGBO, JANA MAZELOVÁ, and DANIEL ŠUTA.
Institute of Experimental Medicine, Academy of Sciences of the Czech Republic, Vídeňská 1083, 142 20 Prague, Czech Republic

1. INTRODUCTION

Inactivation of a part of the brain is a traditional tool used in experimental studies of brain function. Two types of inactivation - temporary or permanent – yield different results. Temporary inactivation is usually performed by local cooling or by the local application of drugs producing a loss of function of the structure for several minutes or hours. Typical drugs used are tetrodotoxin, lidocain or muscimol (similar results may also be achieved by transcranial magnetic stimulation or local cooling). There are only limited compensatory mechanisms available for such short periods of inactivation. Usually a complete recovery of function is expected to follow a temporary inactivation. A permanent loss of function is traditionally caused by mechanical or thermal lesions or with more sophisticated methods based on the use of laser beam or x-rays. The loss of function of the destroyed part of the brain then activates compensatory mechanisms that allow other parts of the brain to take over the function of the lost structure. Compensatory processes developing in the brain after permanent lesion are usually considered as manifestations of brain plasticity. We report here about the results of several experiments in rats in which either temporary (functional) ablations of the auditory cortex (AC) were made by local injections of tetrodotoxin (TTX) at several places in the auditory cortex or the auditory cortex was removed mechanically, either unilaterally or bilaterally.

Temporary inactivation of the auditory cortex was employed as a tool for investigating the role of the descending corticollicular pathway, particularly its influence on the function of the inferior colliculus (IC) (Nwabueze-Ogbo *et al.* 2002; Popelář *et al.* 2003). It was already shown by Neff *et al.* (1975) that bilateral ablation of the auditory cortex is accompanied by defects in the processing of the temporal aspects of an acoustical signal. Therefore, we investigated the influence of AC ablation in two experiments, one focused on changes in gap function (Syka *et al.*, 2002) and the other on changes in the perception of frequency-modulated (FM) signals (unpublished data).

In the rat, as in other mammalian species, the principal targets of descending cortical fibers in the inferior colliculus are the dorsal cortex (DCIC) and, to a lesser extent, the external cortex (ECIC) of the IC (Beyerl, 1978; Druga and Syka, 1984a; 1984b; Faye-Lund, 1985; Winer *et al.*, 1998). Whereas the morphology of the corticofugal auditory pathways is well documented, their role in the processing of auditory signals has not been explained to date. For example, electrical stimulation of the AC was shown to produce facilitation, depression or a combination of both effects on sound-evoked neuronal responses recorded in the IC (Syka and Popelář, 1984; Sun *et al.*, 1996). Studies in the bat (reviewed by Suga and Ma, 2003) revealed that subcortical neurons are subject to corticofugal modulation in the frequency, amplitude and time domains. For example, focal electrical stimulation of the AC produces a shift of the frequency-tuning curves of the IC neurons along the frequency axis, which is a function of the relationship in frequency tuning between the stimulated and recorded neurons. The aim of our study was to investigate the changes in acoustically evoked activity (evoked responses and single or multiple unit activity) in the IC of the rat after functional ablation of the AC by a local intracortical application of TTX.

The auditory cortex is considered to play an important role in the processing of the temporal aspects of acoustical signals. The measurement of gap detection threshold (GDT) offers the possibility to investigate pathologies of temporal resolution ability and to estimate their severity. For such purpose, studies dealing with the influence of experimentally induced impairments in gap detection in animals may be of importance. The aim of such studies is usually to determine the involvement of individual structures of the auditory system in temporal resolution. Only a few papers on gap detection in animals with a cortical lesion have been published so far (Ison *et al.*, 1991; Kelly *et al.*, 1996). The purpose of our study (Syka *et al.*, 2002) was to examine the role of the AC in the detection of gaps in noise. Gap detection threshold was measured in pigmented rats before and after bilateral ablation of the AC.

It is usually supposed that under normal conditions, both hemispheres of the brain are involved in the processing of acoustical signals to the same extent – with the exception of a specific situation in the human brain. In mammals, the auditory cortex plays an important role in the recognition of complex acoustical signals such as frequency-modulated (FM) signals or animal vocalizations (see, for example, Ehret and Geissler in this volume). A special FM-FM area exists in the auditory cortex of the mustached bat (Suga and O'Neill, 1979), the neurons of which detect the echo of FM signals emitted by the animal. FM direction-selective neurons were found in the primary auditory cortex, and recently direction selectivity was reported to be topographically ordered in parallel with characteristic frequency (i.e. with frequency tuning of cortical neurons - Zhang *et al.*, 2003). In contrast to the situation in man, where speech areas are strictly localized to one of the brain hemispheres, in non-human mammals the specialization of hemispheres, i.e. lateralization, has not been observed or is the subject of discussion (for review, see Rogers and Andrew, 2002). The results of a recent study by Wetzel *et al.* (1998) pointed out that the detection of the direction of an FM signal change is less precise in gerbils with lesions in the right AC compared with those having lesions in the left AC. In our study, the results of which are presented in the third part of this paper, we were interested in whether the detection of the direction of FM signal change in rats, trained by an operant conditioning technique, was more compromised in animals with an ablation of the left or right AC or both.

2. THE ROLE OF DESCENDING CORTICOTECTAL PATHWAYS

In contrast to several studies aimed at elucidating the function of the descending auditory corticotectal pathways, in which localized small lesions of the AC were performed (Zhang and Suga, 1997; Zhang *et al.*, 1997), we were interested in the question of what are the effects of a total unilateral ablation of the AC on the function of the inferior colliculus (Nwabueze-Ogbo *et al.*, 2002). Therefore, 30 ng of tetrodotoxin were applied on the surface of the AC in ketamine-xylazine anaesthetized rats. The TTX application resulted in a complete disappearance of electrical activity in the AC that lasted for several hours. Middle latency responses (MLR), recorded with an electrode placed on the surface of the AC, were not possible to elicit by acoustical stimulation. Full recovery of MLR amplitude was observed only 1-2 days post-application. Several types of acoustically evoked electrical responses were recorded in the IC to characterize the effects of decortication. Gross IC evoked responses (IC-ER), recorded from electrodes implanted in the IC, slightly changed their amplitude during cortical inactivation (in 48% of the cases an increase was evident, in 30% a decrease), but the latency of some waves was significantly prolonged. Prolongation was typical for the latency of the N_1 and subsequent waves (Fig. 1). After TTX injection, the latency-intensity function of the IC-ER was shifted to higher values and recovered to pre-TTX value several hours post-injection. The average latency shift varied around 0.5 ms regardless of the decrease or increase in the IC-ER amplitude. In some of the implanted electrodes (insulated nichrome wires, diameter 0.002 inches) multiple unit activity was possible to detect. The frequency tuning of neuronal activity in the proximity of the electrode tip was investigated and a frequency-tuning curve (FTC) was constructed. Total functional ablation of the AC did not induce any change in the FTC. Multiple unit activity was, similarly as in the case of the the IC-ER amplitude, either enhanced, depressed or unchanged. This was evident from shifts in the spike-intensity functions. Even though the number of spikes in the response increased or decreased, the shape of the response, as expressed by the peri-stimulus time histogram, remained essentially unchanged. Functional ablation of the AC resulted in a depression in the firing rate of multiple unit responses in about half of the electrodes, whereas in the other half the number of spikes in response to auditory stimulation increased after intracortical TTX injection. Similarly as in acute experiments, as described further below, functional ablation of the AC resulted in more pronounced changes in the later part of the response whereas the onset part of the response was changed to a lesser extent.

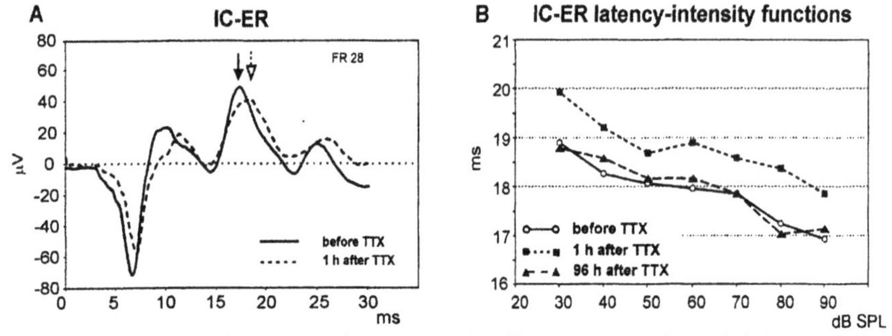

Figure 1. Changes in the latency of click-evoked auditory responses in the inferior colliculus after decortication by the application of tetrodotoxin. A: An example of IC-ER recorded in one animal (FR 28) with the indicated latency shift after AC inactivation. Sound intensity 80 dB SPL. B: IC-ER latency-intensity functions for the P3 wave before and after tetrodotoxin application.

In acute experiments (Popelář *et al.*, 2003) a multichannel 16-electrode probe (produced by the Center for Neural Communication Technology,

University of Michigan) was used for recording electrical activity in the IC instead of implanted electrodes, otherwise the design of the experiment was similar as in chronic experiments. Similarly as in chronic experiments, we did not find any changes in the tuning curves of individual neurones in the IC after TTX application on the AC. Significant differences were, however, found in the firing rate of some of the recorded neurons. Twenty to thirty minutes after TTX injection, many IC neurons decreased or increased their spiking during stimulus presentation without changing the threshold of their response. The proportion of neurons decreasing or increasing their spontaneous and acoustically evoked activity was roughly the same as in chronic experiments (40% and 44%, resp.). Differences between neuronal responses before and after TTX injection were found in other parameters: in rate-level functions and in the shapes of PSTHs. In most cases neurons did not change the type of their rate-level function (in monotonic functions they generally changed only the slope, in saturated types of functions they shifted the level of saturation), but some non-monotonic rate-level functions changed to monotonic ones after TTX injection. Even though the number of spikes in the responses of the majority of neurons changed after AC inactivation, the basic shape of the PSTHs was not significantly altered. However, in about one-third of the recorded neurons, a more pronounced change in the firing rate occurred mainly in the later, sustained part of the PSTH, whereas the onset part of the PSTH was either not affected at all or changed insignificantly. The opposite effect, i.e. a change in the onset part of the PSTH without affecting the later part of the response, was not observed in any of the recorded neurons. The multichannel electrode used in the experiments enabled us to record the activity of several units at the same time and consequently to evaluate their mutual relationship by computing cross-correlograms. 80% of recorded pairs displayed a synchronization of their activity, manifested by a peak in the cross-correlogram occurring a few milliseconds around zero time. Temporal decortication did not change the strength of the correlation in one-third of the evaluated neuronal pairs. A significant decrease in the cross-correlation value was observed in 45% of neuronal pairs. In contrast, an originally small correlation became stronger after TTX injection in 21% of cases. The type of changes in the cross-correlograms after TTX injection were not related to frequency tuning, the position of the neuron in the IC or the type of firing rate changes (i.e. increase or decrease) of either one or both neurons in a pair.

In spite of many experimental results coming mainly from the lab of Nobuo Suga (see e.g. Gao and Suga, 2000; Suga *et al.*, 2002; Suga and Ma, 2003), the function of the descending corticotectal auditory pathways is not completely understood. Yet changes in some parameters of the electrical activity of the IC demonstrated in our experiments are significant: 84% of units changed their overall firing rate after cortical inactivation, and a large number of neuronal pairs changed their functional interaction as indicated in the cross-correlograms. These findings do not contradict data from the Suga laboratory since the two approaches are different: their inactivation with lidocaine or muscimol is usually limited to a small cortical area, and in this way they are able to influence the tuning of a subcortical neuron. When a recorded neuron in the IC is matched in frequency tuning to the inactivated cortical neuron, its response to the best frequency is reduced. Moreover, the responses of other subcortical neurons tuned to different frequencies are increased, and their preferred frequencies are shifted towards those of the inactivated cortical neurons (Zhang *et al.,* 1997). In our case the AC activity is completely abolished, therefore we do not observe changes in the tuning of IC neurons; however, we demonstrate a profound disarray in IC function elicited by such intervention.

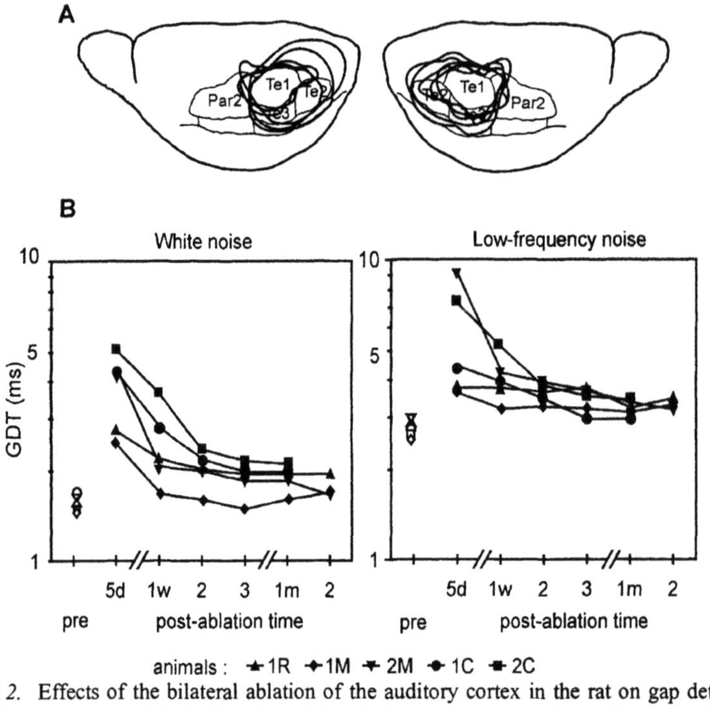

Figure 2. Effects of the bilateral ablation of the auditory cortex in the rat on gap detection thresholds. A: Extent of the lesions in both hemispheres in five experimental animals. B: Pre- and post-ablation GDT values in five animals. (After Syka *et al.*, 2002).

3. TEMPORAL RESOLUTION AND AUDITORY CORTEX ABLATION

In another series of experiments (Syka *et al.*, 2002) pigmented rats (strain Long Evans) were trained by an operant conditioning procedure with food reinforcement to detect the presence of gaps in a continuous noise. The experiment was designed so that 0.5 – 5 s after the animal pressed the starting lever in a modified Skinner box, test signals consisting of five gaps were triggered. Pressing the responding lever during the time window when the five gaps were presented was scored as a hit reaction, and the rat was rewarded with a pellet. GDT was measured using gaps embedded in continuous white noise or continuous low-frequency noise. (upper cut-off frequency 3 kHz) of 70 dB SPL. The threshold of gap detection was first estimated by a method of limits; final GDT values were obtained by the method of constant stimuli using five or six different stimulus durations. The GDT was defined as the duration of a gap corresponding to a 0.5 hit rate corrected for false alarms. After estimating the GDT values, the rats were anaesthetized with a mixture of ketamine and xylazine and the primary auditory cortex in both hemispheres was ablated with a sharp knife under an operating microscope (Fig. 2). GDT was first tested five days after ablation.

The mean values of GDT, measured before cortical ablation when the gaps were embedded in a continuous white noise of 70 dB SPL, were significantly ($P<0.05$) lower than when the gaps were embedded in a low-frequency noise of the same intensity: the values were 1.57±0.07 ms and 2.9±0.34 ms, respectively (Fig. 2). Bilateral ablation of the primary auditory cortex resulted in increased GDT values. On the fifth day after cortical ablation, all rats were capable of gap detection; however, the number of false alarm reaction was high. The values of GDT amounted to 250-300% of the preoperative levels and were on average 4.2±1.1 ms for white noise of 70 dB SPL and 7.4±3.1 ms for low-frequency noise of the same intensity. Seven days after ablation, rats showed an improved performance due to a decreased rate of false alarm reactions. A gradual decrease in GDT was observed over the course of the following 2 – 3 weeks. One month after cortical ablation, gap detection ability had almost recovered. However, GDT values remained slightly higher than before ablation (1.8±0.18 ms for white noise and 3.22±0.15 ms for low-frequency noise). These values were still significantly ($p<0.05$) higher than the values observed before AC ablation. A further decrease of GDT during the following month was not observed.

The results of our experiments suggest that the complete bilateral destruction of the Te1 and the partial destruction of the Te2 and Te3 areas of the auditory cortex in rats substantially reduces not only the temporal resolution abilities of the animal, but also learning and memory mechanisms.

The increased false alarm rate that appeared after cortical ablation indicates a change in the rats' ability to respond correctly to test stimuli. Similar results were obtained by other authors in trained animals (Kelly and Whitfield, 1971; Kelly 1973; Ohl *et al.*, 1999). Lesions in naïve animals before training led, in contrast, to a reduced hit rate and had no effect on the false alarm rate (Kelly, 1973; Ohl *et al.*, 1999).

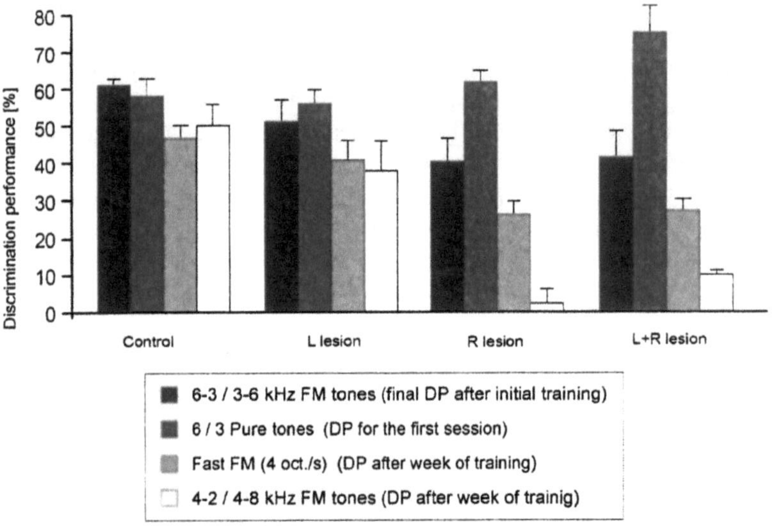

Figure 3. Effects of unilateral and bilateral ablation of the auditory cortex in the rat on discrimination of different acoustical signals. Vertical bars indicate S.D.

4. ROLE OF THE LEFT AND RIGHT HEMISPHERES IN THE PERCEPTION OF FM SIGNALS

In another series of experiments we investigated the ability of pigmented rats (strain Long Evans) to discriminate between different FM stimuli. Normal control rats as well as rats with a unilateral (right or left) or bilateral lesion of the auditory cortex were used in the experiments. Using a shock-avoidance conditioning procedure, thirsty rats were trained to drink in the presence of a rising FM tone, from 3 to 6 kHz (a safe stimulus), and to stop drinking when a falling FM tone, from 6 to 3 kHz (warning stimulus - paired with a mild electric shock), was presented. A cessation of licking in the response period following a warning stimulus was classified as a hit response, while a cessation of licking following a safe stimulus was considered as a false alarm response. The ability to discriminate was quantified as the discrimination performance (DP), which was calculated as the hit rate - false alarm rate, expressed as a percentage. Control animals learned to discriminate between rising and falling FM tones in 11-16 training days with a DP of 60.5% (on average) (Fig. 3). The difference in the final DP after training between the normal and sham-operated rats was not significant (unpaired t-test, $p>0.05$). Rats with an ablation of the auditory cortex were able to learn to discriminate between falling and rising FM tones. However, in comparison with control rats, they had a lower DP and in most cases needed a longer training period. The mean group values of the final DP in rats with a left, right or bilateral lesion of the auditory cortex were 51.2%, 40.5% and 41.7%, respectively (Fig. 3). The results of discrimination training in the group of left-lesioned rats were more homogeneous and more similar to the control group than in the case of right- and bilateral-lesioned animals, both of which displayed a larger deficit in the ability to discriminate between rising and falling FM tones. Study of the spontaneous transfer of the conditioned response to novel pairs of stimuli showed that both control and lesioned rats were able to discriminate between FM tones of a lower (1-2 kHz vs. 2-1 kHz) or higher (8-16 kHz vs.16-8 kHz) frequency range than the FM tones that were used in the initial discrimination training. However, the mean DP values in these test sessions were lower than the final DP values after the initial training. Testing the ability of trained rats to discriminate between a pair of pure tones (a 6 kHz tone as a warning stimulus and a 3 kHz tone as a safe stimulus) has shown that in lesioned rats, DP is significantly higher than the DP reached in previous training with FM tones. When a 3 kHz tone was used as a warning stimulus and a 6 kHz tone as a safe stimulus, DP values were negative. Rising and falling FM tones with a faster modulation rate resulted in a decreased DP both in control and lesioned animals. Additional one week training improved the DP in all animals, however, the improvement was smaller in right-lesioned animals (Fig.3). When the rats had to discriminate FM tones that began at the same frequency, the mean value of DP in all groups was very low. Additional one week discrimination training resulted in a significant improvement in DP only in control and left-lesioned rats.

The results have shown that animals with a lesion of the right auditory cortex are deficient in discriminating the direction of a frequency modulation; however, they are capable of discriminating FM stimuli on the basis of their starting frequencies. Similar results were obtained in Mongolian gerbils that were trained to discriminate pure tones and FM tones before and after bilateral lesions of the auditory cortex (Ohl *et al.*, 1999). In animals with lesions, the discrimination of pure tones was not influenced whereas discrimination of FM tones was impaired. The effects of a lesion were also dependent on the timing of the lesion relative to FM tone discrimination training. In the case of unilateral lesions of the auditory cortex in gerbils (Wetzel *et al.*, 1997), the discrimination performance of the gerbils in a task with rising and falling FM tones was found to be the same as in controls in animals with a left side lesion whereas it was impaired in the case of a right side lesion.

5. CONCLUSIONS

The primary auditory cortex, as a target of the auditory pathway, represents a structure that exerts a strong influence on the processing of acoustical information in other auditory nuclei. The corticothalamic and corticotectal pathways are involved in as-yet not completely understood process that result in plastic changes in the function of IC and MGB neurons. For example, the local inactivation of a small area in the AC of the bat produces amplified IC and MGB responses to a tone (Zhang and Suga, 1996) and a shift in the preferred frequencies of IC neurons towards those of the inactivated cortical neurons (Zhang *et al,* 1997). The complete inactivation of the auditory cortex by tetrodotoxin in our experiments produced changes in the timing of the acoustical signal processing in the IC as reflected in the prolongation of the waves of the IC evoked responses and the reversible enhancement or reduction of the responses of individual neurons to a sound. Changes in the functional interaction of IC neurons after AC ablation with tetrodotoxin, revealed by the cross-correlation technique, support the assumption that feedback from the auditory cortex plays an important role in the function of the IC. Further details about the function of this feedback remain to be elucidated.

As already pointed out by Neff *et al.* (1975), bilateral ablation of the auditory cortex results in defects in the localization of a sound source in space and defects in detecting the temporal features of a sound. Ison et al. (1991) demonstrated in rats an inability to detect a gap of 15 ms duration following functional decortication by KCl application on the cortex. Kelly *et al.* (1996) reported that bilateral destruction of the auditory cortex in ferrets resulted in an increase in GDT from 10 ms to 20 ms. In our case bilateral ablation of the auditory cortex in the rat resulted in a significant prolongation of GDT from approximately 2 ms to 5-7 ms soon after ablation. However, the GDT recovered to near preablation levels during retraining, thus demonstrating a high degree of plasticity in the central auditory system.

Finally, our experiments revealed that some degree of hemispheric specialization in acoustical signal processing exists not only in man but also in other mammals. In agreement with studies performed in the Mongolian gerbil (Wetzel et al., 1997), the results of our experiments clearly demonstrate that the right hemisphere in the rat plays an important role in discriminating between rising and falling FM tones, whereas even a bilateral ablation of the auditory cortex does not impair the discrimination of pure tones. Thus, we can confirm even in rodents the key data from experiments performed thirty years ago in cats and monkeys (see Neff et al., 1975) with an important addition: emerging knowledge about the different roles of the hemispheres in processing specific acoustical signals.

ACKNOWLEDGEMENTS

The research was supported by grants 309/01/1063 and 309/04/1074 from the Grant Agency of the Czech Republic and NK 6454-3 and NR 8113-4 from the Ministry of Health of the Czech Republic.

REFERENCES

Beyerl, B.D., 1978, Afferent projections to the central nucleus of the inferior colliculus in the rat. *Brain. Res.* **145**: 209-223.

Druga, R., and Syka, J., 1984a, Ascending and descending projections to the inferior colliculus in the rat. *Physiol. Bohemoslov.* **33**: 31-42.

Druga, R., and Syka, J., 1984b, Neocortical projections to the inferior colliculus in the rat. (An experimental study using anterograde degeneration techniques). *Physiol. Bohemoslov.* **33**: 251-3.

Ehret, G., and Geissler, D.B. 2004, Communication-call representation in the mouse auditory cortex: perception vs. recognition. This volume, pp. 85-96.

Faye-Lund, H., 1985, The neocortical projection to the inferior colliculus in the albino rat. *Anat. Embryol. (Berl).* **173**: 53-70.

Gao, E. and Suga, N., 2000, Experience-dependent plasticity in the auditory cortex and the inferior colliculus of bats: role of the corticofugal system. *Proc. Natl. Acad. Sci. U. S. A.* **97**: 8081-8086.

Ison, J.R., O'Connor, K., Bowen, G.P., and Bocirnea, A., 1991, Temporal resolution of gaps in noise by the rat is lost with functional decortication. *Behav Neurosci.* **105**:33-40.

Kelly, J.B., 1973, The effects of insular and temporal lesions in cats on two types of auditory pattern discrimination. *Brain. Res.* **62**: 71-87.

Kelly, J.B. and Whitfield, I.C., 1971, Effects of auditory cortical lesions on discriminations of rising and falling frequency-modulated tones. *J. Neurophysiol.* **34**: 802-816.

Kelly, J.B., Rooney, B.J., and Phillips, D.P., 1996, Effects of bilateral auditory cortical lesions on gap-detection thresholds in the ferret (Mustela putorius). *Behav. Neurosci.* **110**: 542-50.

Neff, W.D., Diamond, I.T., and Casseday J.H., 1975, Behavioral Studies of Auditory Discrimination: Central Nervous System. In: *Handbook of Sensory Physiology, Vol. V/2 Auditory System* (W.D. Keidel and W.D. Neff, eds.), Springer-Verlag, Berlin, pp. 307-400.

Nwabueze-Ogbo, F.C., Popelář, J., and Syka, J., 2002, Changes in the acoustically evoked activity in the inferior colliculus of the rat after functional ablation of the auditory cortex. *Physiol. Res.* **51**: S95-S104.

Ohl F.W., Wetzel, W., Wagner, T., Rech, A., and Scheich, H., 1999, Bilateral ablation of auditory cortex in Mongolian gerbil affects discrimination of frequency modulated tones but not of pure tones. *Learn. Mem.* **6**: 347-362.

Popelář, J., Nwabueze-Ogbo, F.C. and Syka, J., 2003, Changes in neuronal activity of the inferior colliculus in rat after temporal inactivation of the auditory cortex. *Physiol. Res.* **52**: 615-628.

Rogers, L.J. and Andrew, R.J., 2002, *Comparative Vertebrate Lateralization*. Cambridge University Press, Cambridge, UK

Suga, N. and Ma, X., 2003, Multiparametric corticofugal modulation and plasticity in the auditory system. *Nat. Rev. Neurosci.* **4**: 783-794.

Suga, N. and O'Neill, W.E., 1979, Neural axis representing target range in the auditory cortex of the mustache bat. *Science* **206**: 351-353.

Suga, N., Xiao, Z., Ma, X. and Ji, W., 2002, Plasticity and corticofugal modulation for hearing in adult animals. *Neuron* **36**: 9-18.

Sun, X., Chen, Q.C., and Jen, P.H., 1996, Corticofugal control of central auditory sensitivity in the big brown bat, Eptesicus fuscus. *Neurosci. Lett.* **212**: 131-4.

Syka, J. and Popelář, J., 1984, Inferior colliculus in the rat: neuronal responses to stimulation of the auditory cortex. *Neurosci. Lett.* **51**: 235-240.

Syka, J., Rybalko, N., Mazelová, J., and Druga, R., 2002, Gap detection threshold in the rat before and after auditory cortex ablation. *Hear. Res.* **172**: 151-159.

Wetzel, W., Ohl, F.W., Wagner, T., and Scheich, H., 1998, Right auditory cortex lesion in Mongolian gerbils impairs discrimination of rising and falling frequency-modulated tones. *Neurosci Lett.* **252**: 115-8.

Winer, J.A., Larue, D.T., Diehl, J.J. and Hefti, B.J., 1998, Auditory cortical projections to the cat inferior colliculus. *J. Comp. Neurol.* **400**: 147-174.

Zhang,Y., and Suga, N., 1997, Corticofugal amplification of subcortical responses to single tone stimuli in the musteched bat. *J. Neurophysiol.* **78**: 3489-3492.

Zhang, Y., Suga, N. and Yan, J., 1997, Corticofugal modulation of frequency processing in bat auditory system. *Nature* **387**: 900-903.

Plastic Changes in the Primary Auditory Cortex in Cochlear Implanted Deaf Cats

RAINER KLINKE, ANDREJ KRAL, SILVIA HEID, and RAINER HARTMANN

Institut Für Sinnes- Und Neurophysiologie, Goethe-Universität Frankfurt, D–60590, Frankfurt/M, Germany

1. INTRODUCTION

The congenitally deaf cat (CDC) is a convenient model for (i) deafness in human infants, (ii) for the study of central deficits as a consequence of auditory deprivation and (iii) for central effects of cochlear implantation. Finally (iiii) adult animals of thc this strain can bc used to tests for conceivable cross-modal reorganization of the primary auditory cortex as a consequence of life-long auditory deprivation.

The organ of Corti in these animals degenerates during the first two postnatal weeks (Mair, 1973) a period during which in normal kittens the inner ear starts to take up function. Thus the CDCs have never any hearing experience as also shown from a longitudinal study in a few of our individuals (Heid *et al.*, 1998). The CDC cochlea is devoid of hair cells, bony structures and neuronal elements are preserved, i.e. there is a Scheibe type of dysplasia. The good preservation of primary auditory afferents during the first year of life has to be emphasized (Heid *et al.*, 1998), an aspect in which CDCs clearly differ from pharmacologically deafened normal animals.

Clearly CDCs also present dystrophic changes in the central auditory system (Saada *et al.*, 1996; Heid *et al.*, 1997; 1998; Redd *et al.*, 2000) e.g. reductions in somatic area of the brain stem neurones or reduction in number of dendritic ramifications in cortical neurones in A1 (Wurth *et al.*, 1999).

Nevertheless, the auditory pathway is rudimentarily present in CDCs and allows for a feed-through of neural activity from the spiral ganglion to the auditory cortex (Hartmann *et al.*, 1997; see also Klinke *et al.*, 1999; Popelar *et al.*, 1995). We additionally suppose, experimental proof is missing, that there is spontaneous activity in the primary afferents and thus in the nuclei of the auditory pathway.

2. EXPERIMENTAL DESIGN

For the experiments exclusively CDCs were used the complete deafness of which had been verified at the age of 4 weeks by electrophysiological means.

As a rule the animals were unilaterally implanted with custom-made intracochlear electrodes (left ear) in different ages from 2 to 6.5 months. The electrodes were fed through the skin on the back and connected to a VIENNA-type sound processor carried in a jacket (see Klinke *et al.*, 1999). The coding was compressed-analogue. Electrical stimuli were fed to the most effective intracochlear electrode, stimulation was monopolar. The system therefore only provided the time structure of the acoustic signals, place coding was not made use of. The stimulus currents were set using the pinna reflex as an indicator. After recovery from surgery, the animals were exposed to electrical cochlear stimuli 24 h per day and lived in an acoustically enriched environment. To call their attention to the acoustic world they were also conditioned and rewarded when responding to sound stimuli.

After a period of 2 to 6 months of experience with the implant, the animals were investigated in a final neurophysiological experiment under light isoflurane anaesthesia (Kral *et al.*, 1999). The primary auditory cortices (A1) were exposed. Intracochlear electrical stimulation was applied using single charge-balanced pulses of 200 μs duration, 10 dB above threshold of the most sensitive point (hot spot). Surface potentials, intracortical potentials, single- and multi-unit activities were recorded. From the depth-profiles of intracortical responses current-source-density analyses were calculated. Naive unimplanted CDCs, normal animals deafened pharmacologically within the second week of life and normal hearing cats acutely deafened before neurophysiological testing served as controls.

3. RESULTS

3.1 Behaviour

As already mentioned in the introduction, the pinna-reflex was found to be established upon the first electrical stimulation. The respective neural circuit is thus not depending on experience. After a short period of experience cats implanted in young age made use of the newly acquired sensory channel. Successful hits to the conditioned stimulus were learned in about 2 weeks. The cats soon responded adequately to other sound stimuli, could be awakened by sound (e.g. rattling of keys) and some of the individuals searched actively for sound sources.

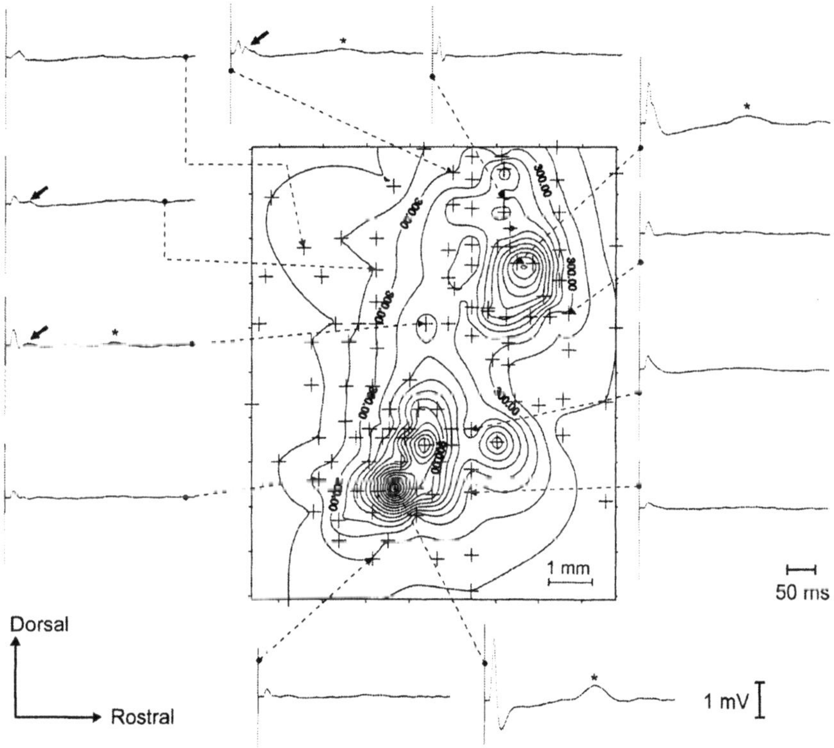

Figure 1. Electrical intracochlear stimulation and map of P_a-amplitudes taken at different locations of A1 and morphology of the potentials in a chronically implanted congenitally deaf cat (from Kral et al., 2002 with permission).

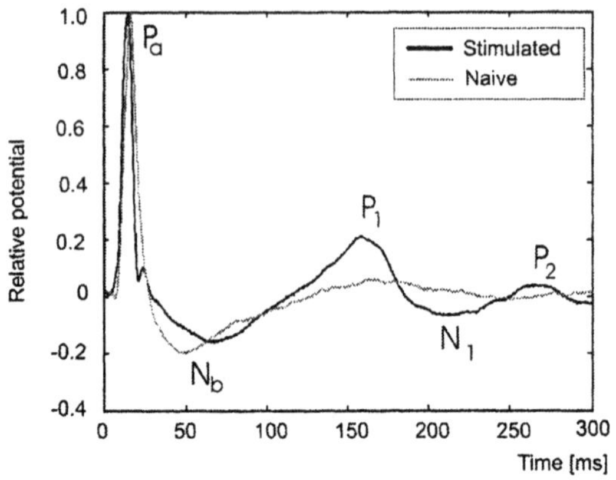

Figure 2. Comparison of surface potentials in a naïve and in a chronically stimulated deaf cat (implantation age 3.5; recording 8.5 months) upon intracochlear electrical stimulation. Notice P_a-amplitudes are normalized to 1.0, i.e. enlarged by approximately factor 3 in naïve cat.

3.2 Gross Cortical Activity

Recordings were taken from some 100 different locations in A1 of the auditory cortex.

In naïve as well as in cochlear implanted CDCs middle latency cortical surface potentials (< 50 ms) can be recorded. While the amplitudes of these middle latency responses (P_a) are larger in chronically stimulated animals by a factor of 3. Fig. 1 gives an example of the morphology of these responses, recorded from different positions of A1 in an animal chronically implanted and stimulated for 5 months. In addition to the middle latency peak P_a also late responses can be found in chronic CDCs (P_1, P_2 at approximately 150 and 300 ms, see Fig. 2. In this figure the recordings are normalized to identical amplitude of P_a. I.e. the potentials of the naïve cat are enlarged). The late waves are not normally found in naïve CDCs of comparable age. Further analyses reveals that γ-activity is to be found in chronically stimulated CDCs as well. It should be mentioned, that the above described differences in amplitude of field potentials of naïve and chronic CDCs are not found in recordings of the respective brain stem potentials. Thus the differences found are brought about by maturation of the auditory thalamus and/or the auditory cortex. The development of late waves and γ-activity indicates that processes of intracortical or thalamo-cortical information processing are established during the acquisition of hearing experience in the implanted animals.

3.3 Cortical Areas to be Activated

If maps are computed showing the cortical area to be activated by intracochlear electro-stimulation and normalized for the size of the growing brain (Kral *et al.*, 2003) an increase of the amplitude of the local field potentials as well as an enlargement of the activated area can be seen, depending of the duration of the electrical stimulation. In addition, the latencies of the different peaks decrease with increasing stimulation period. These changes again show that an increasing population of neurons is recruited for the processing of the peripheral input with increasing "hearing" experience in the chronically implanted CDCs. Although this increase of activated area may be partly due to the circumstance that single channel intracochlear stimulation was applied, it nevertheless shows that it is possible to recruit more and more neurones for the processing of afferent activity. In multichannel intracochlear stimulation most probably each stimulation channel would compete with neighbouring ones, nevertheless finally occupying as many neurons as ever possible.

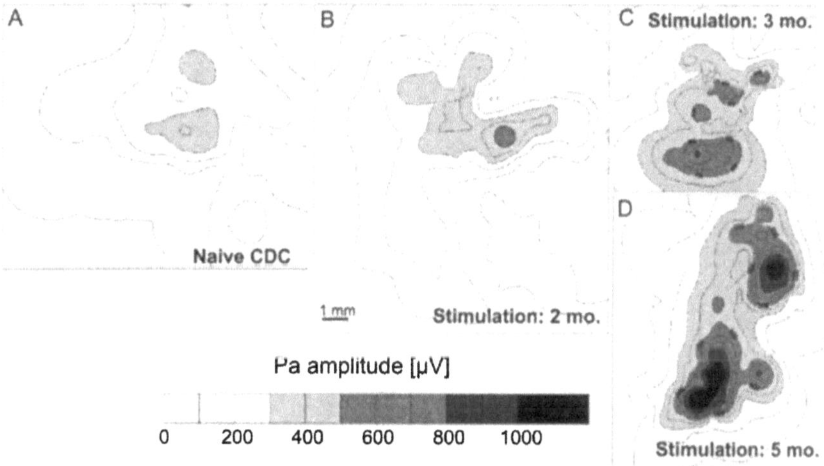

Figure 3. Increase of cortical area activated with increasing duration of stimulation.

While Fig. 3 and the above text refers to the cortex contralateral to the chronically implanted ear, A1 of both the contralateral and ipsilateral side in reference to the stimulated ear can be activated by stimulation through the chronic implant. Even more, if in the final experiment a stimulation electrode is placed in the hitherto unstimulated cochlea both cortices can also be activated from the untrained afferent pathway.

3.4 Synaptic Activity

The current source density analysis (Mitzdorf, 1985) results in current profiles in different layers of nervous tissue, in our case of A1. The profiles are thus layer-specific; the method suppresses remote current generators, a particularly important property! As a matter of fact the currents reveal transmembrane currents and, with some caution, can be looked upon as synaptic currents. Fig. 4 gives an example of such analyses in a naïve CDC, a chronically stimulated one and in a normal hearing cat acutely deafened and stimulated electrically by an intracochlear electrode. Sinks are upwards and shaded. Judged from the presence of sinks in the normal cat (right hand panel) excitations starts more or less simultaneously in layers VI, V, IV and III, then spreads into II and remains strong in the deep portion of layer III. Subsequently sinks appear in layers V and IV. Activity continues up to 300 ms after the stimulus. The initial input to layer IV coincides with the N_a-wave of the surface potential. We further suggest that the sources in layers III/IV following the initial things represent inhibitory currents as during these latencies only weak multi-unit activity was found (Klinke *et al.*, 1999; Kral *et al.*, 2001).

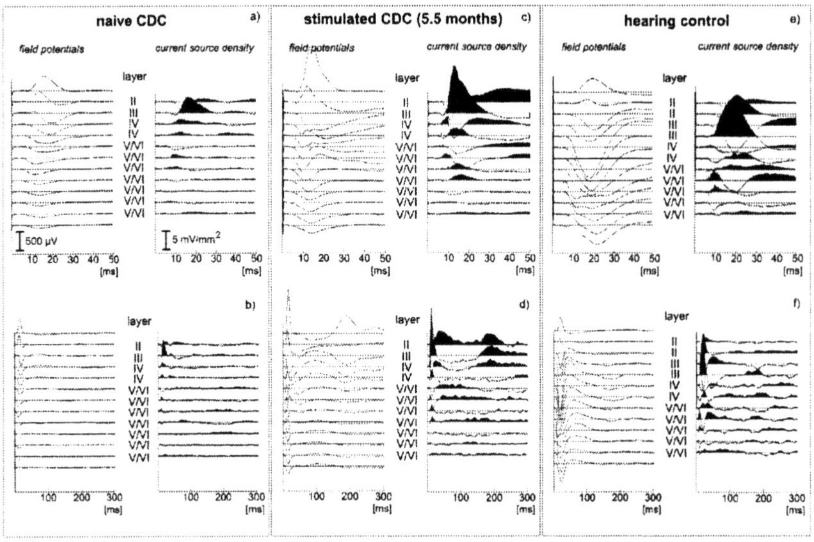

Figure 4. Current source density profiles in naïve, chronically stimulated deaf cats and in a hearing but electrically stimulated control (from Klinke et al., 1999, with permission).

In the naïve CDC (left hand panel) the amplitudes of the currents are much smaller and more or less restricted to the period 50 ms post stimulus (see Kral *et al.*, 2000 and 2003a for further details).

In chronically stimulated CDCs (middle panel) there are stronger currents than in naïve ones. As in hearing controls currents extend up to 300 ms and longer. The sinks may be larger than even in hearing controls, particularly in layer III where neuronal plasticity is believed to the greatest. In short, the current profiles in implanted animals approach the findings in hearing controls. However, in details there are also differences which still wait for explanation. In particular, the sequence of activation in different layers seems to differ. It should be mentioned that further analyses of these currents was undertaken using independent component analyses (Hubka *et al.*, 2003).

3.5 Single- and Multi-Unit Activity

To summarize shortly: Intracochlear electrical stimulation in naïve CDCs evokes only action potentials in the 30 ms latency range in A1. The intensity functions of these discharges are monotonic, the dynamic range is less than 5 dB. In chronically stimulated CDCs, however, a variety of responses types can be found. Most units display a middle latency peak (~30 ms) but after a silent period there is further activity extending into the 150 – 300 ms range. Other units only display long latency responses. The intensity function of units in chronic animals are mostly non-monotonic. They respond over a larger dynamic range (Klinke *ct al.*, 1999; Kral *et ul.*, 2001).

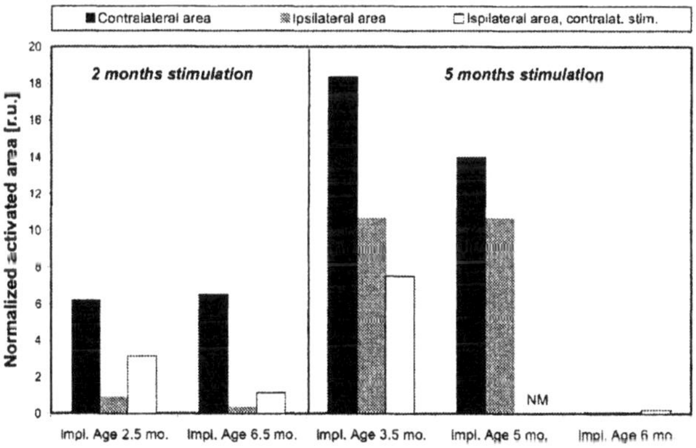

Figure 5. Activated cortical areas with different implantation ages.

3.6 Is there a Sensitive Period?

As already reported, different CDCs were implanted at different ages. The behavioural and neurophysiological changes brought about by "hearing" experience were seen up to an implantation age of 5 months, whereby e.g.

conditioning to sound stimuli was less successful with older implantation age. Implanting CDCs with an age of 6 months and above did hardly lead to a recruitment of A1. An illustration is given in Fig. 5. The figure represents the activated area in A1 of both the contralateral and ipsilateral cortices. In particular the cat illustrated in the right hand panel, which was allowed 5 months for becoming familiar with the implant, did hardly show any activation of the auditory cortex. We conclude that for the auditory cortex of cats the closure of a critical period is around the 5th postnatal month which corresponds to the data provided by Eggermont (1996).

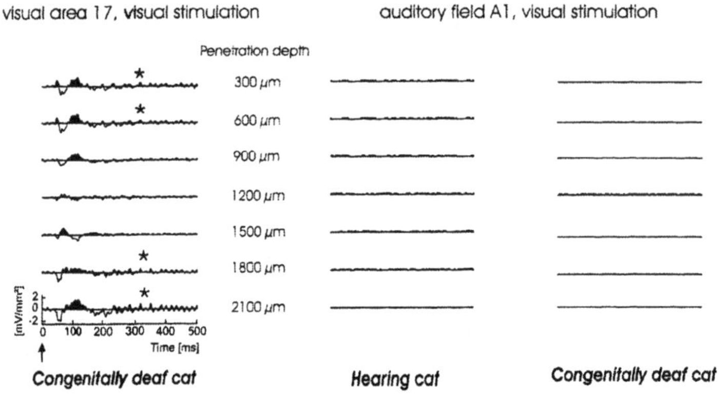

Figure 6. Current source density profiles taken in a deaf cat from the visual cortex (right panel) and in A1 (right panel). For comparison data from a hearing cat are also included (middle panel). Visual stimulation: Phase-reversal gratings.

3.7 No Cross-Modal Input to A1 in CDCs

There are reports on congenitally deaf patients the auditory cortex of whom could be activated by visual stimuli (see Bavelier and Neville, 2002). Rebillard *et al.* (1977, 1980) had reported on visually evoked potentials in the auditory cortex of CDCs. In fact, adult CDCs should be an ideal model to test for cross-modal activation of the auditory system if it is really taken in possession by other modalities in cases of sensory deprivation. However, already Hartmann *et al.* (1997) were unable to reproduce Rebillard's findings. The question was reinvestigated applying current source density analyses and multi-unit-recordings (Kral *et al.*, 2003 b). Phase-reversal-gratings of different orientation and spatial frequencies were used as visual stimuli. This type of visual stimulation activates both the dorsal and ventral visual pathways. In addition moved visual targets were used as stimuli. Also the somatosensory system was activated by mechanical stimuli applied to the whiskers, the face and the forepaws of the CDCs. Microelectrode recordings were taken from the visual cortex (area 17) and auditory (A1). Neither in current source density analyses nor in multi-unit-responses visually evoked local activity could be spotted in A1. An illustration is provided in Fig. 6. The upper panel shows current source density profiles from the visual cortex upon stimulation with phase-reversal-gratings (stimulation reversal frequency 2 Hz). The lower panels show that any evoked activity is missing in both A1 from hearing and congenitally deaf cats. Admittedly local surface potentials correlated with the visual stimuli could be recorded in both CDCs and hearing cats. These potentials were, however, tiny (~50 μV) and were explained by volume conductance. Remember that the current source density analysis removes remote generators and thus is a save method for our purposes. So we maintain that in the CDC model there is no indication of a cross-modal take-over of A1.

4. CONCLUSION

From our data we argue that in CDCs the deprived field A1 of the auditory cortex is not engaged with analyses of other sensory modalities; that early supply with an cochlear implant secures maturation of the auditory cortex and that this supply has to be provided within a critical period, about 5 months in the cat, in order to be successful. The data are in agreement with human data by Scharma *et al.* (2002).

ACKNOWLEDGEMENT

The above investigations were supported by the Deutsche Forschungsgemeinschaft (SFB 269, C1).

REFERENCES

Bavelier, D., and Neville, H. J., 2002, Cross-modal plasticity: Where and how? *Nat. Rev. Neurosci.* **3**: 443-452.

Eggermont, J. J., 1996, Differential maturation rates for response parameters in cat primary auditory cortex. *Auditory Neuroscience* **2**: 309-327.

Hartmann, R., Shepherd, R. K., Heid, S., and Klinke, R., 1997, Response of the primary auditory cortex to electrical stimulation of the auditory nerve in the congenitally deaf white cat. *Hear. Res.* **112**: 115-133.

Heid, S., Hartmann, R., and Klinke, R., 1998, A model for prelingual deafness, the congenitally deaf white cat-population statistics and degenerative changes. *Hear. Res.* **115**: 101-112.

Heid, S., Jähn-Siebert, T. K., Klinke, R., Hartmann, R., and Langner, G., 1997, Afferent projection patterns in the auditory brainstem in normal and congenitally deaf white cats. *Hear. Res.* **110**: 191-199.

Hubka, P., Kral, A., and Klinke, R., 2003, Input desynchronization and impaired columnar activation in deprived auditory cortex revealed by independent component analysis. Satellite Symposium to the 6th IBRO World Congress of Neuroscience, Prague.

Klinke, R., Kral, A., Heid, S., Tillein, J., and Hartmann, R., 1999, Recruitment of the auditory cortex in congenitally deaf cats by long-term cochlear electrostimulation. *Science* **285**: 1729-1733.

Kral, A., Hartmann, R., and Klinke, R., 2003a, Recruitment of the auditory cortex in congenitally deaf cats. In *Rewiring the Cerebral Cortex* (S. Lomber and J. Eggermont, eds.) in press.

Kral, A., Hartmann, R., Tillein, J., Heid, S., and Klinke R., 2000, Congenital auditory deprivation reduces synaptic activity within the auditory cortex in a layer-specific manner. *Cereb. Cortex* **10**: 714-726.

Kral, A., Hartmann, R., Tillein, J., Heid, S., and Klinke, R., 2002, Hearing after congenital deafness: Central auditory plasticity and sensory deprivation. *Cereb. Cortex* **12**: 797-807.

Kral, A., Hartmann, R., Tillein, J., Heid, S., and Klinke, R., 2001, Delayed maturation and sensitive periods in the auditory cortex. *Audiol. Neurootol.* **6**: 346-362.

Kral, A., Schröder, J.-H., Klinke, R., and Engel, A.K., 2003b, Absence of cross-modal reorganization in the primary auditory cortex of congenitally deaf cats. *Exp. Brain Res.* **153**: 605-13.

Kral, A., Tillein, J., Hartmann, R., and Klinke, R., 1999, Monitoring of anaesthesia in neurophysiological experiments. *Neuroreport* **10**: 781-787.

Mair, I. W. S., 1973, Hereditary deafness in the white cat. *Acta Otolaryngol. (Stockh.) Suppl.* **314**: 1-53.

Mitzdorf, U., 1985, Current source-density method and application in cat cerebral cortex: Investigation of evoked potentials and EEG phenomena. *Physiol. Rev.* **65**: 37-100.

Popelar, J., Hartmann, R., Syka, J., and Klinke, R., 1995, Middle latency responses to acoustical and electrical stimulation of the cochlea in cats. *Hear. Res.* **92**: 63-77.

Rebillard, G., Carlier, E., Rebillard, M., and Pujol, R., 1977, Enhancement of visual responses on the primary auditory cortex of the cat after an early destruction of cochlear receptors. *Brain Res.* **129**: 162-164.

Rebillard, G., Rebillard, M., and Pujol, R., 1980, Factors affecting the recording of visual-evoked potentials from the deaf cat primary auditory cortex (AI). *Brain Res.* **188**: 252-254.

Redd, E. E., Pongstaporn, R., and Ryugo, D. K., 2000, The effects of congenital deafness on auditory nerve synapses and globular bushy cells in cats. *Hear. Res.* **147**: 160-174.

Saada, A. A., Niparko, J. K., and Ryugo, D. K., 1996, Morphological changes in the cochlear nucleus of congenitally deaf white cats. *Brain Res.* **736**: 315-328.

Sharma, A., Dorman, M. F., and Spahr, A. J., 2002, Rapid developnet of cortical auditory evoked potentials after early cochlear implantation. *Neuroreport* **13**: 1365-1368.

Wurth, N., Heid, S., Kral, A., and Klinke, R., 1999, Morphology of neurons in the primary auditory cortex (AI) in normal and congenitally deaf white cats – study of DiI labelled cells. Proc. *27th Göttingen Neurobiol. Conf.* **27**: 318.

Input Desynchronization and Impaired Columnar Activation in Deprived Auditory Cortex Revealed by Independent Component Analysis

PETER HUBKA[*], ANDREJ KRAL[#], and RAINER KLINKE[#]
[*]*Institute of Pathological Physiology, Comenius University, Sasinkova 4, SK-811 08 Bratislava, Slovak Republic; and* [#]*Institut für Sinnes- & Neurophysiologie, J.W. Goethe Universität Frankfurt am Main, Theodor-Stern-Kai 7, D-60590 Frankfurt, Germany.*

1. INTRODUCTION

Congenital auditory deprivation leads to abnormalities in the cortical processing in adult subjects (Klinke *et al.*, 1999; Raggio and Schreiner, 1999; Kral *et al.*, 2000; for comparable human data see Ponton *et al.*, 2000; Sharma *et al.*, 2002) indicating that auditory input is necessary for proper development of the auditory system. When auditory system was not stimulated during development, auditory inputs could not be properly processed and adequate feature extraction could not be achieved. To study functional effects of deprivation, congenitally deaf cats and normal hearing controls were electrically stimulated and activation of primary auditory cortex was recorded. In previous studies, deficits in the auditory cortex of congenitally deaf cats were most pronounced in the infragranular layers and at longer poststimulus latencies (Klinke *et al.*, 1999; Kral *et al.*, 2000). Functional cortical deficits underlying abovementioned impairments are not well understood yet.

2. MATERIAL AND METHODS

Six adult congenitally deaf cats (6, 10, 18, 20, 21, 29 month of age) were used, 3 normal adult hearing cats served as controls (10, 14 and 21 months of age). The animals were electrically stimulated by charged-balanced biphasic pulses (200 µs/phase, 2 Hz repetition rate). Field potentials (FP) were recorded in 78 recording positions in different cortical layers (6 tracks, 13 recording depth in 300 µm steps, see Fig. 1).

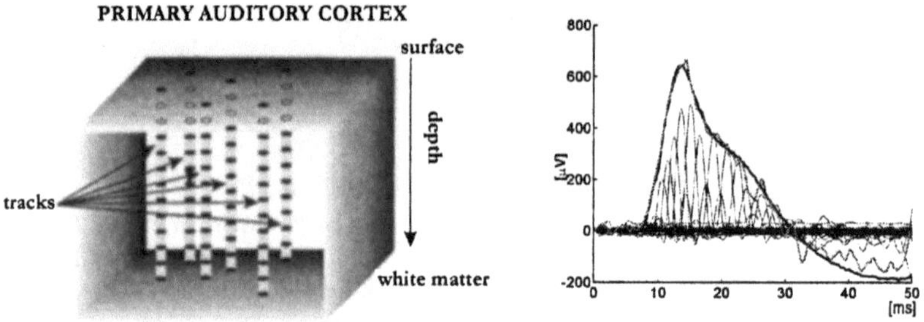

Figure 1. Left panel: Schematical view on recording points (black ellipses) in the region with the largest surface response in the primary auditory cortex.
Right panel: Relation of recorded FP (thick black line) to IC-FP separated from a set of compound FPs (thin black lines).

Independent component analysis (ICA), a method for extraction of individual signals contributing to their linear mixture, was used in the present study (Hyvärinen, 2000). At present, ICA is a dominant method for solving so-called 'cocktail-party' problems. In the present study, ICA was used to decompose compound FP set into a series of source signals or independent component signals (IC-FP) with their specific cortical profiles.

For each single IC-FP profile, current source density (CSD) profile was computed to acquire information about the transmembrane currents underlying its electrical field distribution in the cortex. Normalized CSD time representations IC-FP are called activation units (AU).

3. RESULTS AND DISCUSSION

The individual activation unit could reflect synchronous activation of a 'synaptic patch' containing many single synapses located in (or close to) the same structure. Significantly shorter duration of single activation units in hearing cats ($p = 0.00018$) indicates more synchronous activation of single synapses within the synaptic patch. On the other hand, longer activation units in deaf cats could be a consequence of less synchronous input into the synaptic patch. It was also shown that duration of excitatory postsynaptic currents shortens during postnatal development (Carmignoto and Vicini, 1992, Wall *et al.*, 2002). We therefore assume that activation units separated from the deprived primary auditory cortex could reflect a persistent immature state of the cortical synapses.

Differences in time course and layer distribution of activation in the cortex were quantified using partial summation of the activation units with peak latencies in a 3 ms time window for the layers I/II, III, IV and V/VI separately. We have revealed missing or delayed superficial input components (layer I/II) and substantially decreased sources (outward transmembrane currents) in the layer III and IV with latencies 9-30 ms in deaf animals (hearing cats – 231.2 ± 151.7 mV/mm^2; congenitally deaf cats – 27.0 ± 12.3 mV/mm^2, $p < 0.05$).

It was shown that the synchronous coupling of inputs into different cortical layers (interlayer synchronization) affects responsiveness of layer V pyramidal neurons (Larkum *et al.*, 1999, Miller *et al.*, 2001). Thus, missing or delayed thalamic inputs into the superficial layers in deaf animals indicate interlayer desynchronisation that could negatively affects further flow of information in the deprived auditory cortex.

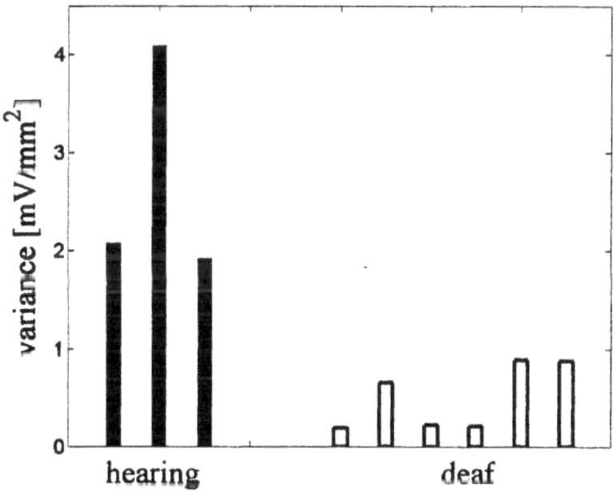

Figure 2. Mean lateral variances of activation unit amplitudes in the same penetration depth of all analysed animals. Lateral variances in hearing cats were significantly higher than variances in the group of congenitally deaf cats ($p < 0.05$).

Lateral relation of activations in the same layer was analysed by computing of variance of activation unit amplitudes in the same penetration depth. Activation of the cortex of the congenitally deaf animals showed significantly smaller intercolumnar variability ($p = 0.024$, see Fig. 2). This result together with substantially decreased outward transmembrane currents in the layer III and IV indicate decreased inhibition in the primary thalamic input layers of the naïve primary auditory cortex.

The decreased inhibition and input desynchronisation leads to a diffuse activation of the primary auditory cortex of congenitally deaf adult cats (Fig.3). These findings could negatively affect coincidence detection, feature extraction and flow of information to the higher auditory cortical regions in naïve animals.

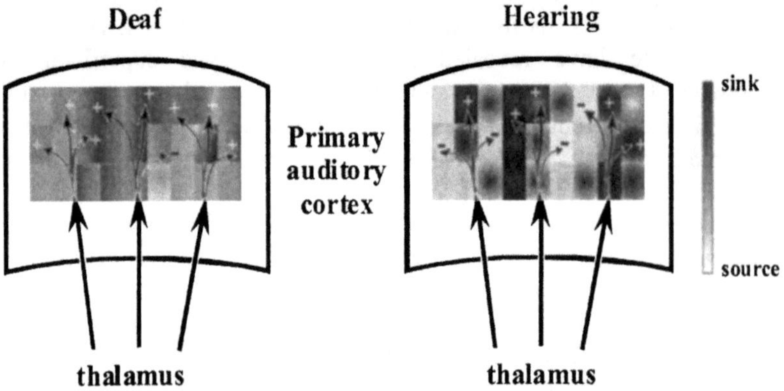

Figure 3. Schematic comparison of the diffuse activation of the primary auditory cortex in adult congenitally deaf cats and well-structured activation in hearing animals.

4. CONCLUSIONS

Presented results show that in adult congenitally deaf cats there is:
a local desynchronization of inputs within adjacent synapses,
an interlayer desynchronization of inputs to different layers in cortex, and
possibly also a decreased inhibition in layer IV and III.

ACKNOWLEDGEMENTS

The present study was supported by DFG (SFB 269/C1) and DAAD.

REFERENCES

Carmignoto, G., and Vicini, S., 1992, Activity-dependent decrease in NMDA receptor responses during development of the visual cortex. *Science* **258**: 1007-1011.

Hyvarinen, A., and Oja, E., 2000, Independent component analysis: algorithms and applications. *Neural Netw* **13**: 411-430.

Klinke, R., Kral, A., Heid, S., Tillein, J., and Hartmann, R., 1999, Recruitment of the auditory cortex in congenitally deaf cats by long- term cochlear electrostimulation. *Science* **285**: 1729-1733.

Kral, A., Hartmann, R., Tillein, J., Heid, S., and Klinke, R., 2000, Congenital auditory deprivation reduces synaptic activity within the auditory cortex in a layer-specific manner. *Cereb Cortex* **10**: 714-726.

Larkum, M. E., Zhu, J. J., and Sakmann, B., 1999, A new cellular mechanism for coupling inputs arriving at different cortical layers. *Nature* **398**: 338-341.

Miller, L. M., Escabí, M. A, and Schreiner, C.E, 2001, Feature selectivity and interneuronal cooperation in the thalamocortical system. *J Neurosci* **21**: 8136-8144.

Ponton, C. W., Eggermont, J. J., Kwong, B., and Don, M., 2000, Maturation of human central auditory system activity: evidence from multi-channel evoked potentials. *Clin Neurophysiol* **111**: 220-236.

Raggio, M. W., and Schreiner, C. E., 1999, Neuronal responses in cat primary auditory cortex to electrical cochlear stimulation. III. Activation patterns in short- and long- term deafness. *J Neurophysiol* **82**: 3506-3526.

Sharma, A., Dorman, M. F., and Spahr, A. J., 2002, A sensitive period for the development of the central auditory system in children with cochlear implants: implications for age of implantation. *Ear Hear* **23**: 532-539.

Wall, M. J., Robert, A., Howe, J. R., and Usowicz, M. M., 2002, The speeding of EPSC kinetics during maturation of a central synapse. *Eur J Neurosci* **15**: 785-797.

Temporal Firing Activities of Auditory Cortical Neurons and Modification of Their Activities by Laser Irradiation

HIROSHI RIQUIMAROUX[*], and YOSKY KATAOKA[#]
[*]*Dept. of Knowledge Engineering and Computer Sciences, Doshisha University, Kyotanabe, Japan; and* [#]*Dept. of Anatomy and Cell Science, Kansai Medical University, Moriguchi, Japan*

1. INTRODUCTION

It has been well known that tonotopy exists in the primary auditory cortex (AI) of mammals at threshold level. However, we showed using Japanese monkeys (*Macaca fuscata*) that there was no simple tonotopy at suprathreshold level (60~70 dB SPL). In other words, neurons in AI responded to totally different frequencies with totally different temporal firing patterns (Riquimaroux *et al.*, 1994; Ishikawa and Riquimaroux, 2001). First, we examined whether what we found in Japanese monkeys at suprathreshold level could be also observed in Mongolian gerbils. Previous works reported the tonotopic organization in AI of Mongolian gerbils (Thomas *et al.*, 1993). We have been developing a focal, reversible inactivation method for neuronal activities by using near-infrared laser irradiation. We found that auditory evoked field potentials were attenuated by the laser irradiation. In in-vitro experiments using hippocampal slices hyperpolarization of membrane potential and decrement in membrane resistance were observed (Kataoka *et al.*, 2001). Then, the second aim of the present work was to examine how complicated temporal neuronal activities in AI would be modified by the laser irradiation.

2. METHODS

2.1 Subjects and Experimental Preparation

Thirteen Mongolian gerbils (about 6 mo. old) were used as subjects. Animals were anesthetized with Ketamin (50 mg/kg, i. m.) and Xylazin (10mg/kg, i. m.). After the animal's head was immobilized, a hole was made in the skull. The hole was located 5.0-7.0 mm lateral and 2.0-4.0 mm rostral to the lambda. The site of the primary auditory cortex was neurophysiologically estimated by referring to previously published data (Thomas *et al.*, 1993).

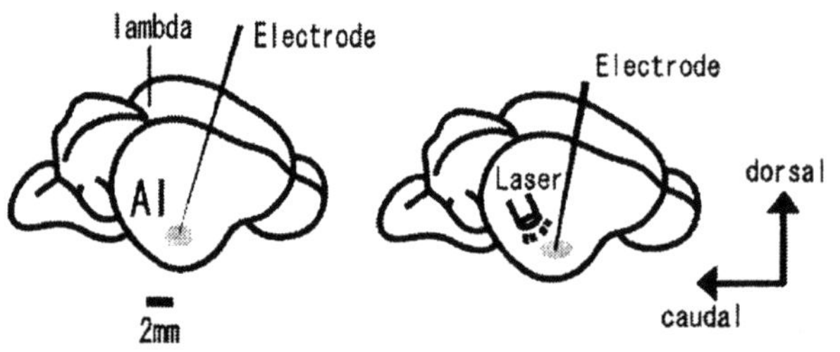

Figure. 1 Recording sites in the primary auditory cortex of the gerbil. These figures show dorso-lateral view of the right hemisphere of gerbil brain. The bar indicates 2mm. Recordings were made from the right auditory cortex (contra-lateral to stimulation side). We irradiated AI area with 1.5 mm diameter centered at electrode penetration.

2.2 Sound Stimuli

White noise bursts and tone bursts were used. White noise bursts were used to search auditory neurons. Duration of sounds was 75 ms including 5 ms rise/fall times. Frequency was varied from 500 Hz to 12 kHz. The frequency range was logarithmically divided into 25 parts. The sound pressure level used was 15 dB above neurons' thresholds. Sounds were presented with random order with inter-stimulus interval of 1.5 s. All stimuli were digitally generated by a PC with sampling rate of 44.1 kHz. The sounds were delivered to the left external ear canal via an ear phone (SONY, MDR-EX70SL) and a sound card (SEK'D, Siena) with D/A rate of 44.1 kHz. A probe tube connected to a microphone (Etymotic, ER-7C) was inserted in the external ear canal close to the tympanic membrane to monitor the sound pressure level. The output of the microphone was used to construct the transfer function. The output from the earphone was adjusted so that the sound pressure level within the frequency range (from 500 Hz thorough 12 kHz) would be flat (± 1 dB).

2.3 Experimental Procedure

All experiments were performed within an electrically shielded sound attenuated chamber. Neuronal activities of the right AI (contra-lateral to the stimulated ear) were recorded by an Elgiloy electrode (1~3 MΩ at 1 kHz; Fig. 1 left). A hydraulic micromanipulator (Narishige, SW-8) was used to move the electrode. The neuronal activities were amplified (WPI, DAM-80), band-passed (from 300 Hz to 3 kHz) and sent to a window discriminator to isolate a single unit activity. Then, the neuronal activities were digitally recorded on the hard disk of PC at a rate of 44.1 kHz for off-line temporal-spectral analysis (Fig. 2). Temporal firing patterns, temporal wave shapes of action potentials and post stimulus time (PST) histograms were monitored on-line. Further, during recordings AI was irradiated by low-power near-infrared laser (830 nm wave length, 3~5 W/cm^2) at the recordings site (1.5 mm diameter) where the electrode penetrated in the middle (Fig. 1 right).

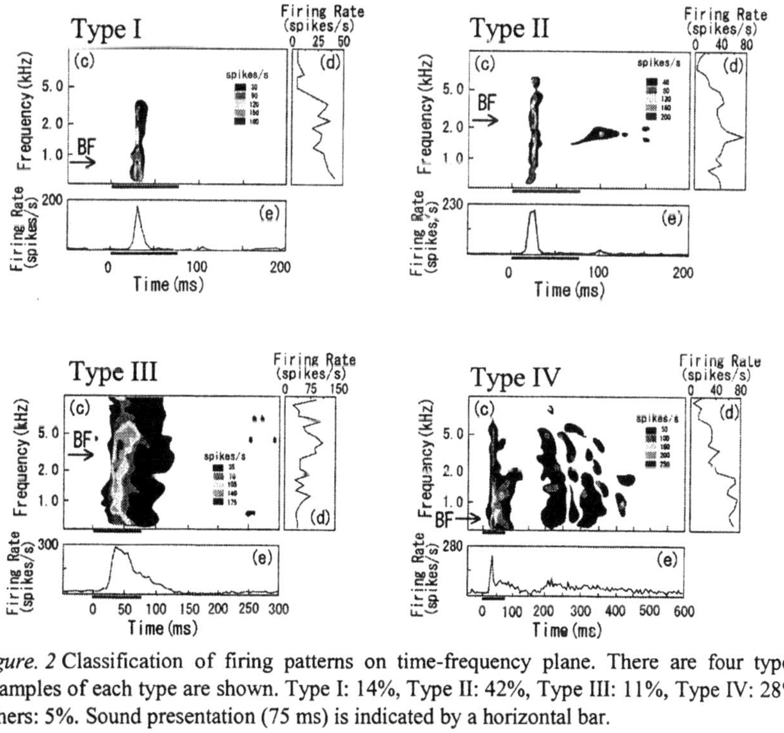

Figure. 2 Classification of firing patterns on time-frequency plane. There are four types. Examples of each type are shown. Type I: 14%, Type II: 42%, Type III: 11%, Type IV: 28%, Others: 5%. Sound presentation (75 ms) is indicated by a horizontal bar.

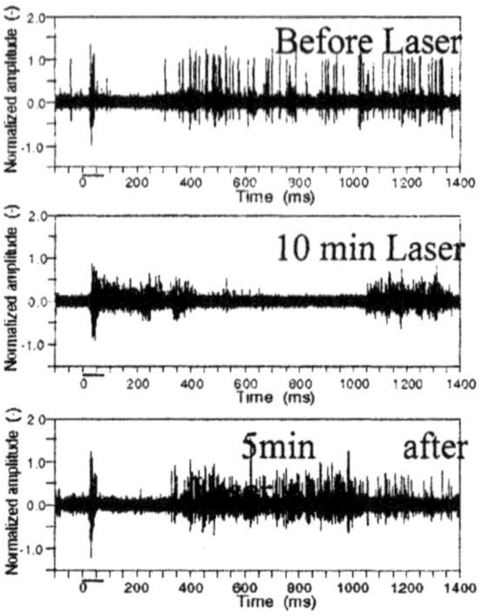

Figure. 3 Comparison between firing patterns before, during and after the near-infrared laser irradiation. Firings were temporally modulated by the irradiation.

3. RESULTS AND DISCUSSIONS

3.1 Temporal Firing Patterns of the Primary Auditory Cortex

We examined temporal firing patterns of 36 AI neurons and classified them into four types (Fig. 2). The first type was a simple onset response, where we could see firings only shortly after the stimulus onset without following firings (Type I, 14%). The second type was a simple onset response followed by firings after an interval (Type II, 42%). The third type was a sustained response which lasted during the duration of stimulus (Type III, 11%). The fourth type was a sustained response followed by late firings (Type IV, 28%). An example of Type IV shown in Fig. 2 presented inhibitory and excitatory periods repetitively appeared after the sustained firings. Further, there were two neurons which did not belong to any of above types ('Others', 5%). Thus, we could find four major types of temporal firing patterns in the auditory cortical neurons of Mongolian gerbils. So, data show that the simple onset response (Type I) was a minority and most neurons demonstrated complicated temporal firing patterns. Further, the present findings indicate that the same AI neuron respond to different frequencies with totally different temporal firing patterns, which was found in AI neurons of Japanese monkeys (Riquimaroux *et al.*, 1994).

3.2 Modification of Neuronal Activities by Near-Infrared Laser Irradiation

We examined how temporal activities of AI neurons were modified by low powered near-infrared irradiation. We confirmed that 2~5 min laser irradiation modified temporal firing patterns of AI neurons and 10~20 min after terminating the irradiation the activities recovered. An example is shown in Fig. 3.

4. SUMMARY

We classified temporal firing characteristics of AI neurons in Mongolian gerbil into four types, where 1) population of simple onset response appeared to be very small (14%) while complex responses occupied large population (86%) and 2) temporal firing pattern to different frequencies became very different which we found in AI of Japanese macaques (Riquimaroux *et al.*, 1994). Irradiation of low-powered near-infrared laser could reversibly modify temporal firing pattern of AI neurons. The laser irradiation could decrease the firing rate in about 5 minutes and recovery of responses took about 10 minutes.

ACKNOWLEDGEMENTS

The authors show great appreciation to K. Sumida and K. Maki for obtaining data. The present research was supported by Special Coordination Funds from the Ministry of Education, Culture, Sports, Science and Technology, by Grant to RCAST at Doshisha University from the Ministry of Education, Culture, Sports, Sciencc and Technology.

REFERENCES

Ishikawa, K., and Riquimaroux, H., 2001, Temporal neuronal properties of frequency responses in macaque auditory cortex. *Trans. Tech. Comm. Psychol. Physiol. Acoust.*, Acoust. Soc. Jpn. H-2001-8.

Kataoka, Y., Tamura, Y., Cui, Y. L., Yamada, H., and Watanabe, Y., 2001, Effects of low-level laser irradiation on membrane potential and membrane resistance in cultured rat hippocampal neurons, *Soc. Neurosci. Abstr.*, **27**: 725.10.

Riquimaroux, H., Takahashi, T., and Hashikawa, T., 1994, Neuronal responses of the primary auditory cortex in Japanese monkey: Tonotopicity revisited, *J. Acoust. Soc. Jpn. (E)*, **15**: 171-177.

Thomas, H., Tillein, J., Heil, P. and Scheich, H., 1993, Functional organization of auditory cortex in Mongolian gerbil (Meriones unguiculatus). I. Electrophysiological mapping of frequency representation and distinction of fields, *Eur. J. Neurosci.* **5**: 882-897.

Neurodynamics in Auditory Cortex During Category Learning

FRANK W. OHL[*,#], WOLFRAM WETZEL[*], MATTHIAS DELIANO[*], HENNING SCHEICH[*], and WALTER J. FREEMAN[#]
[*]*Leibniz Institut für Neurobiologie, Brenneckestr. 6, D-39118 Magdeburg, Germany; and*
[#]*University of California, Berkeley, CA, USA*

1. INTRODUCTION

The research on learning has for historical reasons been divided mainly into behaviourally oriented studies performed on animals and more cognitively oriented studies in humans (Anderson, 2000). Accordingly, studies aimed at the presumed neurophysiological basis of learning have either exploited the full range of cognitive-phenomenological approaches in humans while being methodologically confined mainly to non-invasive imaging techniques, or alternatively, have made use of the better accessibility of the animal nervous system physiology while being restricted in the definition and analysis of the cognitive aspects involved. Consequently, our physiological understanding of learning processes is best for simple behaviours that can be studied in suited animal models (e.g. Kandel, 2001) but declines for cognitively more demanding aspects of learning.

Here we present an animal model, amenable to detailed physiological analysis, of a cognitively demanding learning task, namely the formation of categories (concept formation) and the sorting of novel stimuli into these categories. Both of these aspects are encapsulated in the term 'category learning'.

In the present chapter we will show that learning phenomena having aspects beyond stimulus-response associations are of high relevance for studying the neuronal basis of learning in general, in that they preclude explanation by a broad class of simple neurophysiological models which are otherwise discussed as elemental for physiological theories of learning. Further, we describe an electrophysiological correlate of category-specific processing of stimuli. The paradigm of category learning allows the neurophysiologist to open an objective observation window on a subjective cognitive structure. Finally, we show that the physiological results derived from the categorization paradigm solve the problem of the lacking invariance of certain spatiotemporal patterns in cortical activity that is reported in the literature since the 1950s.

1.1 Relationship between 'Learning' and 'Neuronal Plasticity'

The exact nature of the relationship between learning and neuronal plasticity is highly nontrivial, because the former is a purely psychological concept while the latter is a physiological concept. Conceptual difficulties in inter-relating these two domains are therefore predictably similar to other situations in science where conceptually different levels have to be linked, like the case of statistical physics and thermodynamics, or the mind-body problem.

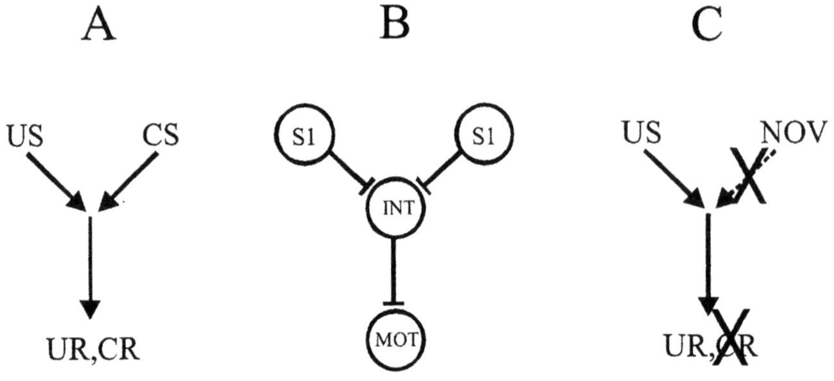

Figure 1. A. General scheme of flow of information before and after Pavlovian conditioning. Before conditioning an unconditioned stimulus (US) will elicit a particular behavior, then referred to as unconditioned response (UR). After conditioning, a previously behaviourally neutral stimulus can elicit this behaviour as a conditioned response (CR) and is then referred to as the conditioned stimulus (CS). B. A straightforward translation of the *flow of information* during Pavlovian conditioning into a *flow of neuronal excitation* within a neuronal substrate. C. The architecture in B cannot explain responses to novel stimuli (NOV), because novel stimuli have not been associated with unconditional triggers and, consequently, cannot be conditioned.

If the role of neuronal plasticity for learning is considered by physiologists, it is usually seen in the capacity for re-routing the flow of excitation through neuronal networks. The case of Pavlovian conditioning, a learning phenomenon that has been intensively studied by physiologists, lends itself to this concept in a very straightforward way: Pavlovian conditioning can be described, and - in fact - has traditionally been defined, as a process by which an initially behaviourally neutral stimulus can later elicit a particular behaviour, when it has previously been paired with a stimulus (US) that unconditionally triggers this behaviour (Fig. 1A). It has proved successful to formulate models of the role of neuronal plasticity for learning which basically consist in a one-to-one translation of this idea into a neuronal substrate (Fig. 1B). In the case of the conditioned gill withdrawal reflex in *Aplysia*, this concept is manifested in the feedforward convergence of – in the simplest case - two sensory neurons on a shared interneuron which in turn projects on a motor neuron. The concept is so straightforward, that its appraisal as a generic element of physiological models of learning has been put forward, as, for example, expressed by the metaphor of a 'cellular alphabet' (Hawkins and Kandel, 1984) or the metaphor of a 'molecular alphabet' (Kandel *et al.*, 1995).

1.2 Category Learning as a Tool to Study Neuronal Correlates of Learning

The claim of the elemental nature of the above sketched neuronal model for learning has been challenged by cognitive science (e.g. Schouten and De Long, 1999) where learning is viewed as a process by which an animal gains information about conditions in its environment that help it to behave in meaningful ways. It should be noted that this perspective purposely includes, among other things, the possibility of meaningful behaviours to novel stimuli. Novel stimuli, however, have not been encountered before, specifically, they have not been associated with unconditional triggers (Fig. 1C). Therefore, a simple feedforward convergence scheme as in Fig. 1A cannot be used as an explanation for such aspects of learning, i.e. aspects that go beyond mere stimulus-response associations. The current chapter focuses on an instant of category learning because category learning particularly emphasizes this aspect of learning, i.e. the meaningful behaviour in response to novel and unfamiliar stimuli, and therefore precludes explanation by a broad class of reductionist schemes. In this sense, the study of category learning is of general relevance to the physiological understanding of learning phenomena in general.

Moreover, the formation of categories is fundamental to cognition. Category learning transcends simple stimulus-response associations, typically studied in systems neurophysiology, in that it involves abstraction from learned particular stimulus features. Category learning goes beyond the information given (Kommatsu, 1992).

2. AN ANIMAL MODEL OF AUDITORY CATEGORY LEARNING - BEHAVIORAL MEASURES

To approach a physiological investigation of category learning we have developed an animal model of an auditory category learning (Wetzel *et al.*, 1998; Ohl *et al.*, 2001). In this paradigm a rodent species, the Mongolian gerbil (*Meriones unguiculatus*), is trained to discriminate the modulation direction of linearly frequency modulated tones, to form the categories of 'rising' and 'falling' frequency modulations and to categorize novel, i.e. previously unheard, stimuli as belonging to one or the other category.

2.1 Generalization and Categorization

Before we turn to a description of the training procedure it seems appropriate to consider the terms 'generalization' and 'categorization' in more detail, as these terms are frequently used in the relevant literature with variable definitions or connotations.

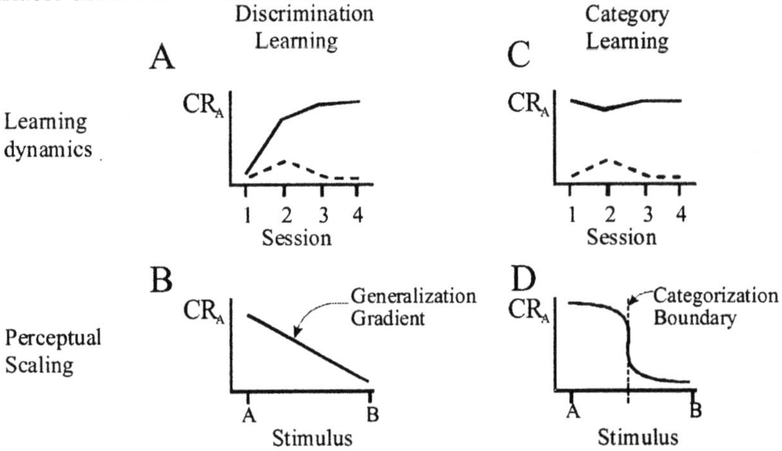

Figure 2. Suitable observables differentiating discriminative conditioning and category learning. Learning curves and psychometric functions after discrimination learning (A,B) and afer category learning (C,D). CRA is typically a measure of the rate of occurrence or of the strength of the response that has been trained to be elicited by stimulus a A. Solid and dashed curves in A and C depict the hit rate and false alarm rate, respectively.

When an animal (including human) is trained to discriminate a stimulus A from a stimulus B the discrimination performance usually shows a more or less gradual development over time as is manifest in the various forms of so-called 'learning curves'. A typical example is depicted by the scheme in Fig. 2A, which displays the temporal evolution of the hit rate and false alarm rate in a GO/(NO-GO) discrimination experiment. Other depictions for learning curves are possible depending on the kind of experiment performed (symmetric choice experiments, detection approaches, etc.) and the selection of behavioural observables (e.g., various transformations of hit rate and false alarm rates as are suitable under certain conditions, explicit measurement and report of changes in response criteria as done in signal detection approaches, etc.). In most cases, however, the registration of a particular behaviour in response to a stimulus and its changing (conditional) rate of occurrence must be assessed at some point in the analysis (Fig. 2A).

The detailed form of the learning curve can be assumed to be dependent on behavioural and perceptual states (vigilance, response criteria, etc.) as well as reflecting the kinetics of a larger number of diverse processes. Such processes involve the unspecific familiarization of a subject with the apparatus and paradigm, changes in sensory or behavioural thresholds, kinetics of learning-related processes like association, memory formation, etc. Some type of 'saturation kinetics' is typically observed. The asymptotic value of the discrimination performance in such an experiment again reflects various influencing parameters like the internal response criteria of the subject, its discrimination thresholds, behavioural strategies of the subject testing endurance of reinforcement rules, etc. Despite the heterogeneous nature of processes contributing to so measured discrimination performance this measure is typically characterized by a certain degree of stimulus specifity. This can be demonstrated by generalization gradients. In a GO/(NO-GO) experiment in which a subject is trained to show the GO reaction in response to stimulus A and the NO-GO reaction in response to stimulus B, for example, a generalization gradient can be measured by varying physical stimulus parameters of A gradually until stimulus A becomes stimulus B. It will express itself as the gradual fall-off in the A-specific response amplitude (seen in the strength of the conditioned response or its frequency of occurrence or both) upon stimulus variation (Fig. 2B).

When a subject has formed categories it can recognize even novel stimuli as representants of the learned categories. A depiction of behavioural variables analogous to a learning curve would therefore indicate a high discrimination performance even in the first training session (Fig. 2C). The experimental demonstration of this behaviour is critical for assessing category learning and is sometimes referred to as the criterion of the 'transfer of learned behaviours to novel stimuli'. Category learning is distinguished from simple or discriminative conditioning learning also by the psychometric functions it produces. Instead of gradual generalization gradients we find sigmoid psychometric functions with a more or less sharp boundary at some location in the stimulus parameter space, called the categorization boundary (Fig. 2D). Categories develop as cognitive constructs; they epitomize subjective hypotheses that are expressible as parcellations of the set of actually perceived or imaginable stimuli, conditions or actions, into equivalence classes of meaning in particular contexts. Transfer of learned behaviours to novel stimuli therefore follows the subjective laws of this parcellation, rather than being guided by (physical) similarity relations between stimuli or stimulus features. These are important criteria for cognitive structures that have to be accounted for by physiological models of learning.

In a nutshell, generalization is a general feature of learned stimulus-cued behaviours reflecting the converse of stimulus specificity, while categorization is a cognitive process based on the parcellation of the represented world into equivalence classes of meaning, valid for an individual in a particular context and in a particular time.

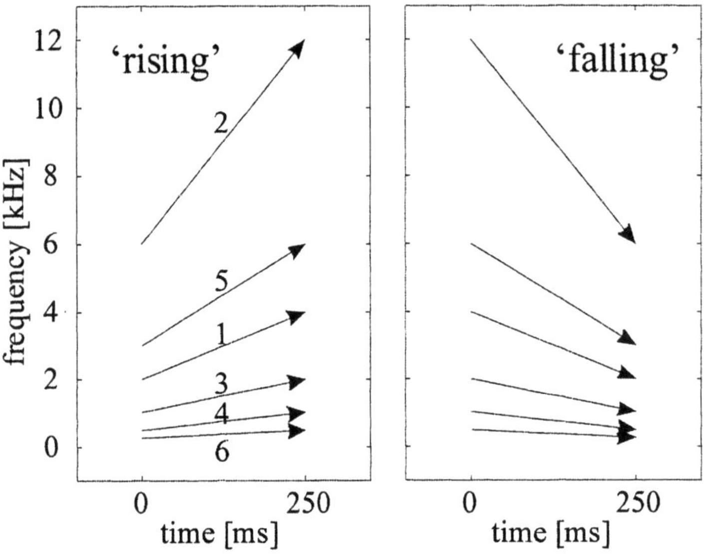

Figure 3. Linearly frequency modulated tones can be varied along various parameter dimensions but can also be grouped into the two categories of 'rising' and 'falling' frequency modulations depending on the direction of change of the instantaneous frequency with time.

2.2 Stimuli, Training Paradigm, Apparatus

The developed animal model for auditory category learning utilizes linearly frequency modulated tones. In linearly frequency modulated tones a number of parameters can be (co-)varied, like their duration, their intensity, the frequency ranges covered, and the (constant) modulation rate, i.e. the rate of change of the tones' instantaneous frequency (Fig. 3). Most of the possible stimuli (all, except for those with zero modulation rate, i.e. pure tones) can be categorized as either 'rising' or 'falling' corresponding to the change of the instantaneous frequency with time.

Gerbils could be trained to form the categories of 'rising' and 'falling' frequency modulations and sort novel stimuli into these categories by a training protocol which invoked a series of training blocks, with a training block defined as a period in time in which the discrimination between one particular rising frequency modulated tone and one particular falling frequency modulated tone was trained. After the learning performance had saturated in a given training block, the next training block (with another pair of a rising and a falling frequency modulated sound) was started. Checking the criterion of transfer of the conditioned response to novel stimuli (at the beginning of a new training block) enabled the experimenter to detect the transition from discrimination learning to category learning (Wetzel *et al.*, 1984; Ohl *et al.*, 2001).

Trainings were carried out in a two-compartment shuttle-box using a GO/(NO-GO) procedure. Crossing a little hurdle between the compartments in a critical time window after stimulus was defined as the GO response. Animals received a mild but aversive electrodermal stimulation through the metal floor grid on misses and false alarms (Wetzel *et al.*, 1998). Care was taken that in each new training block stimuli were well outside the generalization gradients established by previous trainings.

2.3 Results on Behavioral Measures

Animals trained on one or more discriminations between rising and falling frequency modulated tones never generalized to pure tones of any frequency (e.g. start frequency, stop frequency or intermediate frequencies of the frequency modulation) as demonstrated by direct transfer experiments (Ohl *et al.*, 2001, supplementary material) or by measuring generalization gradients for modulation rate (Ohl *et al.*, 2001).

The behavioral analysis showed that gerbils were able to form the concepts of modulation direction, i.e. the categories of rising and falling frequency modulations, and sort novel (not previously learned or heard) frequency modulated tones into these categories, as indicated by both the transfer criterion (Wetzel *et al.*, 1998; Ohl *et al.*, 2001) and the sigmoid psychometric function (Ohl *et al.*, 2001). Notably, different individuals showed the transition from discrimination learning to category learning (cf. Fig. 2) in different training blocks although all were trained with the same sequence of blocks. Also, the transitions occurred abruptly rather than gradually as indicated by the sequence of discrimination performances in the first training sessions (when stimuli were newly introduced) across training blocks. After the transition had occurred, the categorization remained stable, meaning that additional novel stimuli were all correctly categorized.

3. ELECTROPHYSIOLOGICAL MEASURES

A suitable level for studying electrophysiological correlates of perceptual organization is the mesoscopic level of neurodynamics (Freeman, 2001), which defines the spatial scale of phenomena observed and provides focus on electrical phenomena emergent from the mass action of ensembles of some 10^4 to 10^5 neurons. For cortical structures, this level of description is accessible by measurement of the electrocorticogram. We have therefore combined the described category learning paradigm with the measurement of the electrocorticogram using arrays (3×6) of microelectrodes chronically implanted on the dura over the primary auditory cortex. The spatial configuration and interelectrode distance (600 μm) of the recording array were so designed to cover the tonotopic representation of the frequency modulated stimuli used and avoid spatial aliasing of electrocorticogram activity (Ohl *et al.*, 2000a).

3.1 Cortical Representation of Frequency Modulated Sounds in Naïve Animals

The spatial organization of the thalamic input into the auditory cortex can be studied by averaging electrocorticograms across multiple stimulus presentations, yielding the well-known auditory evoked potential (Barth and Di, 1990, 1991). Our studies of pure-tone-induced (Ohl *et al.*, 2000a) and frequency-modulated-tone-induced (Ohl *et al.*, 2000b) auditory evoked potentials in primary auditory cortex, field AI, revealed that their early components (P1 and N1) are topographically organized, i.e. localized at positions within the tonotopic gradient of the field that correspond to the frequency interval traversed by the frequency modulation, while their late components (P2 and N2) are not. On a finer spatial scale, the localization of the early components of rising and falling frequency modulated tones was found to be shifted towards tonotopic representations of the respective end frequencies of the modulations, i.e. towards higher frequencies for rising modulations and towards lower frequencies for falling modulations. These 'tonotopic shifts' (Ohl *et al.*, 2000b) could be explained by the finding that single neurons are usually activated more strongly when the frequency modulation is towards the neuron's best frequency than when it is away from it (Phillips *et al.*, 1985). In the former case, the activation of frequency channels in the neighbourhood of the best frequency of a single neuron are recruited more synchronously than in the latter case, due to the increasing response latency with increasing spectral distance from the neuron's best frequency. If this asymmetry is transferred onto a tonotopically organized array of neurons a tonotopic shift as described will result. Tonotopic shifts have previously been reported in the cortex analogue of the chick (Heil *et al.*, 1992).

3.2 Electrophysiological Correlates of Category Learning

Since physiological correlates of category learning could not be expected to occur time-locked to stimulus presentation we analyzed electrocorticograms recorded during the training with a single trial type of analysis. Instantaneous spatial patterns in the ongoing cortical activity were described by state vectors (Ohl *et al.*, 2001). State vectors were formed from estimates of signal power in 120 ms time windows obtained for each channel. A state vector moved through the state space along a trajectory according to the temporal evolution of the spatial pattern. For each trial, the Euclidean distance (parameterized by time) to a reference trajectory was calculated and termed 'dissimilarity function'. In each case, the reference trajectory was the centroid over trajectories associated with trials associated with stimuli from the respective other category measured in the same training session. I.e., each trajectory associated with a rising frequency modulated tone was compared to the centroid over all trajectories associated falling frequency modulated tones in the same session, and vice versa. Comparison of single trajectories with centroids of trajectories, rather than other single trajectories, ensured, on a statistical basis, that transient increases in the pattern dissimilarity (peaks in the dissimilarity function) were due to pattern changes in the observed trajectory rather than in the centroid. In naïve animals, dissimilarity functions showed a 'baseline behavior' with a sharp peak (2 – 7 standard deviations of baseline amplitude) after stimulus onset. This peak occurred predictably because of the topographically dissimilar patterns (tonotopic shifts) of early evoked responses that rising and falling frequency modulated tones produce (Ohl *et al.,* 2000b). With learning, additional peaks emerged from the ongoing activity, thus labelling spatial activity patterns in single trials with transiently increased dissimilarity to the reference trajectory indicating a potential relevance for representing category-specific information processing. These patterns were therefore termed 'marked states'.

To test whether marked states do in fact represent processing of category-specific information we analyzed the similarity and dissimilarity relations among them in the entire course of the training. While animals were in their discrimination phases (prior to the formation of categories), we observed that dissimilarities between marked states within categories were of the same order of magnitude than between categories. After an individual animal had entered its categorization phase dissimilarities within a category were significantly smaller than between categories (Ohl *et al.*, 2001). This indicated the existence of a metric which reflected the parcellation of stimuli into equivalence classes of meaning. This type of metric is therefore different from the known tonotopic organization reflecting similarity relations of physical stimulus parameters, namely spectral composition, in that it reflects subjective aspects of stimulus meaning, namely its belongingness to categories formed by previous experience.

The spatial organization of the emerging marked states was analyzed in more detail and compared to that of the early evoked activity (also yielding peaks in the dissimilarity function) by a multivariate discriminant analysis, identifying the regions in the recording area which maximally contributed to the dissimilarity between the observed pattern and the reference pattern (Ohl *et al.*, 2003b), or identifiying the regions which contribute most information about the pattern (Ohl *et al.*, 2003a).

4. CONCLUSION

It was possible to develop an animal model of auditory category learning which demonstrated the formation of categories as a process which has three main characteristics in the behavioural data: First, categorization develops abruptly rather than gradually. Second, it develops at a point in time that is specific for each individual subject in its learning history, and is typically different from subject to subject. Third, when categorization has occurred in a subject it remains stable for the rest of the subject's training experiences (unless a change in the reinforcement schedule forces a change in the meaning attributed to stimuli). A process which conforms to these characteristics is sometimes termed 'Aha'-event to indicate a change in the cognitive state of a subject.

The neurophysiological analysis revealed that the process of associating meaning to acoustic stimuli as indicated by an increasing discrimination performance in the behavioural data was paralleled by the emergence of transient activity states in the ongoing cortical activity, that could be identified on the basis of their dissimilarity to patterns found in trials associated with stimuli not belonging to the category. It is noteworthy, that spatial patterns of electrocortical activity in single trials were already observed by Lilly in his toposcopic studies (Lilly and Cherry, 1954). To him it was already apparent that the long lasting dynamics was not just random 'noise', but was better described by 'figures' moving in time and space across the cortical surface. At that time the majority of research programs had already turned to the analysis of averaged data in which such spatiotemporal structures are no longer detectable. The few research programs pursuing analysis of activity patterns in single trials (e.g. DeMott, 1970; Livanov, 1977) had faced a major problem for the interpretation of such patterns: the lack of invariance with the applied stimuli. A large body of data accumulated over the last decades (summarized in Freeman, 2000) showed that such patterns might remain stable when repetitively evoked by sensory stimulation for a certain period in time, but might typically vary with behavioural context, particularly in learning situations when stimuli were associated with particular meanings. This lack of invariance of these patterns with the mere physical parameters of stimuli challenged their interpretation as 'sensory representations'. The observed metastability of the patterns was hypothesized to reflect context aspects of the stimulation as well as the perceptual history of the individual, and it was inferred that such patterns reflect subjectively relevant cognitive structures (for summary see Freeman, 2000, and references therein). The results described above critically confirm this interpretation: The category learning paradigm, first, allows determination of the point in time when a particular cognitive structure (the formation of the categories 'rising' and 'falling') emerges, and second, predicts that the main source of variance in the stimuli (the spectral interval traversed by the frequency modulation) is no longer a relevant feature after a subject's transition to categorization. Consequently, it was found that the dissimilarity between marked states associated with stimuli belonging to the same category was significantly reduced after the transition to categorization, although the physical dissimilarity of the corresponding stimuli was still high, as also reflected in the topographic organization of the stimulus locked peaks in the dissimilarity function and the fact the dissimilarities remained high in individuals that had not yet formed categories. In this sense, the utilized paradigm and analysis strategy provided an objective (for the experimenter) window of a subjective cognitive structure (that of the animal).

ACKNOWLEDGEMENTS

This work was supported by grants from the German Ministry of Education and Research, BMBF, and by grants from the Land Sachsen-Anhalt. We thank Brian Burke and Daniela Labra Cardero (Berkeley), as well as Kathrin Ohl and Thomas Wagner (Magdeburg) for technical assistance.

REFERENCES

Anderson, J. R., 2000, *Learning and memory. An integrated approach*, John Wiley & Sons, New York.

Barth, D. S., and Di, S., 1990, Three-dimensional analysis of auditory-evoked potentials in rat neocortex. *J. Neurophysiol.* **64**: 1527-1636.

Barth, D. S., and Di, S., 1991, The functional anatomy of middle latency auditory evoked potentials. *Brain Res.* **565**: 109-115.

DeMott, D. W., 1970, *Toposcopic studies of learning.* Thomas Books, Springfield, IL.

Freeman, W. J., 2000, *Neurodynamics. An exploration in mesoscopic brain dynamics.* Springer-Verlag, London.

Hawkins, R. D., and Kandel, E. R., 1984, Is there a cell-biological alphabet for simple forms of learning? *Psych. Rev.* **91**: 375-391

Heil, P., Langner, G., and Scheich, H., 1992, Processing of frequency-modulated stimuli in the chick auditory cortex analogue: evidence for topographic representations and possible mechanisms of rate and directional sensitivity. *J. Comp. Physiol. A* **171**: 583-600.

Kandel, E. R., 2001, The molecular biology of memory storage: A dialogue between genes and synapses, *Science* **294**: 1030-1038.

Kandel, E. R., Schwartz, J. H., and Jessell, T. M., 1995, *Essentials of neural science and behaviour.* Appleton & Lange, Norwalk, CT.

Kommatsu, L. K., 1992, Recent views of conceptual structure. *Psychol. Bull.* **112**: 500-526.

Lilly, J. C., and Cherry, R. B., 1954, Surface movements of click responses from acoustic cerebral cortex of cat: leading and trailing edes of a response figure. *J. Neurophysiol.* **17**: 531-537.

Livanov, M. N., 1977, *Spatial organization of cerebral processes.* Wiley, New York.

Ohl, F. W., Scheich, H., and Freeman, W. J., 2000a, Topographic analysis of epidural pure-tone-evoked potentials in gerbil auditory cortex. *J. Neurophysiol.* **83**: 3123-3132.

Ohl, F. W., Scheich, H., and Freeman, W. J., 2000b, Spatial representation of frequency-modulated tones in gerbil auditory cortex revealed by epidural electrocorticography. *J. Physiol. (Paris)* **94**: 549-554.

Ohl, F. W., Scheich, H., and Freeman, W. J., 2001, Change in pattern of ongoing cortical activity with auditory category learning. *Nature* **412**: 733-736.

Ohl, F. W., Deliano, M., Scheich, H., and Freeman, W. J., 2003a, Early and late patterns of stimulus-related activity in auditory cortex of trained animals. *Biological Cybernetics* **88**: 374-379.

Ohl, F. W., Deliano, M., Scheich, H., and Freeman, W. J., 2003b, Analysis of evoked and emergent patterns of stimulus-related auditory cortical activity. *Rev. Neurosci.* **14**: 35-42.

Phillips, D. P., Mendelson, J. R., Cynader, M. S., and Douglas, R. M., 1985, Response of single neurons in the cat auditory cortex to time-varying stimuli: Frequency-modulated tones of narrow excursion. *Exp. Brain Res.* **58**: 443-454.

Schouten, M. K. D., and De Long, L., 1999, Reduction, elimination, and levels: The case of the LTP-learning link. *Phil. Psychol.* **12**: 237-262.

Wetzel, W., Wagner, T., Ohl, F. W., and Scheich, H., 1998, Categorical discrimination of direction in frequency-modulated tones by Mongolian gerbils. *Behav. Brain Res.* **91**: 29-39.

Comparison of Two Rat Models of Aging

Peripheral pathology and GABA changes in the inferior colliculus

JEREMY G. TURNER, and DONALD M. CASPARY
Department of Pharmacology, Southern Illinois University School of Medicine, Springfield, IL 62702

1. INTRODUCTION

Studies in rat inferior colliculus (IC) suggest that a significant change in neurotransmitter function related to the inhibitory amino acid neurotransmitter, gamma aminobutyric acid (GABA), occurs in aged animals. These age-related changes include significant decreases in GABA content, GABA release, GABA neurons, glutamic acid decarboxylase (GAD) enzymatic activity, $GABA_A$&$_B$ receptor binding and changes in the subunit composition of the $GABA_A$ receptor. If similar changes occur within the inferior colliculus of humans, it could affect the ability of elderly listeners to process complex acoustic signals, particularly in the presence of background noise.

Much of this research has been conducted using the inbred albino Fischer 344 (F344). The F344 was developed by M.R. Curtis at Columbia University Institute for Cancer Research in 1920. The F344 strain has been a major rat model of aging throughout the 1980s and early 1990s during which time, the strain began to show increasing tumor pathology and a shortened life span. By 1994 the NIA began to recommend the F1 hybrid cross between the F344 and Brown Norway rat (FBN) as a new and better model for aging research (Figure 1). To continue auditory aging studies and justify the use of the FBN hybrid strain, we felt it necessary to confirm previous F344 observations using the FBN strain. Age-related threshold changes as well as changes in neurochemical measures of pre- and postsynaptic GABA function were compared between two rat models of aging.

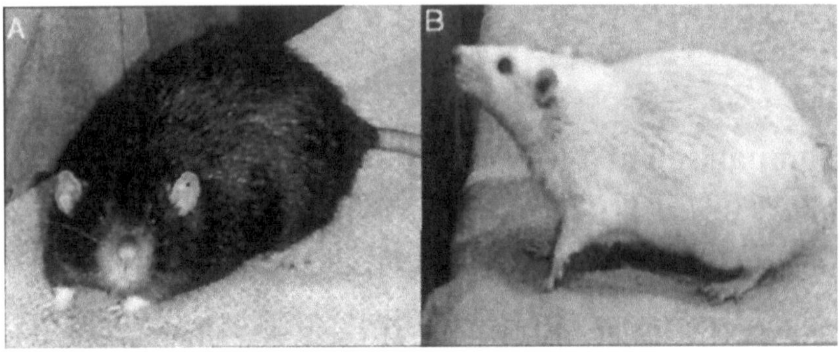

Figure 1. Two rat models of aging. A. The pigmented FBN is the F1 hybrid cross between F344 females and Brown Norway males. FBN's have a median lifespan of 36 months and a maximal lifespan of 44 months. B. F344 is an inbred albino rat. F344's have a median lifespan of 26 months and a maximal life span of 31 months.

2. METHODS

Several methods were used to compare the strains with respect to peripheral pathology and GABA changes in IC. Peripheral pathology was measured using cochleograms to determine the number of missing inner and outer hair cells and auditory brainstem-evoked response (ABR) thresholds. Age-related changes in GABA in the IC were measured using the Fonnum method for GAD, the rate limiting GABA synthetic enzyme, activity. Quantitative receptor binding autoradiography using tritiated GABA ligands and Western blotting immunocytochemistry were used to measure $GABA_A$ subunit protein levels.

Cochleograms: Temporal bones were removed and immersion fixed and processed according to procedures of Sponger *et al.*, 1997 (courtesy of Dr. Richard Salvi). Surface preparations of the organ of Corti were viewed with a differential interference microscope and missing inner (IHCs) and outer hair cells (OHCs) counted over 0.24-mm intervals along the entire basilar membrane. Hair cells were counted as present if the cell body and cuticular plate were intact. IHC and OHC loss was plotted as a percentage of young hair cells missing.

ABRs: ABRs were recorded from the vertex (apex) and subcutaneous electrodes placed in the nose (reference) and neck (ground). Thresholds were determined at 4, 8, 12, 16 and 32 kHz tones (amplification = 500,000x; 512 trials; duration 3 ms, ramp 1ms, rate 20/s).

GAD Activity: GAD activity was measured from punched (F344) or dissected (FBN) samples of the central nucleus of the IC (CIC), using the enzymatic method of Fonnum *et al.* (1977) and Raza *et al.* (1994).

Receptor Binding: Quantitative receptor binding autoradiography was with three different $GABA_A$ receptor ligands (Milbrandt *et al.*,1996). Measures of receptor binding were made using single concentrations of tritiated muscimol, t-butylbicycloorthobenzoate (TBOB) and flunitrazepam, as well as modulation experiments where binding at the TBOB/picrotoxin site was modulated by increasing concentrations of GABA.

Subunit Protein Levels: Subunit selective antibodies were used in a standard Western blotting paradigm to measure $GABA_A$ subunit protein levels in preparations obtained from F344 and FBN rats (Caspary *et al.*,1999).

3. FINDINGS

The two strains displayed subtle differences in patterns of hair cell loss (Figure 2) and ABR thresholds (Figure 3). They also appeared to differ in GABA ligand binding in the IC (Table 1). However, FBN and F344 rats demonstrated similar age-related reductions of GAD activity (Figure 4) and GABA modulation of TBOB binding (Figure 5), as well as similar age-related changes in the subunit makeup of the $GABA_A$ receptor (Figure 6).

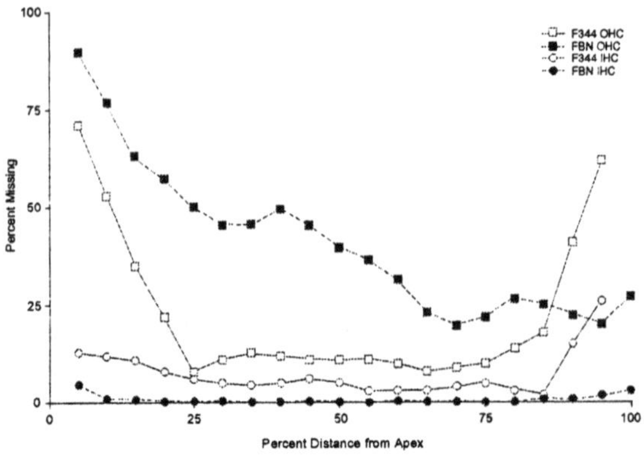

*Figure 2.*Comparisons between *cochleograms* from aged F344 (24 month, n=8) and aged FBN (32 month, n=7) rats showing percent of missing hair cells relative to a young adult rat. The pattern of age-related hair cell loss was different between the two rat strains. Aged FBN rats lost few IHCs, while aged F344 rats displayed small IHC losses (<10%) throughout the cochlea with a pronounced increase in IHC loss near the basal end. F344 rats exhibited U-shaped loss of OHCs with the greatest losses (as high as 70%) confined to apical and basal turns. Low levels of OHC losses were observed throughout the F344 rat cochlea. FBN rats had significant OHC losses starting at the apex, which tapered to moderate losses in basal regions. This pattern of OHC loss resembles the pattern of spiral ganglion cell loss described by Keithley and colleagues (1992) in Brown Norway rat cochleas.

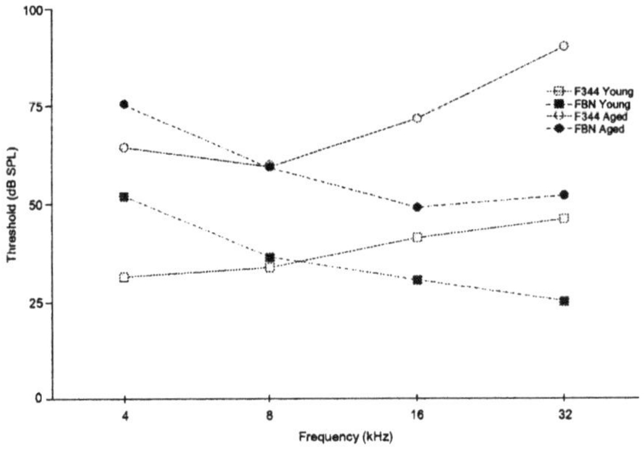

Figure 3. Strain Comparison of Age-Related Threshold Changes. ABR thresholds for young and aged F344 rats (3 month, n=28; 24 month, n=18) were compared to young FBN and aged rats (4 month, n=11; 32 month, n=10). F344 rat thresholds were lower at 4 and 8 kHz than at 16 and 32 kHz. FBN rats displayed a significantly different threshold pattern, with the lowest thresholds at higher frequencies (ANOVA, $p<0.01$). Age affected both strains similarly with near parallel upward threshold shifts. For the FBN strain, age-related threshold shifts did not correlate with the apical pattern of outer hair cell loss. Frequencies below 4.0 kHz were tested in 4 young and 4 aged in both strains (not shown). The age-related threshold shift at low frequencies was not disproportionately greater in the FBN strain.

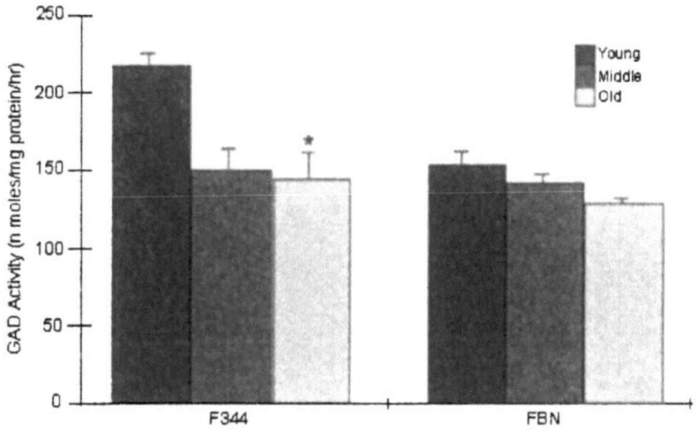

Figure 4. Strain Comparison of Age-Related Alterations of GAD. When compared to young F344 rats, there was a significant ($p<0.05$) age-related reduction in GAD synthetic activity in the CIC of mature (-31%) and aged (-30%) animals (n=5). F344 rat ages were: young, 3-7 mos.; mature, 15-17 mos. and aged, 24-26 mos. Aged FBN rats showed a nonsignificant ($p=0.06$) reduction of GAD synthetic activity when compared to young animals. FBN rat ages were: young, 5 mos. (n=8); mature, 21 mos. (n=7) and aged, 30 mos. (n=8). These data are qualitatively similar for both strains.

F344 Ligand	Young Adult	Aged	% Young
^{3}(H) Muscimol	766.4	866.7	113.1
^{3}(H) Flunitrazepam	942.9	950.7	100.8
^{3}(H) TBOB	1138.8	1058.2	95.9
FBN Ligand			
^{3}(H) Muscimol	657.4	381.3	58%
^{3}(H) Flunitrazepam	726.2	427.5	78.5%*
^{3}(H) TBOB	1402.1	1111.9	79.3%

Table 1. Strain Comparison of Age-related Changes in $GABA_A$ Receptor Binding in Inferior Colliculus. Binding levels are given as fmoles/mg protein. * $p<.05$

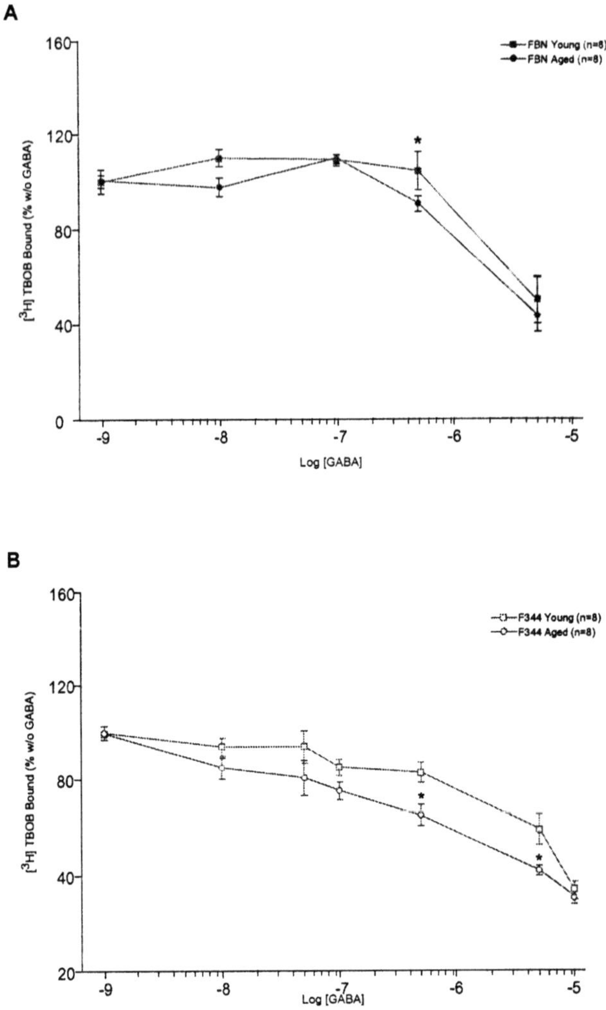

Figure 5. Strain Comparison of Age-Related Binding Changes of GABA modulation at the picrotoxin site. Dose response curves showing GABA modulation of 20 nM TBOB binding. Increasing doses of GABA caused a significant (p>0.05) age-related decrease in binding in the IC of rats in each age group of F344 rats (A) as well as FBN rats (B). In both strains, the GABA dose-response curve was shifted down and to the left in aged animals. This change was indicative of GABA being more efficacious at inhibiting [^{3}H]TBOB binding in the IC of aged rats when compared to young adult rats (significant at GABA concentrations indicated with an *).

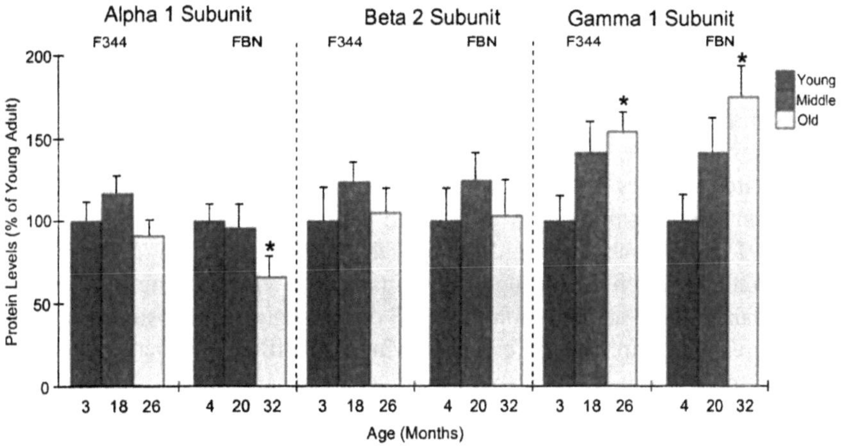

Figure 6. Strain Comparison of Age-related Changes in the Subunit Makeup of the $GABA_A$ Receptor. The effects of aging on levels of the α1, β2 and γ1 $GABA_A$ receptor subunit protein in the IC from two strains of rats (F344 and FBN; numbers on abscissa correspond to age in months). **p*=0.05 for F344 γ1; **p*=0.03 and 0.04 for FBN α1 and γ1, respectively.

CONCLUSIONS

The two strains displayed different patterns of age-related hair cell loss. The FBN rat cochlea displayed aging patterns similar to the paternal Brown Norway pattern previously described by Keithley *et al.* (1992) using spiral ganglion cell counts. Despite the differences in cochleograms, both strains displayed similarly shaped (<20dB) age-related parallel threshold shifts which were highly significant. This was unexpected due to the large apical OHC losses in the FBN strain. This lack of correlation could be due to the inability to accurately and selectively test low frequencies with short 1 ms rise-fall tone-pips.

Both strains showed an age-related decline in the activity of the GABA synthetic enzyme GAD. This change was significant for the F344 rat and nearly significant (p<0.06) in the FBN rat strain. However, while single point binding of several GABA ligands (flunitrazepam, muscimol and TBOB) showed no changes in aged F344 rat IC, significant age-related reductions were observed in the IC of FBN rats for flunitrazepam and trends for age-related reductions were observed for the muscimol and TBOB. Both strains did however display an age-related change in the ability of GABA to modulate binding at the picrotoxin site of the $GABA_A$ receptor.

Collectively, these findings suggest an age-related change in the pharmacology of the $GABA_A$ receptor, likely due to the age-related subunit shift described below. Patterns of $GABA_A$ receptor subunit reorganization were similar in both strains. Both strains displayed significant age-related increases in the $\gamma 1$ subunit and decreases in the a1 subunit. No significant age-related changes in the $\beta 2$ subunit were observed for either strain. These changes in subunit protein levels confirm the *in situ* findings of Milbrandt *et al.*, 1997 and are published in Caspary *et al.*, 1999.

Together these findings suggest that many age-related changes of GABA neurotransmission are common to the IC of both aging model strains in spite of age-related peripheral differences. Minor IC differences between strains appear more quantitative than qualitative. Peripheral differences between strains are significant and cannot be ignored. It is likely that peripheral aging signals central alterations in GABA neurotransmission and that these molecular and/or activity cues may be subtle.

ACKNOWLEDGEMENTS

The authors would like to thank Dr. Richard Salvi for his help with the cochleograms and Dr. Larry Hughes and Lynne Ling for their helpful comments concerning the development of this manuscript. This work was supported by NIH DC00151 to DMC.

REFERENCES

Caspary, D. M., Holder, T. M., Milbrandt, J. C., McKernan, R., and Naritoku, D. K, 1999, Age-related changes in $GABA_A$ receptor subunit composition and function in the rat auditory system. *Neuroscience* **93**: 307-312.

Fonnum, F., Walaas, I., and Iversen, L, 1977, Localization of GABAergic, cholinergic and aminergic structures in the mesolimbic system. *J. Neurochem.* **29**: 221-230.

Keithley, E. M., Ryan, A. F., and Feldman, M. L. Cochlear degeneration in aged rats of four strains. *Hear. Res.* **59**: 171-8.

Lipman, R. D, 1997, Pathobiology of aging rodents: inbred and hybrid models. *Exp. Gerontol.* **32**: 215-28.

Lipman, R. D., Chrisp, C. E., Hazzard, D. G., and Bronson, R. T, 1996, Pathologic characterization of brown Norway, brown Norway x Fischer 344, and Fischer 344 x brown Norway rats with relation to age. *J. Gerontol. A Biol. Sci. Med. Sci.* **51**:B54-9.

Milbrandt, J. C., Albin, R. L., Turgeon, S. M., and Caspary, D. M., 1996, $GABA_A$ receptor binding in the aging rat inferior colliculus. *Neuroscience* **73**: 449-458.

Milbrandt, J. C., Hunter, C., and Caspary, D.M., 1997. Alterations of $GABA_A$ receptor subunit mRNA levels in the aged Fischer 344 rat inferior colliculus. *J. Comp. Neurol.* **379**: 455-465.

Palombi, P. S., and Caspary, D.M., 1996. Physiology of the aging Fisher 344 rat inferior colliculus: Responses to contralateral monaural stimuli. *J. Neurophysiol.* **76**:3114-3125.

Raza, A., Americ, S. P., Milbrandt, J., and Caspary, D.M., 1994, Age-related changes in brainstem auditory neurotransmitters; Measures of GABA and acetylcholine function. *Hear. Res.* **77**: 221-230.

Sponger, V. P., Flood, V., Frisina, R. D., and Salvi, R. J., 1997, Quantitative measures of hair cell loss in CBA and C57BL/6 mice throughout their life spans. *J. Acoust. Soc. Am.* **101**: 3546-3553.

Age-Related Changes in Cochlear Function in Young and Adult Fischer 344 Rats

JIŘÍ POPELÁŘ*, DANIEL GROH*#, and JOSEF SYKA*
* *Institute of Experimental Medicine, Academy of Sciences,* # *ENT Clinic, 2nd Medical Faculty, Charles University, Prague, Czech Republic*

1. INTRODUCTION

Investigating the mechanisms underlying age-related changes in hearing, studied in normal animal populations, requires time consuming and long-lasting observations. Thus, special mutant strains of rodents, characterized by faster aging, are preferentially used for such investigations. Among these animal models, Fischer 344 (F344) rats are a widely used general-purpose inbred rat strain, being particularly favored for cancer research and toxicology. Aged rats develop cochlear degeneration connected with the elevation of hearing thresholds (Keithley *et al.*, 1992; Simpson *et al.*, 1985) as well as neurochemical and neurophysiological changes in the central auditory system (Backoff and Caspary, 1994; Palombi and Caspary, 1996).

The hearing threshold in F344 rats has been previously evaluated by auditory brainstem responses (ABRs) recordings over a limited frequency range, starting at 3-4 kHz and ending at frequencies below 40 kHz (Palombi and Caspary, 1996; Seidman *et al.*, 2000). Thus, the hearing threshold in F344 rats at frequencies below 3 kHz was not quantified until now. In addition, to date, no measurements of transiently evoked otoacoustic emissions (TEOAEs) or distortion product otoacoustic emissions (DPOAEs) in F344 rats have been reported in the literature.

The aim of the present study was to monitor hearing thresholds, assessed on the basis of ABR recordings, and TEOAEs and DPOAEs in F344 rats during the first year of their life (thus extending the previously published report about hearing function development in the first six months of life, Popelář *et al.*, 2003). Similar data were obtained in control groups of age-matched Long Evans rats. The data should be taken into account in interpreting the results of aging studies in the F344 rat strain.

2. RESULTS

Recordings of ABRs, TEOAEs and DPOAEs were made in two groups of the inbred albino strain Fischer 344 (F344) rats. Ten rats were tested at the age of one month, another group of eleven animals was measured repeatedly from the age of three to twelve months. The results were compared with similar measurements obtained in age-matched hooded Long Evans (LE) rats.

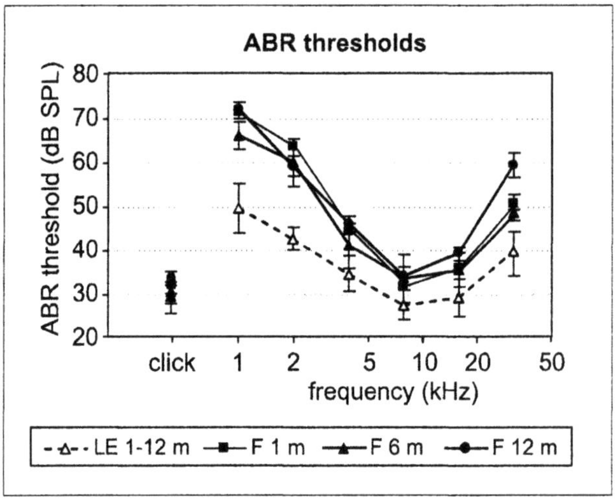

Figure 1. Average hearing thresholds (ABR audiograms) in one-, six- and twelve-month-old F344 rats and in one- to twelve-month-old LE rats. Bars represent ± SEM.

2.1 ABR Audiograms

The hearing thresholds in F344 and LE rats were determined using ABR recordings in anesthetized animals (38 mg/kg b.w. of ketamine and 5 mg/kg b.w. of xylazine). The ABRs were recorded by three subdermal stainless-steel needle electrodes, placed over the vertex (positive) and the right and left mastoids (negative and ground electrodes) of the animal's head. Acoustical stimuli were presented in free-field conditions via a loudspeaker. Hearing thresholds (the ABR audiogram) for clicks (pulse duration 0.1 ms) and tone bursts (burst duration 5 ms, 2 ms rise/fall times) in one octave steps ranging from 1 to 32 kHz were determined visually by reducing the stimulus intensity from the suprathreshold level in 5 dB steps.

The average hearing thresholds in F344 and LE rats are shown in Fig. 1. The average ABR audiograms of one-month-, six-month- and twelve-month-old F344 rats are very similar and are approximately 20 dB higher over the frequency range of 1-2 kHz and 5 to 10 dB higher at frequencies between 4-32 kHz than the average thresholds of LE rats ($p<0.001$ and $p<0.05$, resp.). The average hearing thresholds at 16 and 32 kHz in twelve-month-old F344 rats are significantly higher ($p<0.002$ and $p<0.001$, resp.) than those in six-month-old F344 rats. The ABR threshold to click stimulation was not significantly different between the two strains.

2.2 Transiently Evoked Otoacoustic Emissions

Otoacoustic emissions were recorded using an ILO 96 otoacoustic emission analyzer with an H-probe for neonate ears. TEOAEs were elicited by clicks of 40 μs duration and an intensity of 65-75 dB peak equivalent SPL. Acquisition was made using the non-linear mode and a time window between 1.5 and 10.5 ms.

Weak TEOAEs, exceeding the background noise level by several dB, were present usually at an isolated frequency in only a few one-month-old F344 rats, but no TEOAEs were measurable in older animals. In LE rats of any age, a weak response peaking at around 1.5 kHz was usually present.

Spontaneous otoacoustic emissions were not detected in any ear of F344 or LE rats.

2.3 Distortion Product Otoacoustic Emissions

Cubic $2f_1$-f_2 DPOAEs were recorded using two primary tones, f_1 and f_2 (ratio $f_2/f_1 = 1.22$), presented with f_1 and f_2 primary tone levels of $L_1/L_2 = 60/50$ dB SPL. DP-grams (the function of DPOAE level on increasing stimulus frequency) were recorded with a resolution of 4 points per octave.

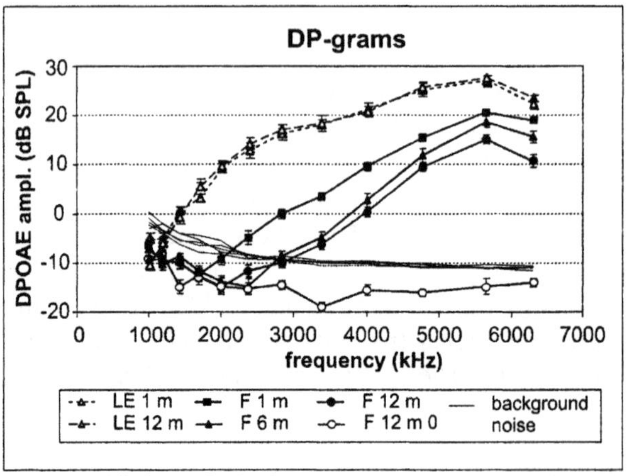

Figure 2. The average $2f_1$-f_2 DPOAE frequency-response curves (average DP-grams) measured in one- and twelve-month-old LE rats (open triangles, interrupted lines) and in one-, six- and twelve-month-old F344 rats (full lines). Bars represent ± SEM.

The DPOAE response amplitudes increased with increasing stimulus frequency, and the shape of the DP-grams was very uniform and typical for the individual rat strains. The average DP-grams of LE and F344 rats are shown in Fig. 2. In one- and twelve-month-old LE rats (open triangles, interrupted lines), the DPOAE frequency-response curves exceeded the background noise level at frequencies above 1-1.5 kHz and reached a plateau (maximal amplitudes of 20-30 dB SPL) at around 5-6 kHz. In one-month-old F344 rats (full squares, full line), the DPOAEs were absent at frequencies below 2-2.5 kHz; at higher frequencies (above 2.5 kHz) the DPOAE frequency-response curve increased rapidly and reached a plateau (maximal amplitudes between 15-25 dB SPL) at similar frequencies as did DPOAEs in LE rats. In six-month- and twelve-month-old F344 rats (full triangels and circles, full lines), the frequency at which DPOAE values exceeded the background noise level was shifted to higher frequencies (above 2.5-3 kHz), and the maximum DPOAE values measured at frequencies around 6 kHz tended to be lower than those observed in LE rats. However, in seven out of eleven twelve-month-old F344 rats, the DPOAEs were present in only one ear whereas the DPOAEs in the opposite ear were not measurable (open circles, full line).

3. DISCUSSION

The aim of this study was to evaluate hearing function in young and adult F344 rats and compare the data with those obtained in age-matched LE rats.

Our previous study revealed only a limited deterioration of the hearing threshold, frequency difference limen and parameters of middle latency responses in very old LE rats (36-months-old) (Rybalko *et al.*, 1998; Rybalko and Syka, 2001; Syka *et al.*, 1996; Syka and Rybalko, 2000). As the first part of this project, the parameters of hearing function were monitored during the first year of life in F344 rats to examine the feasibility of using this rat strain for investigating the mechanisms underlying age-related changes in hearing function.

The results of this study demonstrate that young F344 rats possess significant low-frequency hearing loss in comparison with LE rats of the same age, followed by high-frequency threshold elevation at the age of twelve months. The hearing deficit results in a lack of recordable TEOAEs and in the absence of DPOAEs, mainly at low frequencies. Because changes in average DP-grams were not reflected in similar changes in ABR audiograms, we can speculate that DPOAE recordings may serve as a more sensitive indicator of hearing impairment than ABR thresholds. Since Keithley *et al.* (1992) did not find noticable differences in cochlear morphology between F344 rats and other strains of young rats, we can conclude that hair cell loss cannot be solely responsible for the hearing loss in F344 rats. It seems more probable that the low-frequency defect, which preceeds the later occurring high-frequency hearing loss, can be related to more general genetic mutations present in this rat strain.

The assymetric hearing loss present in seven out of eleven twelve-month-old F344 rats indicates that the progression of degenerative changes can be different in individual ears of an animal. The analysis of the conditions that determine the speed of such progression in individual ears can help to elucidate the general mechanisms underlying presbycusis.

4. CONCLUSIONS

The results of the present study document that a significant low-frequency hearing loss occurrs as early as in one-month-old F344 rats and results in a lack of recordable TEOAEs and in the absence of DPOAEs at low frequencies. The low-frequency defect, which preceeds the later occurring high-frequency hearing loss, is probably not connected with the degeneration of hair cells or specific age-related hearing loss genes, but can be related to more general genetic mutations present in this rat strain.

ACKNOWLEDGEMENTS

Supported by grants No. 309/04/1074 from the Grant Agency of the Czech Republic and No. NR/8113-4 from the Internal Grant Agency of the Ministry of Health.

REFERENCES

Backoff, P.M., and Caspary, D.M., 1994, Age-related changes in auditory brainstem responses in Fischer 344 rats: effects of rate and intensity. *Hear. Res.* **73**: 163-172.

Keithley, E.M, Ryan, A.F., and Feldman, M.L., 1992, Cochlear degeneration in aged rats of four strains. *Hear. Res.* **59**: 171-178.

Palombi, P.S., and Caspary, D.M., 1996. Physiology of the young adult Fischer 344 rat inferior colliculus: responses to contralateral monaural stimuli. *Hear. Res.* **100**, 41-58.

Popelář, J., Groh,D., Mazelova, J., and Syka, J., 2003, Cochlear function in young and adult Fischer 344 rats. *Hear. Res.* **186**: 75-84.

Rybalko, N., Popelář, J., and Syka, J., 1998, Age-related changes in the hearing function in pigmented rats. Proceedings of the 21st Midwinter Meeting, ARO, St. Petersburg Beach, 78.

Rybalko, N., and Syka J., 2001, Susceptibility to noise exposure during postnatal development in rats. *Hear. Res.* **155**, 32-40.

Seidman, M.D., Khan, M.J., Bai, U., Shirwany, N., and Quirk, W.S., 2000, Biologic activity of mitochondrial metabolites on aging and age-related hearing loss. *Am. J. Otol.* **21**, 161-167.

Simpson, G.V., Knight, R.T., Brailowsky, S., Prospero-Garcia, O., and Scabini, D., 1985, Altered peripheral and brainstem auditory function in aged rats. *Brain Res.* **348**, 28-35.

Syka, J., Rybalko, N., Brozek, G., and Jilek, M., 1996, Auditory frequency and intensity discrimination in pigmented rats. *Hear. Res.* **100**, 107-113.

Syka, J., and Rybalko, N., 2000, Threshold shifts and enhancement of cortical evoked rsponses after noise exposure in rats. *Hear. Res.* **139**, 59-68.

Psychoacoustics and Working Memory in Dyslexia

KAREN BANAI, and MERAV AHISSAR
Hebrew University of Jerusalem, ISRAEL

1. INTRODUCTION

Dyslexia affects 5-10% of otherwise normal individuals, even in adulthood, when it is usually manifested in slow reading and difficulties in decoding unfamiliar and pseudo words (Felton, Naylor and Wood, 1990; Pennington *et al.*, 1990; Shaywitz, 1998).

The majority of dyslexics suffer from poor phonological awareness, i.e. difficulties in the explicit manipulation of basic speech sounds, and from poor verbal memory. These deficits impede their ability to develop adequate decoding, namely, quick and accurate mapping of series of visual symbols (written script) to sequences of speech sounds. According to the "core phonological deficit" hypothesis (e.g. Snowling, 2000), both poor verbal memory and poor phonological awareness stem from generally poor phonological representations. Though parsimonious in attributing several manifestations to a common underlying deficit, this hypothesis is not easy to test experimentally since "phonological representations" is an abstract term, which is difficult to probe. One potential consequence, namely that speech perception will be impaired, was found only in the minority of dyslexics (Adlard and Hazan, 1998; Ramus *et al.*, 2003).

An even more general hypothesis was proposed by Tallal and her colleagues (Tallal, 1980; Tallal, Miller and Fitch, 1993), who suggested that the phonological deficit is itself a manifestation of a more fundamental perceptual deficit. According to this hypothesis, dyslexics' processing of rapidly emitted brief stimuli is impaired, whether these acoustic stimuli are verbal or non-verbal. Subsequent behavioral and electrophysiological studies of dyslexics documented difficulties in a range of simple psychoacoustic tasks, consistent with the broad perceptual hypothesis (Behavioral studies: McAnally and Stein, 1996; Witton *et al.*, 1998; Hari *et al.*, 1999; Ahissar *et al.*, 2000; Amitay, Ahissar and Nelken, 2002; Electrophysiology: Nagarajan *et al.*, 1999; Baldeweg *et al.*, 1999). Yet, other studies either failed to document psychoacoustic difficulties in their dyslexic participants (Adlard and Hazan, 1998; Hill *et al.*, 1999), or questioned the relevance of psychoacoustic difficulties to reading (Watson and Miller, 1993; Rosen and Manganari, 2001; Heiervang, Stevenson and Hugdahl, 2002).

One account of this large inter-study variability is that poor psychoacoustic performance characterizes a sub-group of the dyslexic population that was sampled in some studies but not in others, depending on subject-recruitment strategy. The goal of the present study was to determine if such a group exists by recruiting a heterogeneous dyslexic group through various sources and by comprehensively testing the psychoacoustic and cognitive abilities of the dyslexic participants.

2. METHODS

2.1 Participants

One hundred and five native Hebrew speakers (17-30 years old) participated in this study (46 dyslexics and 59 normal readers). Inclusion criteria for all participants were: A. within normal cognitive abilities as measured by the Similarities (≥8) and Block Design (≥7) subtests of WAIS-III (Wechsler, 1997). B. At-least 12 years of formal education. Dyslexics were defined by having: 1. A documented history of reading difficulties throughout school years. 2. Current pseudo-word reading scores of at-least one SD below the control group's average. Control participants were friends of the dyslexic participants with no current or history of reading difficulties.

2.2 Psychoacoustic Tasks

All stimuli were presented to both ears through headphones in a sound-attenuating chamber. Two stimuli were presented in each trial, a fixed reference and a test tone, and the listener had to indicate which was higher, louder or longer. Feedback was given for incorrect responses. Thresholds were measured using a 2 down 1 up adaptive staircase procedure. Each assessment was terminated after 70 trials or 13 reversals. Discrimination threshold (JND) was determined by the arithmetic mean of the last 7 reversals.

Frequency discrimination was measured with a reference tone of 1000 Hz, 70 dB SPL, and 50ms duration; 950 ms separated the two tones in each trial. The initial test frequency was 1200 Hz. All participants completed this task. In ***duration discrimination*** the reference tone was of 1000 Hz, 70 dB SPL and 100ms or 1000ms duration in the first and second conditions, respectively. Initial test durations were of 150 and 1200 ms in these two conditions; 900 ms separated the two tones. This task was completed by 50 controls (40 C-Gs and 10 C-Ps, see division below) and by 37 dyslexic participants (24 D-Gs and 13 D-Ps). In ***intensity discrimination*** the reference tone was of 1000 Hz, 100ms and 30 dB; 900 ms separated the tones. Initial test intensity was 40 dB. This task was completed by 55 controls (45C-Gs, 10 C-Ps) and 44 dyslexic participants (30 D-Gs and 14 D-Ps).

2.3 Reading and Cognitive Tests

2.3.1 Reading and phonological awareness

Oral reading of pointed words (n=48) and pointed pseudo-words (n=36) was assessed using two tests (Deutsch and Bentin, 1996; Shalem and Lachman, 1998). Combined reading scores were defined as the average of accuracy speed Z-scores (relative to the control group's average).

Phonological awareness was tested using the Hebrew version of the Spoonerism task (designed by Ben-Dror & Peleg) in which participants orally hear a word pair and have to swap the initial phonemes of the 2 words

2.3.2 Cognitive tests

Three subtests of the WAIS-III (Wechsler, 1997) were administered: Similarities was used to assess verbal reasoning, Block Design was used to assess spatial reasoning, and Digit Span was used to measure verbal short-term memory. This test has two parts. In the forward part participants should repeat a list of digits in order of presentation and the score characterizes the capacity of verbal memory. In the backward part participants should repeat digits in reverse order. This part is more complex and measures verbal working memory (Gathercole, 1999).

Raven's Standard Progressive Matrices (Raven, Court and Raven, 1992) was also administered as a test of non-verbal cognitive abilities. This test consists 60 items and the score presented is the total number of correct items.

3. RESULTS & DISCUSSION

3.1 Psychoacoustic Processing

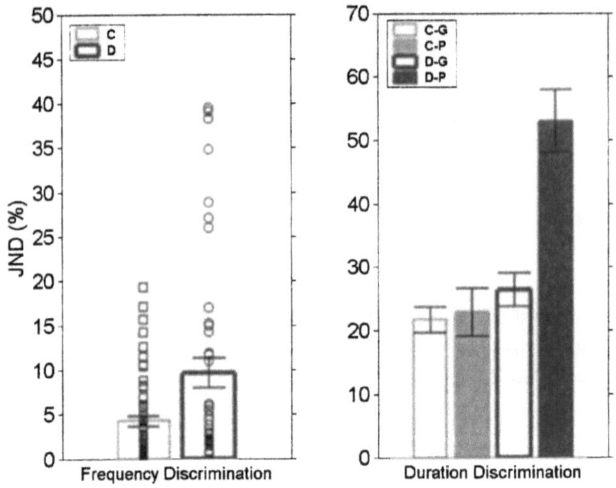

Figure 1. Auditory Processing

Group mean and individual subject JNDs on frequency discrimination are shown in Figure 1 (left). As a group, dyslexics had significantly elevated JNDs compared to controls (9.7% compared to 4.2%; t=3.1, p<.005). However, variability in the dyslexic group was large (range: 0.5 - 39.5%) and as can be seen in Figure 1, the distribution of JNDs was bimodal. Based on this bimodality we divided the dyslexic group into 2 subgroups: dyslexics with low frequency JNDs (<7%, n=31, D-Gs) and dyslexics with high frequency JNDs (n=15; D-Ps). Controls were divided into 2 groups according to the same criterion (C-G, n=48 and C-P, n=11 for controls with low and high JNDs). An interesting consequence of this frequency-discrimination based division was a concomitant cognitive division. Though the resulting 2 dyslexic subgroups (D-P and D-G) did not differ in their reading abilities, they differed in their performance on WAIS III subtests. A similar difference characterized the 2 control subgroups (C-P and C-G).

Thus, this division resulted in two dyslexic subgroups, each with a cognitive matched control subgroup (based on WAIS III subtests), as shown in Table 1. The main question we now asked was whether the 2 dyslexic groups also differed in other abilities. The analysis was conducted using a series of 2-way ANOVAs with Tukey post hoc comparisons between D-Gs and D-Ps and between D-Ps and C-Ps. Subgroup means and significance values for the post hoc tests are shown in Table 1.

Table 1. Subgroup means (SD)

	Controls		Dyslexics		Post-hoc p values DP vs.	
	CG	**CP**	**DG**	**DP**	**CP**	**DG**
Cognitive						
Similarities	13.4 (2.1)	12.6 (1.8)	14.0 (1.8)	12.0 (2.8)	.87	.02
Block Design	12.7 (2.9)	10.5 (1.8)	12.7 (3.1)	10.2 (2.5)	.99	.03
Digit Span	11.7 (2.9)	8.6 (1.7)	9.0 (2.7)	6.5 (2.3)	.21	.02
Raven	56.1 (2.9)	52.1(2.9)	55.0 (3.8)	48.6 (6.1)	.02	.001
Reading						
Accuracy (% correct)						
Words	97.9 (3.3)	96.4 (3.9)	86.6 (8.4)	87.1 (8.2)	.001	.99
Pseudo words	89.0 (9.4)	85.1 (10.0)	59.9 (17.0)	58.1 (15.8)	.001	.97
Rate (w/min)						
Words	102 (23)	96 (28)	54 (20)	44 (12)	.001	.17
Pseudo words	59 (16)	52 (16)	29 (9)	27 (9)	.001	.98
Spelling (% correct)	98.6 (4.1)	98.6 (4.2)	77.8 (18.2)	80.0 (19.0)	.005	.95
Spoonerism (errors)	1.5 (1.7)	2.4 (1.3)	5.9 (5.4)	9.5 (6.1)	.001	.03
Psychoacoustic tasks (JND %)						
Frequency	2.4 (1.8)	12.4 (3.8)	3.2 (1.6)	22.9 (11)	<.001	<.001
Short Duration	21.7 (12.7)	22.9 (12.0)	26.4 (12.7)	53.1 (17.9)	.001	.001
Long Duration	11.5 (5.7)	12.2 (6.5)	15.0 (7.9)	24.9 (10.5)	.001	.001
Intensity JND (dB)	2.7 (2.1)	2.6 (1.3)	3.0 (1.4)	4.0 (2.6)	.33	.39

As seen in Figure 1 (right) and in Table 1 the two dyslexic groups also differed in another auditory discrimination task - discriminating the duration of two pure tones, both around 100 and around 1000 ms. In contrast to the dyslexics, the two control subgroups did not differ in duration discrimination. Thus, among dyslexics but not among controls, poor frequency discrimination was a marker of broader difficulties in performing psychoacoustic comparisons.

3.2 Reading and Verbal Memory

While both dyslexic subgroups differed from both control subgroups on the reading and spelling measures used in this study, they did not differ from each other (see Table 1).

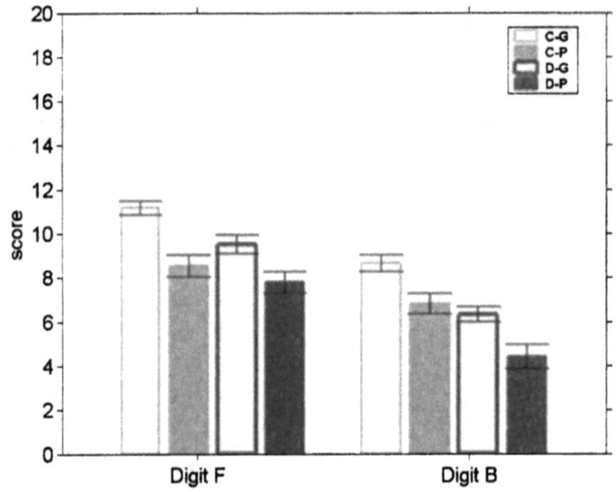

Figure 2. Verbal Memory

This pattern of results is puzzling with respect to the relevance of auditory processing abilities to reading. On one hand we found dyslexics, but no controls, with extremely poor psychoacoustic performance. On the other hand, these individuals were not poorer readers than dyslexics with adequate psychoacoustics. A possible explanation is that the source of reading difficulties among D-Gs and D-Ps is different. To test this alternative we examined whether the two dyslexic subgroups differ in other abilities that are necessary for adequate reading - verbal memory and phonological awareness.

Figure 2 shows the relative performance of the 4 subgroups on the two Digit Span subtests. On the standard score, as expected, dyslexics (as a group) were poorer than controls ($F_{(1,105)} = 15.5$, $p<.001$). There was also a main effect for frequency discrimination ($F_{(1,105)} = 21.9$ $p<.001$) but no interaction.

Figure 2 shows average scores on the two subtests of digit span. Digit Forward (Figure 2, left), measures the number of verbal elements that can be retained in short-term memory. Digit Backwards (Figure 2, right), measures complex span, when manipulation of elements is required. D-Ps were significantly poorer than C-Ps and D-Gs only on the latter subtest ($p = .04$ in the two post hoc Tukey HSD tests), indicating that the crux of their verbal memory difficulties resides in their poor ability to manipulate elements rather than in retaining them. Interestingly, although D-Gs were poorer than their matched controls (C-G) on both subtests of Digit Span, they were not poorer than the other control subgroup (C-P) on either of them. Yet both control subgroups are good readers. Thus, though D-Gs' working memory is somewhat poorer than expected, it does not seem to constitute the main impediment to their reading ability.

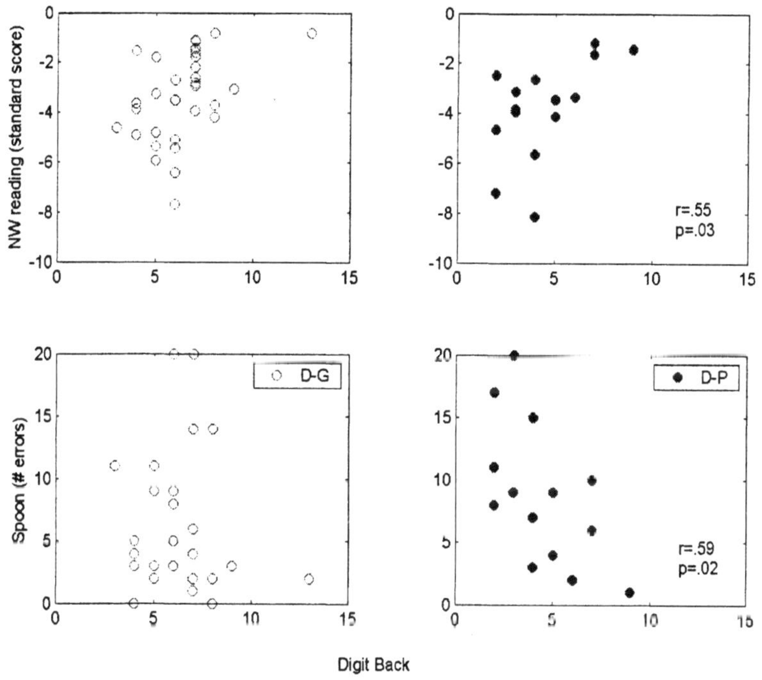

Figure 3. The relationship between working memory and phonological ability

D-Ps were poorer than D-Gs on phonological awareness (see Table 1). Spooncrism and reading scores in this group (but not in the other subgroups) were significantly correlated with working memory scores (measured with Digit Backward; r=.59, p=.02 for spoonerism and r=.55, p=.03 for reading, see Figure 3).

This pattern of results suggests that among D-Ps only, poor working memory (their working memory is poorer than that of any of the other groups) may be the main obstacle for adequate phonological processing and reading. Thus while both dyslexic groups exhibit deficits in phonological processing and poor reading the etiologies of their deficits may be different.

3.3 Cognitive Abilities

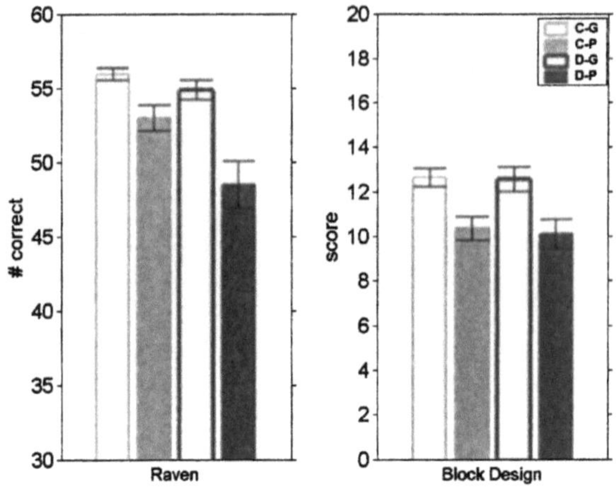

Figure 4. Cognitive abilities with high (Raven) and lower working memory load.

If the core deficit of the individuals in the D-P subgroup is indeed poor working memory, they are expected to perform poorly not only on phonological processing tasks but on cognitive tasks with high working memory load as well. On the other hand, they are expected to perform other cognitive tasks normally. To test this we compared performance in the 2 cognitive tasks described above: Block Design and Raven Matrices. These two tasks measure similar reasoning abilities and are highly correlated in the general population. However, adequate performance of Raven's Matrices (but not of Block Design) requires implicit use of working memory operations (Carpenter, Just and Shell, 1990). Thus, D-Ps are expected to perform worse than their controls (C-Ps) on Raven but not on Block Design. Figure 4 shows that this was indeed the case (see also Table 1). D-Ps and C-Ps were matched on Block Design, but D-Ps were significantly poorer on Raven's Matrices than any of the other subgroups, similarly to their working memory deficit. This is an important finding since Raven's Matrices and similar tests are used wildly in educational and psychometric testing to test analytic reasoning and concept formation. Yet, these tests may put D-Ps at a considerable disadvantage not because their analytic or reasoning abilities are generally poor but because they cannot relate simultaneously to all the different sub-parts of the problem they have to solve, as a result of their deficient working memory.

4. SUMMARY

Our findings suggest a functional division to 2 dyslexic subgroups with different etiologies, one group, who was the focus the current study, suffers mainly from generally impaired working memory and is consequently characterized by more general learning difficulties. Psychoacoustic processing was found an efficient marker for classification of the 2 subgroups, probably because adequate performance of the psychoacoustic tasks we used heavily relies on working memory. Our novel classification may have practical implication as it stresses the need for different intervention techniques for each subpopulation.

Our findings further suggest that working memory for verbal and non-verbal materials share common processing mechanisms. As working memory scores are good predictors of academic success (e.g. Gathercole and Pickering, 2000) this finding may explain why some of the previous studies failed to find auditory processing deficits in their samples that were typically recruited in universities (e.g. Hill *et al.*, 1999).

5. REFERENCES

Adlard, A., and Hazan, V., 1998, Speech perception in children with specific reading difficulties (dyslexia). *Q. J. Exp. Psychol. A* **51**: 153-177.

Ahissar, M., Protopapas, A., Reid, M., and Merzenich, M.M., 2000, Auditory processing parallels reading abilities in adults. *Proc. Natl. Acad. Sci. U. S. A.* **97**: 6832-7.

Amitay, S., Ahissar, M., and Nelken, I., 2002, Auditory processing deficits in reading disabled adults. *J. Assoc. Res. Otolaryngol.* **3**: 302-320.

Baldeweg, T., Richardson, A., Watkins, S., Foale, C., and Gruzelier, J., 1999, Impaired auditory frequency discrimination in dyslexia detected with mismatch evoked potentials. *Ann. Neurol.* **45**: 495-503.

Carpenter, P. A., Just, M. A., and Shell, P., 1990, What one intelligence test measures: a theoretical account of the processing in the Raven Progressive Matrices test. *Psychol. Rev.* **97**: 404-31.

Deutsch, A., and Bentin, S., 1996. Attention factors mediating syntactic deficiency in reading-disabled children. *J. Exp. Child. Psychol.* **63**: 386-415.

Felton, R. H., Naylor, C. E., and Wood, F. B., 1990, Neuropsychological profile of adult dyslexics. *Brain Lang.* **39**: 485-97.

Gathercole, S. E., 1999, Cognitive approaches to the development of short-term memory. *Trends Cogn. Sci.* **3**: 410-419.

Gathercole, S. E., and Pickering, S. J., 2000, Working memory deficits in children with low achievements in the national curriculum at 7 years of age. *Br. J. Educ. Psychol.* **70**: 177-94.

Hari, R., Saaskilahti, A., Helenius, P., and Uutela, K., 1999, Non-impaired auditory phase locking in dyslexic adults. *Neuroreport* **10**: 2347-8.

Heiervang, E., Stevenson, J., and Hugdahl, K., 2002, Auditory processing in children with dyslexia. *J. Child Psychol. Psychiatry* **43**: 931-8.

Hill, N. I., Bailey, P. J., Griffiths, Y. M., and Snowling, M. J., 1999, Frequency acuity and binaural masking release in dyslexic listeners. *J. Acoust. Soc. Am.* **106**: L53-8.

McAnally, K. I., and Stein, J. F., 1996, Auditory temporal coding in dyslexia. *Proc. R. Soc. Lond. B Biol. Sci.* **263**: 961-5.

Nagarajan, S., Mahncke, H., Salz, T., Tallal, P., Roberts, T., and Merzenich, M. M., 1999, Cortical auditory signal processing in poor readers. *Proc. Natl. Acad. Sci. U. S. A.* **96**: 6483-8.

Pennington, B., Van Orden, G., Smith, S., Green, P., and Haith, M., 1990, Phonological processing skills and deficits in adult dyslexic readers. *Child Dev.* **61**: 1753-1778.

Ramus, F., Rosen, S., Dakin, S. C., Day, B. L., Castellote, J. M., White, S., *et al.*, 2003, Theories of developmental dyslexia: insights from a multiple case study of dyslexic adults. *Brain* **126**: 841-65.

Raven, J., Court, J., and Raven, J., 1992, *Progressive Matrices and Vocabulary Scales.* Oxford Psychologists Press, Oxford.

Rosen, S., and Manganari, E., 2001, Is there a relationship between speech and nonspeech auditory processing in children with dyslexia? *J. Speech Lang. Hear. Res.* **44**: 720-36.

Shalem, Z., and Lachman, D., 1998, *Diagnostic Battery for Reading Process in Hebrew.* Nitzan: the Israeli association for the advancement of children with learning disabilities.

Shaywitz, S., 1998, Dyslexia. *N. Engl. J. Med.* **338**: 307-12.

Snowling, M. J., 2000, *Dyslexia.* Blackwell, Oxford.

Tallal, P., 1980, Auditory temporal perception, phonics, and reading disabilities in children. *Brain Lang.* **9**: 182-98.

Tallal, P., Miller, S., and Fitch, R. H., 1993, Neurobiological basis of speech: a case for the preeminence of temporal processing. *Ann. N. Y. Acad. Sci.* **682**: 27-47.

Watson, B. U., and Miller, T.K., 1993, Auditory perception, phonological processing, and reading ability/disability. *J. Speech Hear. Res.* **36**: 850-63.

Wechsler, D., 1997. *WAIS-III - Administration and Scoring Manual.* The Psychological Corporation, San Antonio.

Witton, C., Talcott, J. B., Hansen, P. C., Richardson, A. J., Griffiths, T. D., Rees, A., *et al.*, 1998, Sensitivity to dynamic auditory and visual stimuli predicts nonword reading ability in both dyslexic and normal readers. *Curr. Biol.* **8**: 791-7.

Frequency and Intensity Discrimination in Dyslexia

GAL BEN-YEHUDAH, KAREN BANAI, and MERAV AHISSAR
Hebrew University of Jerusalem, ISRAEL

1. INTRODUCTION

Developmental dyslexia is a specific reading disability, in which individuals do not acquire proficient reading skills despite sufficient cognitive abilities and education (Shaywitz, 1998). Dyslexia typically persists into adulthood when difficulties are characterized by slow reading, poor spelling and impaired phonological processing (Pennington *et al.*, 1990). Although it is commonly accepted that dyslexic individuals suffer from poor phonological awareness, many also perform poorly in simple auditory and visual tasks (e.g., Wright *et al.*, 2000; Stein, 2001; Ramus *et al.*, 2003).

Many previous studies reported that decreasing the inter-stimulus interval (ISI) made auditory tasks particularly difficult for dyslexic individuals (e.g. Tallal, 1980; Helenius *et al.*, 1999). However, results were mixed. In frequency discrimination tasks, several studies found more difficulties using longer intervals (Ahissar *et al.*, 2000; France *et al.*, 2002). Others reported difficulties under a broad range of intervals (Ahissar *et al.*, 2000; Walker *et al.*, 2002; Amitay *et al.*, 2002). Some studies found no difficulties, though typically they did not use brief intervals (Watson, 1992; Hill *et al.*, 1999).

One explanation for the conflicting reports on dyslexics' psychoacoustic performance is the heterogeneous nature of this disorder. Perhaps one dyslexic sub-population with broader learning difficulties (described by Banai and Ahissar; see chapter in this book) has auditory deficits not specific to temporal constraints, whereas another sub-population has auditory deficits limited to brief ISIs. This latter population was the focus of the current study. We therefore excluded individuals whose scores in cognitive tests were average or below from participating in this study.

In previous studies, randomly changing interval duration increased task difficulty (e.g. Tallal, 1980; France *et al.*, 2002). Thus, it is hard to dissociate whether dyslexics' difficulties with brief ISIs stemmed from a general impairment revealed under difficult conditions, or from a deficit specifically associated with temporal constraints. In the current study, we assessed frequency and intensity discriminations. Task difficulty was dissociated from interval duration by gradually decreasing the ISI and measuring thresholds separately for each interval. Thus, in each assessment frequency/intensity level was changed adaptively while the ISI was fixed. We chose frequency discrimination since many studies found high thresholds in dyslexia using intermediate ISIs (e.g. Ahissar *et al.*, 2000). Intensity discrimination was chosen because dyslexics' performance on this task was not impaired for similar intervals (Watson, 1992; Nicolson *et al.*, 1993; Ahissar *et al.*, 2000).

2. METHODS

Participants: Fifty-six adult native Hebrew speakers participated in this study, 23 dyslexic (22.1±3.6 years) and 33 normal readers (22.1±3.3 years). Dyslexics had a documented history of reading disability and a current non-word reading score (a composite z-score of reading rate and accuracy) of at least one standard deviation below controls' average. None of the controls had a history of reading difficulties. Control and dyslexic groups were matched on age and cognitive abilities. Participants had normal hearing level and gave informed consent to take part in this study.

Cognitive, reading-related and attention tests: Cognitive measures were subtests of the WAIS-III (Wechsler, 1997): Similarities (verbal reasoning), Block Design (non-verbal visual spatial reasoning) and Digit Span (verbal memory); and Raven's Standard Progressive Matrices, a general reasoning test shown to be a good predictor of academic achievement (Raven *et al.*, 2000). Note that all of the participants had above average cognitive abilities (Table 1). On both Similarities and Block Design scaled scores were 8 or above and on Raven's Matrices scores were at least in the 50th percentile and typically above (based on percentile norms for adults in the USA).

Reading-related measures included oral reading of single non-words and of an academic level passage, spelling, naming speed of letters and numbers (RAN) and phonological awareness (Spoonerism). For a detailed description of these tests see Banai and Ahissar's chapter in this book. Visual attention was assessed with the standard CPT-II test (Conners, 2000). The outcome of this test is an index, which indicates confidence level in an attention deficit/hyperactivity (ADHD) classification. An index above 60% indicates high confidence, whereas an index between 40-60% is inconclusive.

The performance of control and dyslexic groups on these tests is summarized in Table 1, for participants who completed the frequency discrimination task. Participants that completed the intensity discrimination task had a similar performance profile. The dyslexic group had poor reading and phonological skills, above average cognitive abilities, and a somewhat poorer ability to sustain attention, compared with the control group. This profile is characteristic of adult dyslexics (Pennington *et al.*, 1990).

Table 1. Group means (± SD) for participants who completed the frequency discrimination task. Significant differences are noted for a two-tailed t-test, (*) $P < 0.05$ and (**) $P < 0.01$.

	Control (n=24)	Dyslexic (n=20)
Age (years)	21.4 (3.6)	22.8 (2.6)
Cognitive tests:		
Block Design (scaled score)	12.8 (2.8)	12.8 (2.9)
Similarities (scaled score)	13.8 (2.3)	13.7 (1.9)
Digit Span (scaled score)	10.4 (2.5)	9.0 (2.2)
Raven Matrices (# correct)	56.3 (2.6)	55.6 (1.8)
Reading-related tests:		
Non-word (% correct)	89.5 (6.6)	60.4 (17.5)**
Passage (words/min)	130 (17.6)	89 (23.7)**
Spelling (% correct)	98.9 (3.5)	81.9 (15.9)**
RAN-letters (lett/second)	2.5 (0.4)	2.2 (0.5)*
RAN-numbers (num/second)	2.9 (0.4)	2.5 (0.5)**
Spoonerism (% correct)	92.5 (7.7)	71.0 (24.4)**
Attention test:		
ADHD confidence index (%)	30.9 (16.5)	46.2 (11.7)**

Stimuli and Experimental Design: Most of the participants completed both frequency and intensity discrimination tasks. Participants indicated which interval contained the higher/louder tone, in a 2-alternative temporal forced choice paradigm. Feedback was given for incorrect responses. In both tasks, thresholds were measured with a 2 down/1 up adaptive staircase procedure. Each assessment was terminated after 70 trials or 13 reversals. Discrimination threshold (JND) was determined by the arithmetic mean of the last 7 reversals. Thresholds for each ISI were measured in separate blocks. A training stage preceded data collection in each block.

Frequency discrimination was assessed with a fixed reference tone of 1 kHz and an initial test frequency of 1.4 kHz. Both tones were 50 ms in duration and 70 dB SPL. Thresholds were measured in the following order: 1, 1, 0.7, 0.5, 0.2, 0.1 s; 15 minute break; 1, 2 s. Twenty dyslexic and 24 control participants completed this task. *Intensity discrimination* was assessed with a fixed reference of 30 dB SPL and an initial test level of 40 dB SPL. Both tones were 1 kHz and 100ms in duration. Thresholds were measured in the following order: 1, 0.7, 0.5, 0.2, 0.1 s; 15 minute break; 1, 2 s. Eighteen dyslexic and 20 control participants completed this task.

3. RESULTS

Figure 1 shows average frequency (A) and intensity (B) discrimination thresholds measured at a range of ISIs (0.1–2 s). On both tasks dyslexics' JNDs gradually increased at shorter ISIs. For frequency discrimination, a repeated-measures 2-way ANOVA showed significant main effects of group ($F_{1,38}$=14.9, $P<0.01$) and ISI ($F_{5,190}$=7.4, $P<0.01$). There was also significant interaction between group and ISI ($F_{5,190}$=4.4, $P<0.01$) resulting from the consistent increase in dyslexics' JNDs as ISI was shortened, compared with controls' similar JNDs across all ISIs. In the intensity discrimination task, the dyslexic JNDs had a similar dependency on ISIs. In this task, however, controls' performance was also somewhat poorer with shorter ISIs, though their slope was much shallower. A 2-way ANOVA for intensity discrimination showed significant main effects of group ($F_{1,36}$=4.9, $P<0.05$) and ISI ($F_{5,180}$=17.3, $P<0.01$). The interaction was marginally significant ($F_{5,180}$=2.2, P=0.052).

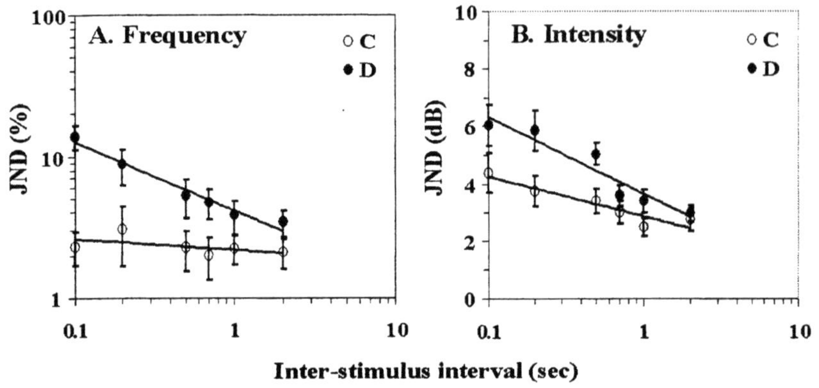

Figure 1. Groups' mean thresholds (mean±SEM) as a function of ISI are shown for frequency (A) and intensity (B) discrimination tasks (dyslexics in filled and controls in empty symbols).

Dyslexics' frequency JNDs measured at ISI of 0.7 s were significantly correlated (Spearman's ranked correlation, r_s) with verbal memory (Digit Span subtest; r_s=-0.73, P<0.01), passage reading rate (Passage; r_s=-0.54, P<0.05) and verbal reasoning abilities (Similarities; r_s=-0.49, P<0.05). A significant correlation was also found between rapid naming of numbers (RAN) and frequency JNDs measured at an ISI of 2 s (r_s=-0.56, P<0.05). Phonological awareness (Spoonerism), however, was not correlated with any of the frequency or intensity JNDs. Among controls, frequency JNDs were not correlated with any of the cognitive subtests. Though, interestingly, controls' frequency JNDs at 0.7 s ISI were significantly correlated with non-word reading accuracy (r_s=-0.42, P<0.05).

4. DISCUSSION

Dyslexics' ability to discriminate between sequentially presented brief tones was impaired, but only when processing time was limited by short ISIs (for frequency < 1 s and for intensity < 0.7 s). At sufficiently long intervals (~ 1-2 s), dyslexics' psychoacoustic performance was similar to that of normal readers. Our finding that dyslexics' frequency and intensity discriminations have a similar dependency on ISI supports the specificity of their deficit to temporal constraints. A correlational analysis showed that dyslexics with higher frequency JNDs at shorter ISIs (500-700 ms) also had poorer verbal memory and slower reading and naming rates. Dyslexics with higher intensity JNDs at longer ISIs (1-2 s) had poorer decoding abilities. On the other hand, Phonological awareness was not correlated with either frequency or intensity JNDs measured at any interval.

These findings are consistent with previous reports of abnormal auditory (Hari and Kiesila, 1996; Helenius *et al.*, 1999) and visual (Hari *et al.*, 1999) performance in dyslexia when stimuli were separated by short intervals. Hari and Renvall (2001) suggested that "sluggish attentional shifts" in dyslexia could account for this abnormal performance. Though the hypothesis of sluggish attention does not clarify whether stimulus encoding or retrieval processes are affected, it does predict that at sufficiently long ISIs dyslexics' performance would be normal, as found for our test group. Thus, in our study, dyslexics' impaired psychoacoustic performance at short ISIs may stem from slower shifts of attention from the first stimulus (that has to be identified) to the second stimulus. The finding that dyslexics' psychoacoustic performance was correlated with their reading rate but not with phonological awareness, suggests that the deficit revealed in our study contributes to their disability, but it is probably not the only source of their reading difficulties.

Taken together, the pattern of psychoacoustic deficits of our dyslexic participants seems to stem from slower shifts of attention between sequentially presented stimuli, rather than from a low-level perceptual impairment. This attentional impairment may underlie their slower reading and naming rate and somewhat poorer memory span.

NOTE

The frequency discrimination task used in this study was administered to a group of dyslexic adults with average cognitive abilities. The findings of this follow-up study have been recently published (Ben-Yehudah, G., Banai, K., and Ahissar, M., 2004, *Neuroreport* 15: 627-631). This expanded study compares the frequency discrimination abilities of dyslexics with average cognitive abilities to those of dyslexics with excellent cognitive abilities (i.e. the dyslexic group described in this chapter). This comparison points to different etiologies underlying similar reading difficulties in the two dyslexic groups.

REFERENCES

Ahissar, M., Protopapas, A., Reid, M., and Merzenich, M. M., 2000, Auditory processing parallels reading abilities in adults. *Proc. Natl. Acad. Sci. U. S. A.* **97**: 6832-7.

Amitay, S., Ahissar, M., and Nelken, I., 2002, Auditory processing deficits in reading disabled adults. *J. Assoc. Res. Otolaryngol.* **3**: 302-320.

Conners, C., 2000, *Conners' Continuous Performance Test (CPT II), Technical guide and software manual.* MHS, Toronto.

France, S. J., Rosner, B. S., Hansen, P. C., Calvin, C., Talcott, J. B., Richardson, A., *et al.*, 2002 Auditory frequency discrimination in adult developmental dyslexics. *Percept Psychophys.* **64**: 169-179.

Hari, R., and Kiesila, P., 1996, Deficit of temporal auditory processing in dyslexic adults. *Neurosci. Lett.* **205**: 138-140.

Hari, R, and Renvall, H., 2001, Impaired processing of rapid stimulus sequences in dyslexia. *Trends Cognit. Sci.* **5**: 525-532.

Hari, R., Valta, M., and Uutela, K., 1999, Prolonged attentional dwell time in dyslexic adults. *Neurosci. Lett.* **271**: 202-4.

Helenius, P., Uutela, K., and Hari, R., 1999, Auditory stream segregation in dyslexic adults. *Brain* **122**: 907-913.

Hill, N. I., Bailey, P. J., Griffiths, Y. M., and Snowling, M. J., 1999, Frequency acuity and binaural masking release in dyslexic listeners. *J. Acoust. Soc. Am.* **106**: L53-8.

Nicolson, R. I., Fawcett, A. J., and Dean, P., 1995, Time estimation deficits in developmental dyslexia: evidence of cerebellar involvement. *Proc. R. Soc. Lond. B Biol. Sci.* **259**: 43-47.

Pennington, B., Van Orden, G., Smith, S., Green, P., and Haith, M., 1990, Phonological processing skills and deficits in adult dyslexic readers. *Child Dev.* **61**: 1753-1778.

Ramus, F., Rosen, S., Dakin, S. C., Day, B. L., Castellote, J. M., White, S., *et al.*, 2003, Theories of developmental dyslexia: insights from a multiple case study of dyslexic adults. *Brain* **126**: 841-65.

Raven, J., Raven, J. C., and Court, J. H., 2000, *Manual for Raven's progressive matrices and vocabulary scales.* Oxford Psychologists Press, Oxford.

Shaywitz, S., 1998, Dyslexia. *N. Engl. J. Med.* **338**: 307-12.

Stein, J., 2001, The magnocellular theory of developmental dyslexia. *Dyslexia* **7**: 12-36.

Tallal, P., 1980, Auditory temporal perception, phonics, and reading disabilities in children. *Brain Lang.* **9**: 182-98.

Walker, M. M., Shinn, J. B., Cranford, J. L., and Givens, G. D., 2002, Auditory temporal processing performance of young adults with reading disorders. *J. Speech Lang. Hear. Res.* **45**: 598-605.

Watson, B. U., 1992, Auditory temporal acuity in normally achieving and learning-disabled college students. *J. Speech Hear. Res.* **35**: 148-56.

Wechsler, D., 1997, *WAIS-III - Administration and Scoring Manual.* The Psychological Corporation, San Antonio.

Wright, B. A., Bowen, R. W., and Zecker, S. G., 2000, Nonlinguistic perceptual deficits associated with reading and language disorders. *Curr. Opin. Neurobiol.* **10**: 482-486.

Speech Perception in Noise among Learning Disabled Teenagers

HANNA PUTTER-KATZ[*,#], KAREN BANAI[*], and MERAV AHISSAR[*]
[*]*Hebrew University of Jerusalem; and* [#]*The Speech and Hearing Center, Tel Hashomer, Israel*

1. INTRODUCTION

Many studies reported deficits in speech perception among a large proportion of the individuals with learning disabilities (LDs) (e.g Tallal, 1980; Eliott *et al.*, 1989; Kraus *et al.*, 1999; Baran, 2002). These deficits are enhanced in the presence of background noise (Nabelek and Picket, 1974; Bellis, 1996; Chermak and Musiek, 1997; Cunningham *et al.*, 2001). Yet, the functional relations between LDs' poor speech perception and their academic difficulties are not well understood (Chermak *et al.*, 1989). Some studies suggested that their perceptual deficits underlie their comprehension difficulties for language presented through the auditory modality (adults- Blalock, 1982; adolescents- Jerger *et al.*, 1987; and children- Ferre and Wilber, 1986). Keith (1981) suggested that poor language comprehension impedes the development of meta-linguistic skills.

The goal of the present study was to examine the auditory-phonetic and auditory-linguistic processing in learning disabled teenagers, with and without reading deficits, in comparison to age and education matched controls. We measured reading, language and working memory skills as well as threshold intensity for perceiving speech in noise. Speech perception in noise was measured using pseudo-words in order to minimize the effects of language and verbal memory on performance.

2. METHODS

2.1 Participants

Fifty 7th grade female native Hebrew speakers participated in this study, 19 controls from a regular public school (13.3±0.4 years) and 31 LDs (13.2± 0.4 years) from a private school for individuals with mild learning difficulties. All participants had within normal range intelligence scores according to the educational authorities and to standard tests administered as part of the school admission procedure. One student who had a history of hearing difficulties was excluded.

2.2 Speech Identification in Noise

Bisyllabic pseudowords were recorded by a female native Hebrew speaker using Tucker-Davis Technologies (TDT, Gainesville, FL) system III processor and a TDT system II microphone module. The recorded word files were normalized for intensity (using SoundForge™) and duration (to 800ms; using Praat stretching algorithm). The masking noise was generated by summing tones at: 500,530,550,580,600,620,680 and 700Hz, and multiplying the sum with a sinusoidal envelope at 4Hz. The total noise RMS was 83 dB SPL. All stimuli were presented binaurally through headphones (Sennheiser HD-265) in a quiet room. Subjects were asked to repeat one bisyllabic pseudo-word per trial. The intensity of the pseudo-words was adapted in a 3 down/1 up staircase procedure (converging on ~80% correct) while the noise level remained fixed. Each assessment terminated after 85 trials or 15 reversals. Identification threshold (JND) was the arithmetic mean of the last 10 reversals.

2.3 Language, Reading and Working Memory Tests

Four language measures, which required production or comprehension of morphological forms, were administered. These measures are highly sensitive for detecting mild language difficulties in the later school years

(Ravid, 2003). Two tasks tested comprehension and production of genitive inflection. The comprehension tasks required the analysis of inflected form into its components, e.g., *metosxa* → *matos shelxa* (your airplane). The production tasks required merging of an analytic form, e.g., *ha-pe shelo* as *piv* (his mouth). Two additional tasks tested the acquisition of derived nominals. The Machine Test involved completing a compound whose head was the bound form *mechonat* (machine) and which required a derived nominal as the modifier (Ravid and Avidor, 1998). Responses were scored for accuracy. All subjects also performed the Hebrew version of the vocabulary subtest of the WISC-R95 (Wechsler, 1974).

Phonological awareness was tested using a Hebrew version of the Spoonerism task (20 items) in which listeners were orally presented with a word pair and were asked to swap the initial phonemes of the two words (e.g. white pig-→ pite wig). In the process of introducing the task, all participants successfully performed its simpler sub-steps (phoneme deletion/ replacement).

Rapid Automatized Naming (RAN) was tested for digits and letters (Spreen and Strauss, 1991).

Verbal memory was assessed by the *Digit Span subtest* (WISC-R95; Wechsler, 1974). It contains 2 subtests: Digit Forward, which measures the number of accurately repeated digits, and Digit Backward which measures the number digits repeated in reverse order.

Oral reading of pointed Hebrew words and non-words was assessed using a list of 24 pseudo-words and a list of 24 words (Deutsch and Bentin, 1996).

2.3.1 Division to Subgroups

We divided the LD class into 2 groups: reading disabled (RDs; n=19) and others (LDs; n=11). Classification was based on a combined phonological score (accuracy of reading pseudo-words and words and accuracy in phonological awareness), calculated with respect to the average of a control class of the same grade from a regular school. RDs were those whose combined phonological score was lower than controls' average by 1.5 standard deviations or more.

3. RESULTS

3.1 Language, Reading and Working Memory Performance

As shown in Table 1, RDs performed significantly poorer than LDs and Controls, not only in reading, verbal memory and phonological awareness tasks, but also in language tests. In contrast, their thresholds for speech perception in noise were not poorer than those of the other 2 groups.

Table 1. Reading disabled, learning disabled and control groups means & (SD)

	RD (n=19)	**LD** (n=11)	**Controls** (n=19)	p[5]
Standard Cog tasks	(Scaled Scores)			
Block Design	9.6 (2.8)	10.7 (2.2)	13 (1.4)	.000
Digit Span	7.7 (1.9)*	9.7 (2.3)	10.5 (3)	.002
Vocabulary	7.4 (1.9)	7.4 (1.9)	10.4 (2.3)	.000
Reading (% correct)				
Words	86.7 (7.3)*	96 (3.7)	95 (6)	.000
Pseudo words	57 (14.3)*	89 (8.7)	83 (12)	.002
Rate (w/min)				
Words	73.7 (26)	99 (28)	79 (22)	.02
Pseudo words	33 (11)*	49 (11)	48 (15)	.001
Phonological processing				
Spoonerism	463 (121)*	273 (95)	222 (108)	.000
Spoonerism[2]	10 (4.4)*	16.6 (3)	15.9 (4)	.000
RAN-digits[1]	22.9 (4.9)	20.9 (3.4)	20 (3.3)	.09
RAN-letters[1]	25.9 (4.8)	24 (3.4)	21.8 (6)	.04
Language				
Machine test-prod[3]	62.3 (22)*	77.6 (7)	89.7 (10)	.000
Genitive nouns-comp.[3]	86 (11.9)*	94.5 (6.8)	96.4 (4.7)	.002
Genitive nouns-prod.[3]	75.6(17.5)*	90.1 (7)	91.1 (11)	.002
Derived nominals-comp[3]	82.6 (11)	86 (5.2)	88 (22.4)	.006
Speech in noise threshold[4]	61.7(3.9)	60.3 (4.3)	60.3 (4.6)	.75

[1]duration in seconds; [2]# correct; [3]% correct ; [4]dBSPL; [5]one way ANOVA; Tasks in which Tukey post hoc indicated poorer performance of RDs compared to both LDs with no reading deficits and controls are marked by (*).

3.2 Speech Identification Thresholds in Noise

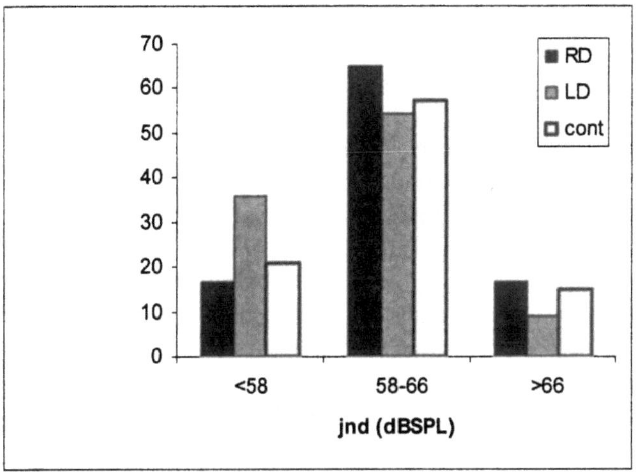

Figure 1. Distribution (% of subjects) of intensity threshold for speech perception in noise among RDs, LDs and Controls.

The distribution of speech identification thresholds in the 3 groups is shown in Figure 1. The distribution of these groups did not differ, just as their average thresholds did not differ. This finding is particularly surprising given that RDs' performance in all language and phonological tasks was consistently poorer than that of the other 2 groups (Table 1).

4. SUMMARY AND DISCUSSION

Our main finding is that RDs' threshold levels for perceiving speech in noise were not different than those of other LDs and from controls'. It indicates that RDs' phonological and language deficits do not stem from poor oral language perception due to increased sensitivity to background noise. Our finding at first seems at odds with previous reports. For example, Chermak *et al.* (1989) found that word identification performance of learning disabled adults was depressed relative to controls' in the presence of speech spectrum noise and linguistic maskers. They concluded that these results suggest that LD subjects are more susceptible to noise. However, having used real mono-syllablic words, perhaps their findings reflect controls' better linguistic knowledge rather than better perception. Elliot *et al.* (1979) also found better identification of monosyllabic nouns in noise among children achieving normal school progress compared to children with learning problems. However, they suggested that the poorer performance of the children with learning problems can be attributed to the less developed language competence of these children. We measured thresholds in noise using pseudo-words in order to minimize such effects.

Most previous studies that reported RDs' impaired speech perception manipulated temporal aspects of the stimulus, rather than noise level. Thus, robustness of categorical boundaries was assessed by manipulating transition rates (e.g. Kraus, 1999; Rosen and Manganari, 2001) or interval (Voice Onset Time) durations (e.g. Elliott *et al.,* 1989). Though these deficits may be interpreted as stemming from a generally noisier perceptual system (e.g. Hartley and Moore, 2002) temporal compression may be special. It may pose a particular perceptual difficulty for this population (Tallal *et al.,* 1993, 1996; Ahissar *et al.,* 2001; Amitay, *et al.*, 2002). In the current study, we found that noise per-se does not hamper RDs' perception more than it does so to other LDs and to controls. The effect of manipulating temporal parameters was not assessed.

Deciphering whether RDs' phonological and language deficits result from a perceptual impairment (e.g. increased sensitivity to compression) or from higher-level deficits (e.g. poor working memory, see Banai and Ahissar in this volume) requires assessments of speech perception under a greater variety of speech manipulations.

REFERENCES

Ahissar, E., Nagarajan, S., Ahissar, M., Protopapas, A., Mahncke, H., and Merzenich, M. M., 2001, Speech comprehension is correlated with temporal response patterns recorded from auditory cortex. *Proc. Natl. Acad. Sci. USA* **98**: 13367-13372.

Amitay, S., Ahissar, M., and Nelken, I., 2002, Auditory processing deficits in reading disabled adults. *J. Assoc. Res. Otolarygol.* **3**: 302-320.

Baran, J. A., 2002, Managing Auditory Processing Disorders in Adolescents and Adults. *Seminars in Hearing* **23**: 327-335.

Bellis, T. J., 1996, *Assessment and management of central auditory processing disorders in the educational setting: from science to practice.* Singular, San Diego, CA.

Blalock, J. W., 1982, Persistent auditory language deficits in adults with learning disabilities. *J. Learn. Disabil.* **15**: 604-609.

Chermak, G. D., Vonhof, M. R., and Bendel, R. B., 1989, Word identification performance in the presence of competing speech and noise in learning disabled adults. *Ear Hearing* **10**: 90-93.

Chermak, G. D., and Musiek, F. E., 1997, *Central auditory processing disorders: new perspectives.* Singular, San Diego, CA.

Cunningham, J., Nicol, T., Zecker, S. G., Bradlow, A., and Kraus, N., 2001, Neurobiologic responses to speech in noise in children with learning problems: deficits and strategies for improvement. *Clin. Neurophys.* **112**: 758-767.

Deutsch, A., and Bentin, S., 1996, Attention factors mediating syntactic deficiency in reading-disabled children. *J. Exp. Child. Psychol.* **63**: 386-415.

Elliott, L. L., Connors, S., Kille, E., Levin, S., Ball, K., and Katz, D., 1979, Children's understanding of monosyllabic nouns in quiet and in noise. *J. Acoust. Soc. Am.* **66**: 12-21.

Elliot, L. L., Hammer, M. A., Scholl, M.E., 1989, Fine-grained auditory discrimination in normal children and children with language-learning problems. *J. Speech. Hear. Res.* **32**: 112-119.

Ferre, J. M., and Wilber, L. A., 1986, Normal and learning disabled children's central auditory processing skills: An experimental test battery. *Ear Hear.* **7**: 336-343.

Hartley, D. E. H., and Moore, D. R., 2002, Auditory processing efficiency deficits in children with developmental language impairments. *J. Acoust. Soc. Am.* **112**: 2962-2966.

Jerger, S., Martin, R. C., Jerger, J., 1987, Specific auditory perceptual dysfunction in a learning disabled child. *Ear Hear.* **8**: 78-86.

Keith, R. W., 1981, *Central Auditory and Language Disorders in Children.* College-Hill Press, San Diego.

Kraus, N., Koch, D. B., McGee, T. J., Nicol, T. G., and Cunningham, J., 1999, Speech-sound discrimination in school-age children: psychophysical and neurophysiologic measures. *J Speech Lang. Hear. Res.* **42**: 1042-1060.

Nabelek, A. K., and Pickett, J. M., 1974, Reception of consonants in a classroom as affected by monaural and binaural listening, noise, reverberation, and hearing aids. *J. Acoust. Soc. Am.* **56**: 628-639.

Ravid, D., 2003, *Later lexical development in Hebrew: Derivational morphology revisited.* Trends in Language Acquisition Research, Benjamines, Amsterdam:.

Ravid, D., and Avidor, A., 1998, Acquisition of derived nominals in Hebrew: Developmental and linguistic principles. *J. Child. Lang.* **25**: 229-266.

Rosen, S., and Manganari, E., 2001, Is there a relationship between speech and nonspeech auditory processing in children with dyslexia? *J. Speech Lang. Hear. Res.* **44**: 720-736.

Spreen, O., and Strauss, E., 1991, *A compedium of neuropsychological tests: Administration, norms and commentary.* Oxford University Press, New York.

Tallal, P., 1980, Auditory temporal perception, phonics, and reading disabilities in children. *Brain Lang.* **9**: 182-98.

Tallal, P., Miller, S., and Fitch, R. H., 1993, Neurobiological basis of speech: a case for the preeminence of temporal processing. *Ann. N. Y. Acad. Sci.* **682**: 27-47.

Tallal, P., Miller, S. L., Bedi, G., Byma, G., Wang, X., Nagarajan, S. S., Schreiner, C., Jenkins, W. M., and Merzenich, M. M., 1996, Language comprehension in language learning impaired children improved with acoustically modified speech. *Science* **271**: 81-83.

Wechsler, D., 1974, *Wechsler Intelligence Scale for Children (Revised)- Manual R-95 edn*, The Psychological Corporation, San Antonio.

How Can The Neural Encoding and Perception of Speech Be Improved?

TRENT NICOL[*], and NINA KRAUS[*,#]
[*]*Auditory Neuroscience Laboratory, Department of Communication Sciences, and* [#]*Departments of Neurobiology & Physiology and Otolaryngology, Northwestern University, Evanston, IL, USA*

1. INTRODUCTION

A major focus of our Auditory Neuroscience Laboratory is gaining an understanding of the physiology underlying speech encoding. Imprecise encoding of speech sounds at the neural level contributes to communication problems in adverse listening conditions. Populations such as the hearing impaired, the elderly, and individuals with auditory-based learning disabilities suffer from poor speech perception even under favourable listening conditions. Unfamiliar sounds and sound combinations present perceptual difficulties for people struggling to learn a foreign language.

In order to improve speech encoding, and ultimately perception, there are two broad courses of action. The first is training. The auditory system has a remarkable ability to reorganize in response to training in order to cope with impoverished—or unfamiliar—signals. Second, steps can be taken to increase the clarity of the signal, shifting the burden of improving encoding outward.

This report, first, will review some work on our lab demonstrating neural plasticity resulting from training in normal, adult populations. Next, physiological changes seen in learning-disabled children following a commercial auditory training program will be examined. Finally, taking a broader definition of plasticity, the neural alterations effected by modifications to speech sounds themselves will be addressed.

2. BACKGROUND

The impact of discrimination training on speech-evoked cortical responses has been examined in our laboratory from multiple angles. The plasticity of the N1/P2 complex—signalling an acoustic event, and the mismatch response—signalling a change between two acoustic events, has been investigated. Particular attention was paid to the time course of neural plasticity and to its generalization to similar acoustic stimuli.

The mismatch response is an attention-independent metric of the physiological detection of acoustic change. Thus, it is useful in the study of training-associated auditory discrimination abilities. If a previously dormant neural population newly reacts to a formerly undetectable difference, it can be inferred that the training regimen has had an effect even in the absence of overt behavioural improvement. Additionally, the N1/P2 complex has been demonstrated to reflect acoustic properties present in complex stimuli such as speech. In particular, voiceless consonant-vowel speech syllables such as /ta/ have been demonstrated to elicit discrete responses to the aspiration of the unvoiced consonant and to the onset of the voiced vowel. Thus, the N1/P2 complex lends itself to the study of training-related discrimination improvements in this type of stimulus. A review of several studies observing altered physiology after directed speech-sound training follows.

2.1 N1/P2 Response Plasticity

Two pre-voiced /ba/ variants, with voice-onset times of –20 and –10 msec, were perceived as /ba/ by ten monolingual English speaking young adults. Subjects were successfully trained, over the course of nine days, to identify the two syllables as /mba/ and /ba/, respectively. In order to determine whether the N1/P2 complex demonstrated changes corresponding to the newly learned temporal cue, cortical responses were acquired to the two /ba/ variants before and after the training regimen. The N1/P2 complex, in response to both stimuli, increased in amplitude following training. Moreover, there appeared to be a relationship between the magnitude of the amplitude increase and the extent of improvement demonstrated behaviourally (Tremblay *et al.*, 2001).

2.2 Mismatch Response Plasticity

A series of mismatch response plasticity investigations was undertaken. First, to establish whether attention-independent neural encoding of sound structure is mutable, thirteen adult subjects underwent speech-sound discrimination training (Kraus *et al.*, 1995). Prior to training, the subjects' discrimination of two similar /da/ stimuli was at chance. Following six one-hour training sessions, the subjects' discrimination significantly improved. Using the same two /da/s to evoke a physiological response, the mismatch response increased in both duration and amplitude in nearly all subjects following the training regimen—demonstrating evidence of cortical neural plasticity.

In a similar design, this time with /ba/ variants as stimuli, a more thorough investigation of the temporal relationship between behavioural improvement and physiological change was undertaken (Tremblay *et al.*, 1998). The mismatch response was measured following each training session in ten subjects. Evidence of neural plasticity—increases in mismatch response duration and magnitude, decrease in onset latency—preceded behavioural improvement in half the subjects. In the remainder, it was concurrent; in no case did the changed behaviour precede the physiological manifestation.

Next, generalization of learning to stimuli other than those used in training was investigated (Tremblay *et al.*, 1997). Nine adult English-speaking subjects were trained to discriminate between and label a pair of pre-voiced bilabial stop consonants—a relevant cue in some languages, but not English. Mismatch responses were collected to the pair before and after nine 20-minute labelling training sessions, over five days. Also assessed behaviourally and physiologically, but not trained, was the ability to discriminate a pair of pre-voiced alveolar stop consonants sharing the same voicing time distinction as the trained bilabial pair. Both behavioural and physiological gains were observed in the untrained alveolar pair as well as the trained bilabial pair.

2.3 Laterality of Physiological Plasticity

In addition to the training-related cortical physiology changes outlined above, physiological responses demonstrate topographical reorganization. With right-ear stimulation, cortical speech-evoked P1 and N1 responses demonstrated a larger post-training increase on the right side (Tremblay and Kraus, 2002). The mismatch response, however, demonstrated a larger magnitude increase on the left (Tremblay *et al.*, 1997).

Thus, several studies have demonstrated that the neural encoding of sound structure is mutable by training. In addition, these pre-consciously recorded changes in neural encoding are evident prior to their behavioural expression and the occurrence of training generalization also is manifested physiologically. Different patterns of laterality in various physiological responses are discernible. Thus, there is considerable evidence of plasticity in the neural encoding of sound structure in auditory cortex. Does plasticity occur subcortically?

With the establishment of this foundation, we have the basis to refocus our attention, outside of the realm of normal adults, on to the more pressing concern of the language-impaired child. The twin goals of improved reading and better perception and understanding of spoken speech are paramount in the treatment of children with language-based learning disabilities. The ability to monitor the brain's response as training improves these skills will enable a better understanding of the mechanisms involved in language acquisition. In addition, because a chief component of auditory training programs is the selective enhancement of speech signals—for example lengthening of consonant-vowel formant transitions, directed studies examining the impact of such enhancements on the neural encoding of speech can be undertaken. In combination, these approaches enable us to assess both the efficacy of the program as a whole and the cue-enhancement technique in isolation.

3. COMMERCIAL AUDITORY TRAINING

A comprehensive project studying physiological response differences between normal children and children with a variety of auditory-based learning disabilities has been ongoing in our laboratory. Differences in neural responses to speech syllables have been established between children with learning problems and normal controls, and relationships between behavioural speech-sound perception abilities and auditory physiology have been demonstrated. Having established norms for school-aged children on a variety of speech-evoked cortical and subcortical responses, we have a good metric against which to evaluate the mutability arising from auditory training in children with learning problems (Kraus *et al.*, 1996; Kraus *et al.*, 1999).

A unique opportunity to investigate neural plasticity to speech sounds in such children has arisen due to the recent popularity of commercial computer-based auditory training programs (Diehl, 1999; Merzenich *et al.*, 1996; Morrison, 1998; Tallal *et al.*, 1996). Children undergoing this type of training comprise a heterogeneous population. This provides an opportunity to study neural plasticity among individuals with a wide variety of learning problems, and of practical importance, to examine physiological patterns in children for whom training may result in a variety of outcomes, thus leading to a potential objective metric presaging probable success.

Twenty-seven subjects with auditory-based learning problems were tested on a number of behavioural and physiological measures prior to and following independently administered *Earobics* (*Cognitive Concepts, Inc.,*

Evanston, IL, USA) training. *Earobics* consists of directed exercises that incorporate phoneme discrimination, auditory memory, sequencing, attention, rhyming and sound blending skills aimed at improving phonological awareness and ultimately reading. The behavioural and physiological test battery was administered to monitor changes in and relationships among these tests resulting from training. The test battery also was administered twice to fifteen non-trained controls spanning a similar time interval.

The behavioural battery included subtests of the Woodcock-Johnson Psycho-Educational Battery (Woodcock and Johnson, 1977; Woodcock and Johnson, 1989) and the Wide Range Achievement Test (Wilkinson, 1993). Listening tasks included sentence perception in noise and just noticeable differences for various consonant-vowel syllable continua in quiet and noise. The physiological battery consisted of subcortical and cortical response recording to /da/, both in quiet and noise, and mismatch responses to a near-threshold /da-ga/ pair in quiet.

Following training, in addition to improving on a number of behavioural tests associated with speech discrimination, comprehension and phonetic awareness, the experimental group demonstrated some physiological changes not seen in the controls. Responses of both cortical and subcortical origin were altered by training.

Cortical P2/N2 responses took on a more mature-looking pattern in quiet following training. Children with more mature cortical responses were most likely to improve on portions of the behavioural test battery. Furthermore, in noise, the amount of response degradation in comparison to the response in quiet, measured by inter-response correlation, was diminished (Hayes *et al.*, 2003). Figure 1 illustrates cortical responses in noise, recorded before and after training (top). Subjects whose inter-response correlations were low prior to training (left) demonstrated sharp increases in correlation values (bottom). Subjects with high inter-response correlations (right) prior to training maintained their strong response in noise following training. The cortical mismatch response was altered as well. Response topography shifted to a more left-dominant pattern following training.

The subcortical response to /da/ consists of a transient response to stimulus onset followed by a sustained frequency-following response to the harmonic aspects of the vowel (King *et al.*, 2002). Training affected the two subcortical response components differently. The transient onset was unaffected by training; both latency and amplitude of this response were stable. However, the subjects whose onset response latencies were most delayed irrespective of test session demonstrated the most post-training improvement on cortical physiology and behavioural measures of auditory perception—potentially revealing an important screening tool. The sustained response, unlike the onset, did demonstrate evidence of modification following training. Quiet-to-noise inter-response correlations increased. Viewed in conjunction with the aforementioned improved cortical inter-response correlations, a consequence of training appears to be improved neural timing in noise in both the auditory brainstem and cortex.

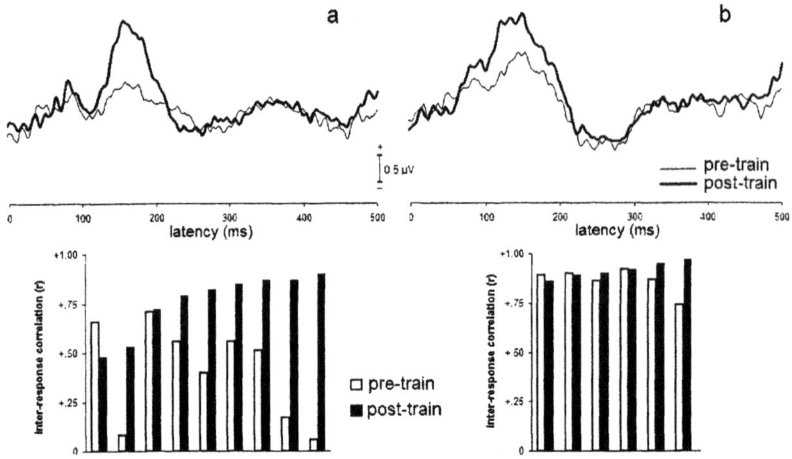

Figure 1. Grand average cortical response to /da/ in noise and individual inter-response correlations. Subjects with poor pre-train inter-response correlations (a) showed a marked morphological change (top) in their response that accompanied their increased correlations (bottom). Subjects whose inter-response correlations were high prior to training (b) demonstrated little change in response morphology and correlations.

Thus, commercial auditory training appears to have a measurable effect on speech-evoked physiology. In particular, the inter-response correlations, both in subcortical and cortical physiology, indicate a movement toward greater precision in neural population timing. Moreover, the source of the improvement appears to lie in enhanced responses to speech signals that are masked with noise.

4. STIMULUS CUE-ENHANCEMENT

In school-age children, certain cortical speech-evoked responses and speech perception tasks are especially affected by the introduction of background noise to the stimuli (Cunningham *et al.*, 2001; Wible *et al.*, 2002). For some learning-disabled children, naturally produced clear speech enhancements are sufficient to improve sentence perception in noise performance equal to normal controls (Bradlow *et al.*, 2003), and speech enhancement techniques are a critical component of auditory training programs (Merzenich *et al.*, 1996). Thus, a line of research has been pursued that investigates physiological response modifications arising from manipulation of the speech signal itself rather than response modifications arising from remediation.

In order to investigate enhanced speech on a syllable level, two cue-enhancement strategies were applied to a 40-step /ada-aga/ stimulus continuum that was characterized as "conversational." First, the stop-gap duration was increased by 80 ms, and second, the amplitude of the consonant burst was increased by 10 dB. A third continuum variant combined both enhancement strategies. The vowel following the stop was not modified. These modifications are common in naturally produced "clear" speech (Picheny *et al.*, 1986).

In nine children with auditory-based learning problems (LP) and nine normal-learning (NL) controls, just noticeable differences were established on the conversational continuum presented in both quiet and continuous background noise. The groups were equivalent in their ability to discriminate among members of the continuum in quiet, but the LP children were significantly poorer at the task than the controls when background noise (+5 dB signal-to-noise ratio) was added. Both cue enhancement strategies afforded behavioural improvement in noise. In particular, the continuum utilizing enhanced burst amplitude restored the LP subjects' performance to a level equal to the controls (Figure 2a).

A similar pattern was seen in evoked response patterns[1]. In quiet, the cortical P2/N2 response complex to the conversational stimulus was equivalent in amplitude between the two groups. However, the addition of background noise, while diminishing the response amplitude in both groups, suppressed the response to a greater degree in the LP children. Likewise, subcortical responses that were alike between groups in quiet were differentially affected upon the addition of background noise to the conversational stimulus. In noise, both transient (later in LPs), and sustained (reduced high-frequency phase locking; poorer stimulus-to-response correlations in LPs) components of the response were significantly different from the controls.

[1] Responses were recorded to the endpoint /da/ rather than /ada/. In assessing the behavioural response enhancements effected by the two cue enhancement strategies, it was found that enhancing the consonant was more effective than elongating the stop gap. Thus, it was decided to concentrate on the former for subsequent physiology recordings, enabling the removal of the leading vowel.

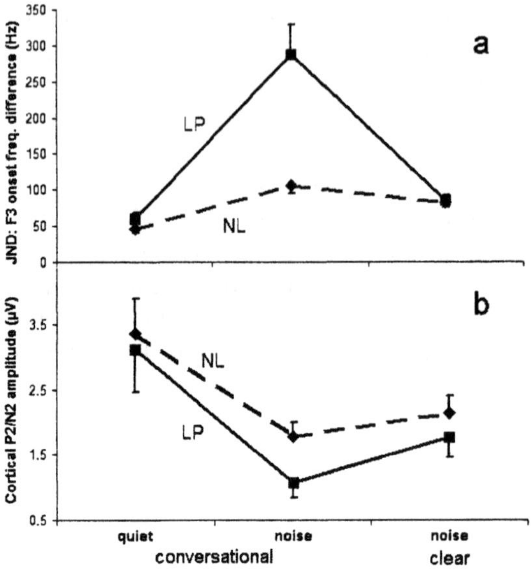

Figure 2. Behavioural discrimination scores (a) and cortical physiology (b). With conversational speech, LP subjects are more affected than normal controls by noise in both behaviour and physiology. Cue-enhanced, clear speech returns their behaviour and physiology to normal. Adapted from Cunningham *et al.* (2001).

Physiological responses to the cue-enhanced stimulus in noise were more robust for the LP subjects. Their cortical P2/N2 response amplitudes were no longer significantly different from the controls' (Figure 2b). The subcortical onset response resumed a normal latency. However, the later, sustained, component of the subcortical response did not rebound with cue enhancement. That the sustained frequency-following response was unchanged by the stimulus cue-enhancement is unsurprising. Because the vowel was unaltered acoustically by cue-enhancement, it served as a useful control; helpful in confirming that the changes seen in responses to the sound onset were due to cue-enhancement alone, and were not attributable to changes in subject state or earphone placement over time.

An accompanying study (Cunningham *et al.*, 2002), using a guinea pig model to investigate near-field response patterns in the inferior colliculus, medial geniculate nucleus and auditory cortex, demonstrated that both the disruption of background noise the benefit afforded by cue-enhancement was greatest at the cortex.

The behavioural improvement offered by stimulus cue-enhancement appears to have a direct link to more accurate neural representation of acoustic events—both cortically and subcortically.

Recent technological advances have permitted the elongation and amplification of certain speech cues to occur almost instantaneously. Consumer products are in development that process radio broadcasts into cue-enhanced speech in real time.

5. CONCLUSION

Accurate speech-sound encoding and perception depends on precise timing of neural events. A deficit in speech-sound perception may be cognitive in origin, but in many instances, imprecise afferent encoding is its source. It has been demonstrated that physiological responses signalling the proper encoding of speech are measurable, and deficits have been discerned in both the auditory cortex and in subcortical regions.

Importantly, this preconscious encoding of sound is plastic. Auditory training has long been used to improve perception. With physiological recording, it is possible to track where neural reorganization has occurred, providing us with insight into the nature of auditory system plasticity as well as an unbiased gauge of training success.

Training and stimulus cue enhancement's impact on optimising neural timing to acoustic sound structure leads to speculation about their impact on learning other types of sounds. The malleability of encoding of sound within the auditory pathway suggests approaches that could be applied more generally in other instances where improved perception of sound is desirable, such as learning music or foreign languages.

ACKNOWLEDGEMENTS

We are grateful for the scientific contributions of Erin Hayes, Bradley Wible, Jenna Cunningham, Kelly Tremblay, Cathrine Warrier and Cynthia King.

This research was supported by NIH NIDCD R01-01510.

REFERENCES

Bradlow, A. R., Kraus, N., and Hayes, E., 2003, Speaking clearly for children with learning disabilities: Sentence perception in noise. *J. Speech Lang. Hear. Res.* **46**: 80-97.

Cunningham, J., Nicol, T., King, C. D., Zecker, S. G., and Kraus, N., 2002, Effects of noise and cue enhancement on neural responses to speech in auditory midbrain, thalamus and cortex. *Hear. Res.* **169**: 97-111.

Cunningham, J., Nicol, T., Zecker, S. G., and Kraus, N., 2001, Neurobiologic responses to speech in noise in children with learning problems: Deficits and strategies for improvement. *Clin. Neurophysiol.* **112**: 758-767.

Diehl, S., 1999, Listen & Learn? A software approach review of Earobics. *Language, Speech and Hearing Services in Schools* **30**: 108-116.

Hayes, E., Warrier, C. M., Nicol, T., Zecker, S. G., and Kraus, N., 2003, Neural plasticity following auditory training in children with learning problems. *Clin. Neurophysiol.* **114**: 673-684.

King, C., Warrier, C. M., Hayes, E., and Kraus, N., 2002, Deficits in auditory brainstem encoding of speech sounds in children with learning problems. *Neurosci. Lett.* **319**: 111-115.

Kraus, N., Koch, D. B., McGee, T. J., Nicol, T. G., and Cunningham, J., 1999, Speech-sound discrimination in school-age children: psychophysical and neurophysiologic measures. *J. Speech Lang. Hear. Res.* **42**: 1042-1060.

Kraus, N., McGee, T., Carrell, T., King, C., and Tremblay, K., 1995, Central auditory system plasticity associated with speech discrimination training. *J. Cogn. Neurosci.* **7**: 27-34.

Kraus, N., McGee, T. J., Carrell, T. D., Zecker, S. G., Nicol, T. G., and Koch, D. B., 1996, Auditory neurophysiologic responses and discrimination deficits in children with learning problems. *Science* **273**: 971-973.

Merzenich, M. M., Jenkins, W. M., Johnston, P., Schreiner, C., Miller, S. L., and Tallal, P., 1996, Temporal processing deficits of language-learning impaired children ameliorated by training. *Science* **271**: 77-81.

Morrison, S., 1998, Computer applications: Earobics Pro. *Child Language Teaching and Therapy* **14**: 279-284.

Picheny, M. A., Durlach, N. I., and Braida, L. D., 1986, Speaking clearly for the hard of hearing. II: Acoustic characteristics of clear and conversational speech. *J. Speech Hear. Res.* **29**: 434-446.

Tallal, P., Miller, S. L., Bedi, G., Byma, G., Wang, X., Nagarajan, S. S., Schreiner, C., Jenkins, W. M., and Merzenich, M. M., 1996, Language comprehension in language-learning impaired children improved with acoustically modified speech. *Science* **271**: 81-84.

Tremblay, K., Kraus, N., Carrell, T. D., and McGee, T., 1997, Central auditory system plasticity: generalization to novel stimuli following listening training. *J. Acoust. Soc. Amer.* **102**: 3762-3773.

Tremblay, K., Kraus, N., and McGee, T., 1998, The time course of auditory perceptual learning: neurophysiological changes during speech-sound training. *Neuroreport* **9**: 3557-60.

Tremblay, K., Kraus, N., McGee, T. J., Ponton, C. W., and Otis, B., 2001, Central auditory plasticity: changes in the N1-P2 complex after speech-sound training. *Ear Hear.* **22**: 79-90.

Tremblay, K. L., Kraus, N., 2002, Auditory training induces asymmetrical changes in cortical neural activity. *J. Speech Lang. Hear. Res.* **45**: 564-572.

Wible, B., Nicol, T. G., and Kraus, N., 2002, Abnormal neural encoding of repeated speech stimuli in noise in children with learning problems. *Clin. Neurophysiol.* **113**: 485-494.

Wilkinson, G., 1993, *Wide Range Achievement Test-3*. Jastak Assoc., Wilmington, DE.

Woodcock, R., Johnson, M., 1977, *Woodcock-Johnson Psycho-Educational Battery. Tests of Cognitive Ability*. DLM Teaching Resources, Allen, TX.

Woodcock, R., Johnson, M., 1989, *Woodcock-Johnson Psycho-Educational Battery-Revised. Tests of Cognitive Ability*. DLM Teaching Resources, Allen, TX.

Hemispheric Processing of Prosody

VENU BALASUBRAMANIAN*, and LUDO MAX#
*Speech-Language Pathology and Audiology, Seton Hall University, South Orange, New Jersey, USA; and #Department of Communication Sciences, University of Connecticut, Storrs, Connecticut, USA

1. INTRODUCTION

Recent psycholinguistic studies of sentence comprehension support the view that listeners can successfully use prosodic cues to disambiguate ambiguous sentences (Cutler *et al.*, 1997). Prosodic cues such as intonation, pauses at phrase boundary, and increased word durations of phrase reportedly facilitate disambiguation of sentences with surface structure ambiguity (Lehiste, 1973; Lehiste *et al.*, 1976). These multiple prosodic cues of sentences are said to be perceived together to facilitate sentence structure identification (Beach, 1991). Lesion studies have mostly targeted the use of single prosodic elements such as stress, word duration and intonation in language comprehension (Blumstein and Cooper, 1974; Mills and Rollman, 1979). These studies have yielded contradictory results regarding the role of cerebral hemispheres in processing prosody (Baum and Pell, 1999; Joanette *et al.*, 1990).

In addition, natural language production and comprehension usually involves use of multiple prosodic cues. Neurolinguistic and neuropsychological studies appear to support the view that processing of complex prosodic cues would tap into the processing potentials of both right and left hemisphere (Wunderlich *et al.*, 2003). Such findings would predict that damage to either hemisphere will interfere with the comprehension of sentences with surface structure ambiguity (SSA). The present study aims at examining the patterns of responses of right hemisphere damaged (RHD), left hemisphere damaged (LHD) and normal controls in comprehending sentences with SSA (e.g., 'The woman told her baby stories'). Although the current study examines lesion data to determine the role of cerebral hemispheres in processing multiple prosodic cues, the role of neural plasticity needs to be recognized in the interpretation of the data (Karbe *et al.*, 1998; Thompson, 2000; Thulborn *et al.*, 1999).

2. METHOD

Subjects. There were three groups of subjects: a control group of 16 normal adults, a group of 12 aphasics, and a group of 11 right hemisphere damaged adults. The age range for all subjects was 50-70 years. All subjects were right-handed native speakers of English. All brain damaged subjects were at least three months post onset and had only one clinically significant cerebro vascular accident. All aphasic subjects demonstrated a mild to moderate form of aphasia as revealed by their test performance on standardized aphasia tests such as Western Aphasia Battery. There were no indications of significant loss of hearing. In addition, there were no reports of limb apraxia, hemianopsia and visual agnosia for the brain damaged patients. Brain damaged subjects were selected from among the patients attending a well known rehabilitation hospital. Control subjects were also from the same facility with no known history of neurological and/or language disorder.

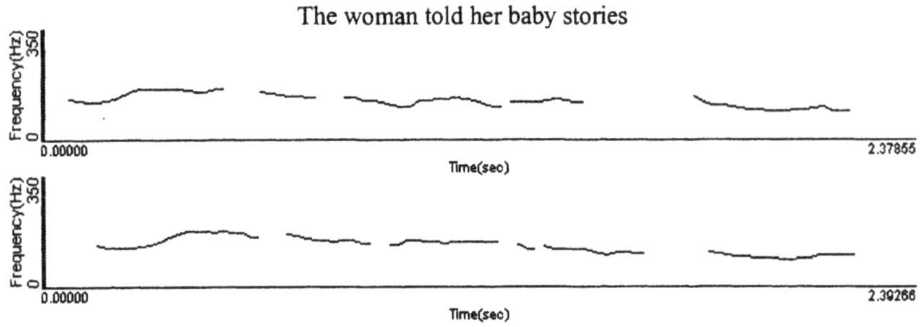

Figure 1. Fundamental frequency traces for all voiced segments in two different productions (i.e., different prosody) of a target sentence: 'The woman told her baby stories'.

Materials. The experimental stimuli included 10 sentences with surface structure ambiguity spoken by a 35-year-old male native speaker of American English (e.g., 'They fed her dog biscuits'). Since the two meanings of a sentence with surface structure ambiguity (SSA) can only be indicated by two different patterns of prosodic features, each sentence was recorded twice with appropriate prosodic cues. A total of 20 sentence stimuli, two for each sentence, were recorded (Balasubramanian, 1987). Acoustic analysis of sentence stimuli confirmed the presence of multiple prosodic cues (See Figures 1 and 2). For each SSA sentence, a set of four pictures was drawn. Of the four line drawings for a given sentence, two of them truly represented the two meanings of the sentence. The remaining two were foils.

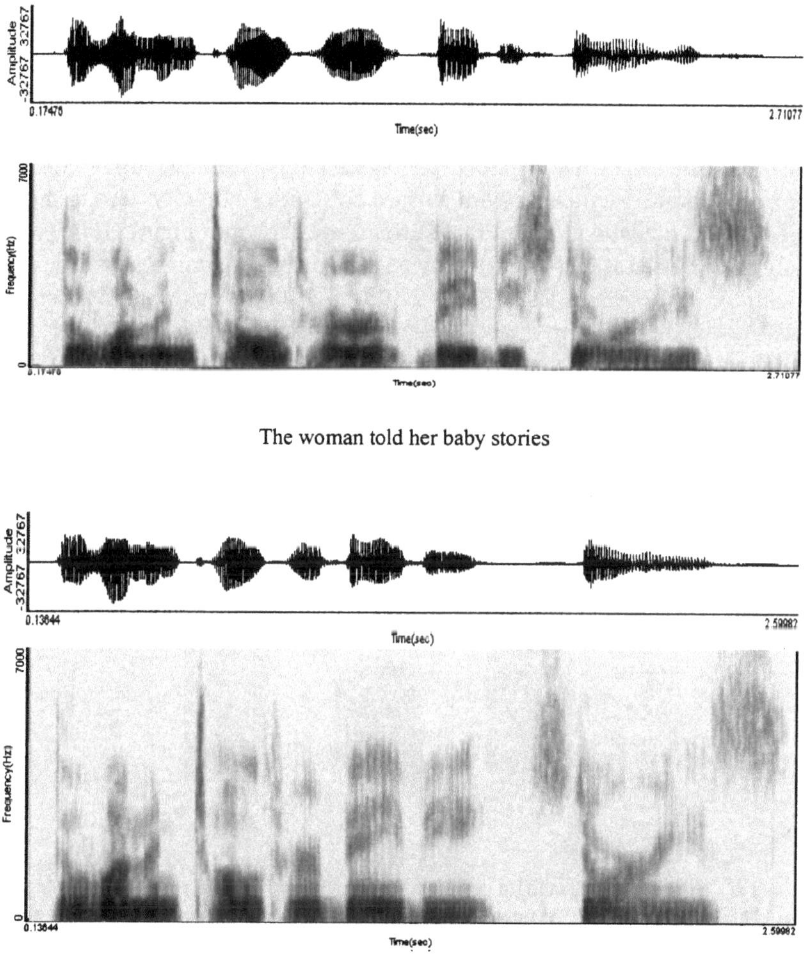

Figure 2. Spectrographic analysis of a sentence with surface structure ambiguity.

Procedures. Each sentence was auditorily presented to the subjects. The subject was to point to the appropriate picture that represented the meaning conveyed by a specific prosodic pattern. Each correct response was given one point.

Four line drawings were placed before the subject as he/she was auditorily presented with the target sentence. The subject was to point to one of the four pictures that matched the particular version of the SSA sentence that was presented. Subjects' responses were categorized into correct (1 point) or incorrect responses (0 point).

3. RESULTS

This interim report on the ongoing comprehensive study offers preliminary information pertaining to the performance of normal controls, aphasics, and right hemisphere damaged patients on a task requiring comprehension of sentences with surface structure ambiguity. Group means and standard deviations for all three groups are shown in Figure 3. One-way analysis of variance (ANOVA) showed a statistically significant difference among the three groups [$F(2, 35) = 22.05$, $p = .000$]. Tukey HSD post-hoc tests showed that performance of the right hemisphere group was statistically significantly different from that of the aphasic group ($p = .000$) but not from the normal controls ($p = .067$). Nevertheless, the group mean for the right hemisphere group was descriptively lower than that for the control group, and this comparison did approach statistical difference.

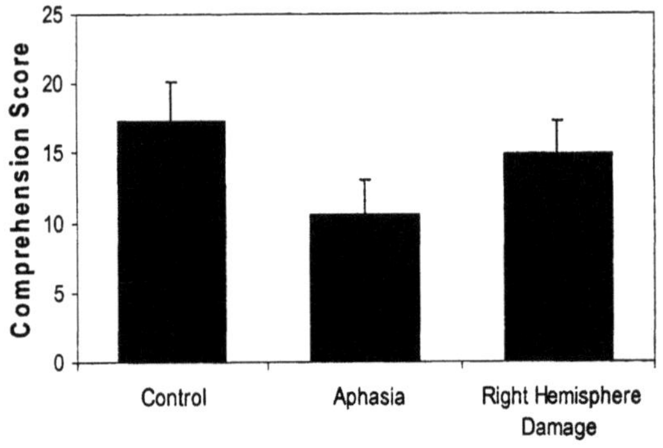

Figure 3. Group means and standard deviations for scores on the Surface Structure Ambiguity Task. Data are shown for a group of normal control adults, adults with aphasia, and adults with right hemisphere damage.

4. DISCUSSION

The findings of the present study are in agreement with that of previous studies with regard to the role of the left hemisphere in processing prosody. Although the right hemisphere damaged subjects were not statistically different from the controls, the mean was lower and the difference almost significant. Such findings appear to offer further support for the emerging views regarding the role of right hemisphere in prosodic comprehension (Baum and Pell, 1999; Wunderlich *et al.*, 2003).

Nearly, all lesion studies appear to derive the interpretation of processing deficits, or lack of such deficits, on the basis of performance patterns of left versus right brain damaged cases on prosodic tasks. What is commonly omitted is the role of plasticity of the uninjured hemisphere(s) in carrying out the experimental tasks. There is a growing body of literature that attests to the significant role played by neural plasticity in the recovery and reorganization of language in patients who had suffered stroke (Karbe *et al.*, 1998; Thompson, 2000; Thulborn *et al.*, 1999). The findings of the current study could be explained using the neural plasticity concept. The performance of aphasics, following left hemisphere damage, may reflect the limited reorganization or compensatory support offered by the right hemisphere in processing linguistic prosody. Conversely, the performance of the right hemisphere damaged subjects may reflect good compensatory reorganization of left hemisphere for prosodic processing. Thus, the significant difference in performance patterns of the two experimental groups could be explained in terms of neural plasticity. In order to gain further insight into the exact role of each hemisphere in processing prosodic structures of language, several different research paradigms (lesion method, different neuroimaging methods) will have to be used, preferably using the same subjects, and the evidence that converge from such methods might offer more dependable data on this issue.

REFERENCES

Balasubramanian, V., 1987, *Comprehension of ambiguous sentences in brain damaged patients: A study of some aspects of 'modularity theory'*. University Microfilms International, Ann Arbor, Michigan.

Baum, S. R., and Pell, M. D., 1999, The neural bases of prosody: insights from lesion studies and neuroimaging. *Aphasiology* **13**: 581-608.

Beach, C. M., 1991, The interpretation of prosodic patterns at points of syntactic structure gambiguity: evidence for cue trading relations. *J. Mem. Lang.* **30**: 644-663.

Blumstein, S., and Cooper, W. E., 1974, Hemispheric processing of intonation contours. *Cortex* **10**: 146-158.

Cutler, A., Dahan, D, and Van Donselaar., 1997, Prosody in the comprehension of Spoken language: A literature review. *Lang Speech* **40**: 141-201.

Joanette, Y. Goulet, P. and Hannequin, D., 1990, *Right hemisphere and verbal communication.* Springer-Verlag. New York.

Karbe, H., Thiel, A., Weber,-Luxenburger, G., Herholz, K., Kessler, J., and Heiss, W. D., 1998, Brain plasticity in post-stroke aphasia: What is the contribution of the right hemisphere? *Brain Lang.* **64**: 215-30.

Lehiste, J., 1973, Phonetic disambiguation of syntactic ambiguity. *Glossa* **7**: 107-122.

Lehiste, J., Oliver, J. P., and Streeter, L., 1976, Role of duration in disambiguating syntactically ambiguous sentences. *J. Acoust. Soc. Amer.* **60**: 1199-1202.

Mills, L., and Rollman, G. B., 1979, Left hemisphere selectivity for processing duration in normal subjects. *Brain Lang.* **7**: 320-335.

Perkins, J. M., Baran, J. A., and Gandour, J., 1996, Hemispheric specialization in processing intonation contours. *Aphasiology* **10**: 343-362.

Thompson, C. K., 2000, Neurobiology of language recovery in aphasia. *Brain Lang.* **71**: 245-8.

Thulborn, K. R., Carpenter, P. A., and Just, M. A., 1999, Plasticity of language-related Brain function during recovery from stroke. *Stroke* **30**: 749-54.

Wunderlich, A., Ziegler, W., and Geigenberger, A., 2003, Implicit processing of prosodic information in patiens with left and right hemisphere strokes. *Aphasiology* **17**: 861-879.

Auditory Cortex Processing Streams: Where Are They and What Do They Do?

ROBERT J. ZATORRE*, and PASCAL BELIN#
*Montreal Neurological Institute, McGill University; and #Department of Psychology, University of Montreal

1. INTRODUCTION

How is the human auditory cortex functionally organized? This basic question has received a good amount of recent attention after years of relative neglect. Inspired in part by several decades of advancement in the visual domain, auditory scientists have been emboldened to attempt to make sense of the cortical organization that underlies auditory cognition. This undertaking has been helped enormously by efforts to bring together primate neuroanatomical and neurophysiological data with human neuroimaging data; in many respects, the papers within this volume are exemplary of this progress.

A natural concomitant of progress is a healthy degree of dissent: nobody yet has the key to answering the basic question posed above, and several different answers have been proposed. The purpose of the present contribution is to summarize some of the relevant findings and to present a framework for thinking about how they might be interpreted. Rather than making assertions about the nature of this organization, in this contribution we wish to propose alternatives to some of the conventional ideas about the problem that we hope may prove useful in moving the field along.

2. TWO PATHWAYS: A PARALLEL TO VISION?

A major contribution to understanding the organization of the auditory nervous system has come from the proposition that there exist distinct processing pathways, located in segregated regions of the cortex, which carry out computationally different tasks (Kaas *et al.*, 1999; Rauschecker and Tian, 2000). Among the most often cited of these models is the one that conceptualizes two main pathways, one dealing with position in space, and the other dealing with object feature processing (Rauschecker, 1998); one reason this model has face validity is that it parallels the organization of visual cortical processing streams, which have been described in terms of the "what/where" dichotomy, such that a ventral stream is specialized for object processing while a dorsal stream is specialized for spatial processing (Ungerleider and Haxby, 1994). Another reason this model has gained support is because of advances in understanding the anatomical connectivity of auditory cortical areas, which strongly suggest a distinction into at least posterior and anterior branches (Rauschecker *et al.*, 1997), and further suggest a hierarchical and parallel arrangement such that information processing proceeds from core to belt and thence to parabelt regions (Kaas *et al.*, 1999). These pathways eventually project to distinct targets within frontal-lobe regions (Romanski *et al.*, 1999).

Although there is considerable evidence from anatomical data for the dual-pathway idea, the functional characterization of these pathways is less certain. Several investigators have adduced neurophysiological evidence favoring the idea that there is a relatively direct mapping between the functional characteristics of the two visual pathways and the two auditory pathways, with ventral and dorsal streams being responsible for object and spatial processing, respectively, in both modalities (Recanzone *et al.*, 2000; Tian *et al.*, 2001). A number of authors have also suggested that patterns of metabolic or hemodynamic change measured with functional neuroimaging techniques support this distinction (Alain *et al.*, 2001; Maeder *et al.*, 2001; Weeks *et al.*, 2000). However, there are alternative ways to think about the fundamental organizational principles involved, which have resulted in useful debates (Belin and Zatorre, 2000; Cohen and Wessinger, 1999; Middlebrooks, 2002; Romanski *et al.*, 2000).

In particular, let us deconstruct the view that a functional parallel exists between visual and auditory computations and their respective cortical specializations. Rather than think in terms of the stimuli to be processed, it may be fruitful instead to consider what computations each pathway would need to carry out. The visual "Where" pathway is concerned with the position and movement of objects in space, which in turn are signaled by the distribution of light on the sensory epithelium (retina). In the auditory system, in contrast, stimulation along the sensory epithelium (basilar membrane) does not represent spatial position, but rather relates to spectral energy distribution. In the auditory domain, of course, space is computed largely based on discrepancies in interaural time and intensity information, along with spectral filtering provided by the pinnae. Thus, if there is a similarity between fundamental neural processes in the visual and auditory domains, the computations required for auditory spatial processing might be more akin to those involved in stereoscopic vision, since in both cases a spatial dimension must be derived based mostly on differences in the stimulation arriving at two distally placed receptor surfaces. This idea has not yet been pursued by auditory neuroscientists, but it may be useful to consider how stereoscopic information is integrated with the "Where" pathway in vision, and then see if this would be a helpful model for how binaural information contributes to cortical processing of auditory space. Conversely, if we are to pursue the analogy between visual spatial computations and their corresponding auditory processes, then this would require that the auditory processes within this dorsal pathway relate to analysis of the distribution of spectral energy, rather than to binaural cues (Belin and Zatorre, 2000). In other words, if the two systems share a fundamental similarity in their neural operations, then space is to vision as spectrum is to audition.

3. POSTERIOR PATH: SENSITIVITY TO SPECTRAL CHANGES OR TO SPATIAL FEATURES?

There is, in fact, a reasonable amount of evidence beginning to accumulate from neuroimaging studies to support the view that analysis of spectral change may involve areas of belt or parabelt cortex lateral and/or posterior to primary zones. For example Thivard *et al.* (Thivard *et al.*, 2000) suggested this interpretation for their finding that synthetic diphtong stimuli, which contain formant transitions (spectral energy changes), yielded cerebral blood flow increases in posterior auditory cortex when compared to steady-state, vowel-like stimuli. Other investigators have also found similar activation patterns in posterolateral auditory cortex when comparing complex tones containing frequency modulation to constant tones (Hall *et al.*, 2002), as well as when comparing amplitude-modulated to constant tones (Hart *et al.*, 2003). Zatorre and Belin (Zatorre and Belin, 2001) studied cerebral blood flow increases elicited by pure-tone sequences which varied parametrically in the number of frequencies sampled within a one-octave range, and reported a correlation with this variable in the lateral portion of Heschl's gyrus bilaterally, with a stronger relationship on the right than the left side (See Fig 1. for illustration of these locations). A similar effect was found in a recent fMRI study in our lab (Hyde *et al.*, 2003) in which the parametric variation involved degree of frequency excursion in a simple melody-like pattern: as the size of the frequency intervals increased, BOLD signal also increased in a location posterior to Heschl's gyrus, once again with an asymmetry favoring the right side. Using isochronous melodies made with iterated noise bursts, Patterson *et al.* (Patterson *et al.*, 2002) observed a similar posterior cortical region of fMRI increase when comparing these sequences with monotone sequences, and also found a right sided asymmetry. This study, however, also reported activity increases in anterior areas in this same contrast, as did a prior study using similar materials (Griffiths *et al.*, 1998). Another recent study (Krumhansl and Zatorre, 2003) contrasted complex melodies that included both rhythmic and pitch changes to sequences that were identical in rhythmic structure to the target melodies, but consisted of only a single repeated pitch. This comparison also yielded a similar result of a BOLD signal increase in a region lateral and posterior to Heschl's gyrus, again stronger on the right than the left.

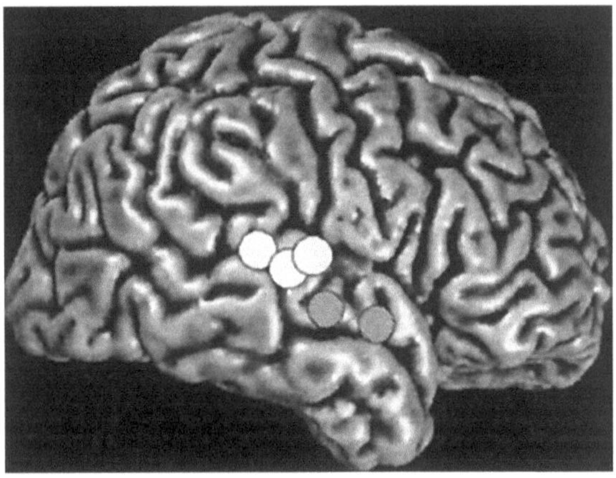

Figure 1. Summary of brain imaging results described in the paper. The location of hemodynamic peak activations in several studies, indicated by circles, are shown on a lateral view of the right hemisphere. Locations are approximate since only anterior-posterior and dorsal-ventral axes are taken into account in this projection. White circles indicate studies in which frequency modulation or change in spectral energy distribution was contrasted with steady-state stimuli; note their posterior distribution. The shaded circles indicate studies focusing on voice or object processing; they are positioned more anteriorly. See text for additional details.

These findings thus seem to converge on the idea that a lateral/posterior stream may be involved in processing changes in spectral energy distribution over time. But does this exclude the possibility that similar or adjacent regions might be involved in spatial processing? The answer is not necessarily, since multiple functions may depend on similar substrates; besides, neuroimaging evidence may not be sufficiently precise anatomically to delineate adjacent areas that are functionally distinct. Additionally, there is the possibility that more than one posterior path may exist: a more medial one sensitive to interaural features (as suggested by many of the animal studies), and a more lateral one sensitive to spectral changes.

However, the existence of posterior temporal cortical areas sensitive to auditory space is not strongly supported from neuroimaing data. A number of imaging studies have been specifically designed to seek activation in the putative posterior or dorsal stream. In most of these studies, a contrast is performed between a condition in which stimuli are perceived to come from distinct locations in space versus a control condition in which stimuli are perceived to come from a single location. What is notable is that many of these studies report no hemodynamic change in auditory cortical regions (Alain *et al.*, 2001; Bushara *et al.*, 1999; Griffiths *et al.*, 2000; Maeder *et al.*, 2001; Weeks *et al.*, 2000; Zatorre *et al.*, 1999), despite explicitly looking for such a change; instead, what most of them do observe is strong recruitment of parietal-lobe areas during active task conditions. These findings have been interpreted by some authors as evidence for a posterior "where" pathway (Alain *et al.*, 2001; Maeder *et al.*, 2001; Weeks *et al.*, 2000), but according to the anatomical evidence, such a system should engage posterior auditory cortices within the superior temporal gyrus (STG), not parietal areas. A more parsimonious explanation of the activity in parietal regions is that they relate to more general aspects of spatial processing, involving sensory motor integration (Andersen, 1995), since tasks used in these studies typically require that spatial information be used in conjunction with other sensory and motor systems in order to arrive at the correct behavior.

A direct test of this hypothesis was provided in a recent PET study from our own lab (Zatorre *et al.*, 2002), in which subjects were tested using a pseudo free-field apparatus which permits physical (as opposed to virtual) spatial presentation within the scanner. Activation of parietal areas was indeed seen, but only when subjects performed active localization tasks (such as using a joystick to locate the sound), in agreement with other studies (Bushara *et al.*, 1999); moreover, cerebral blood flow in this region predicted behavioral accuracy across subjects, suggesting a direct role for this area in sensory-motor processes (such as converting from head-centered to world-centered coordinates, for example). In contrast, no change in auditory cortical areas within the STG could be detected when comparing multiple sound locations to single locations.

Although binaural properties per se do not seem to recruit any specific auditory cortical area in the studies just cited, there are several conditions under which posterior STG areas have been identified as responsive to spatial manipulations. These include moving stimuli (Baumgart *et al.*, 1999; Pavani *et al.*, 2002; Warren *et al.*, 2002), as well as multiple distinct sounds presented simultaneously from several positions (Zatorre *et al.*, 2002, exp 2). These results would therefore seem to support the original idea of a posterior temporal pathway dedicated to spatial analysis. While this interpretation certainly remains viable, another way to conceptualize these data is that the regions recruited in fact represent a computation that takes into account the spectrotemporal analysis needed either to update auditory representations as they move in space, or to disambiguate multiple stimuli. In either case, this computation would be expected to consist of a deconvolution of the sound pattern with the head-related transfer function provided by the filtering properties of the pinnae and head. Such an argument has been made explicitly in a model described by Griffiths and Warren (Griffiths and Warren, 2002). If something like this operation is what is taking place, it would be compatible with the view that posterior regions are not sensitive to space as such, but rather to time-varying spectral changes, as proposed above.

4. ANTERIOR/VENTRAL PATH: WHAT IS AN AUDITORY OBJECT?

To return to the original analogy between visual and auditory functional pathways, we now have to address the issue of what the functional role of the putative anterior/ventral pathway would be. In the visual domain, ventral regions are thought to subserve object processing, which is operationalized to mean sensitivity to features that distinguish one object from another, such as color, form, texture, and so forth. To think about this issue in audition, we believe it is useful to make a fundamental distinction between an auditory pattern and an auditory source. The latter constitutes the carrier for the former. In order to process the information contained within a pattern, by our definition, it is necessary to compute the relationships between elements within the pattern over time, independently of the features which characterize the object producing the sound (i.e., the source). In contrast, an orthogonal computation must take place to identify the source: of relevance for source analysis would be the invariant features which are unique to the source object, independently of the particular sound pattern which the source may be producing at any given time (Fig. 2).

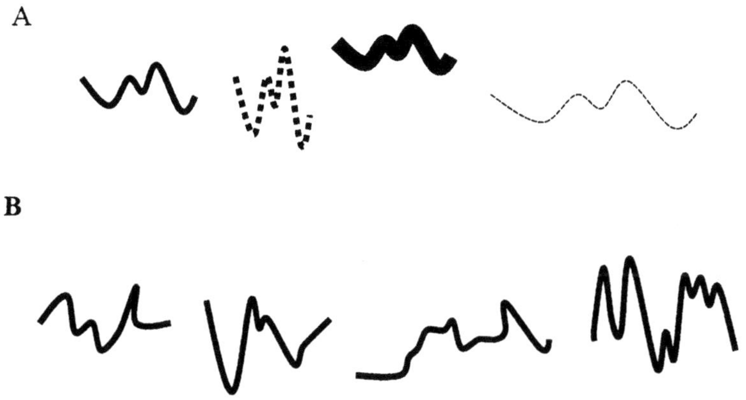

Figure 2. Schematic representation of how the difference between auditory patterns and the source producing them might be conceptualized. The drawings represent a visual analogue of a sound. In the top line (A) the physical features making up the individual items are different, but they all share the same pattern; that is, the relative relationships are constant in the proportions of the figure, so that they are all exemplars of a single pattern. The corresponding auditory example might be a melody played by various instruments, or a speech sound produced by different voices. In (B) the physical features across items are the same, but the pattern is different. That is, the items originated from the same source, but the pattern produced is different in each case. The corresponding auditory example would be a single instrument producing different melodies, or a single voice speaking different words.

The most obvious and intuitive example of this distinction is the difference between a speech utterance and a speaker's voice. Any spoken sound contains information about both what is being said and about who is saying it. But decoding the contents of the utterance requires that the sound pattern be analyzed for the spectrotemporal relationships that specify certain phonemes; this process must at some level be independent of the speaker characteristics, or else speech comprehension would fail when confronted with a novel speaker. Conversely, to identify a particular speaker, the pattern of sound being produced is not directly relevant; more important is extracting the stable features that allow one to identify who the speaker is, irrespective of what is being said. Note that this latter computation is not time-sensitive, but instead would rely on abstracting the appropriate features, such as vocal tract size, spectral shape, etc., that are necessary to be sensitive to the sound source. This is not to say, however, that the processing of speech and voice are absolutely unrelated; indeed, it is known that they affect one another to some extent (Pisoni, 1997). Thus, it is likely that functional interactions take place between these two systems at some stage.

The data reviewed earlier pertaining to processing of stimulus spectrotemporal features should thus be seen as relevant for what we are calling pattern processing. These computations would not, in our view, be carried out primarily within the putative anterior/ventral pathway sensitive to object processing. Instead, the proposed anterior pathway would be relevant for analysis of features which would help to characterize and distinguish one sound source from another. In order for this idea to be validated one needs experimental evidence, and so far the available evidence is relatively limited; but this may be because relatively few studies have been designed to seek out such evidence. One general problem is that any sound will likely contain both pattern and source information. So an experiment in which subjects are presented with auditory stimuli is likely to engage both pathways, and this is indeed often the case with both speech and nonspeech stimuli (Griffiths *et al.*, 1998; Zatorre *et al.*, 1992). Moreover, many studies have not made the distinction we propose, and have treated any auditory pattern (such as a melody or speech sound) as an "auditory object."

There are a few relevant data that would seem to support the contention that anterior and/or ventral areas of auditory cortex are sensitive to object-related features. For example Maeder and colleagues (Maeder *et al.*, 2001) reported significant anterior temporal-lobe hemodynamic signal increases in a task where subjects had to detect the presence of an animal cry amongst a background of other sounds. This finding would seem to support the idea that such anterior regions are involved in object recognition; however there is also an element of figure-ground segregation in this study which could play a role. Also, the comparison condition in this study involved localizing white noise bursts, so that the differences in activity observed could have been driven by either stimulus features or task demands.

Belin *et al.* (Belin *et al.*, 2002; Belin *et al.*, 2000) have reported that regions within the superior temporal sulcus (STS) respond specifically to human vocal sounds (speech or nonspeech), arguably because these areas contain neurons sensitive to those features of voices which distinguish them from other sound-producing sources. To the extent that the STS areas involved could be considered as a part of an auditory ventral processing stream, these findings would be consistent with the model described above. However, in these studies a number of distinct areas were identified, some of which were located in anterior and middle portions of the STS, and others in posterior STS and also STG regions, thus not conclusively supporting the anterior stream proposed here.

In order to address precisely this point, Belin & Zatorre (Belin and Zatorre, 2003) contrasted conditions relating to speech content vs speaker characteristics. The paradigm borrowed from similar studies carried out in vision (Grill-Spector and Malach, 2001), in which adaptation is thought to occur to repetitions of a particular type of stimulus feature. In the present application, Belin & Zatorre presented a sequence of consonant-vowel-consonant stimuli in two different ways: (A) either the same speaker spoke each of twelve different syllables, or (B) each of twelve speakers spoke the same syllable. Across the entire experimental run, the same set of 144 stimuli was used, but distributed differently, such that if adaptation occurs in speaker-sensitive cortical regions, one would expect to observe it as a signal increase in the contrast of conditions B vs A, above (or as a decrease in A vs B, which amounts to the same). When this analysis was performed, a region of fMRI signal change was detected within the anterior portion of the right STS, as predicted by the idea that neurons within this pathway are sensitive to the acoustic features which characterize a given speaker (Fig. 1). When those features are varied because different speakers are speaking (even though the same syllable is produced), then there is relatively more activity than in the condition where the same speaker is producing a sequence of different syllables, presumably because then feature-sensitive neurons are adapted. This finding is related to findings in other studies indicating that cortical areas in the temporal pole may be involved in discriminating individual speakers (Imaizumi *et al.*, 1997). A similar right STS region was also identified in another recent study in which listeners were asked to recognize a target voice within a sentence, as contrasted with recognizing a target word (von Kriegstein *et al.*, 2003).

The findings just discussed add a few more data points to the model under consideration, but clearly much more evidence is necessary to obtain a clearer picture of the functional characteristics of these anterior auditory cortical areas. A neuroimaging study recently carried out in our lab (Zatorre *et al.*, in press) gives additional evidence in this respect. Our aim was to provide a test of the hypothesis that anterior and/or ventral regions of temporal neocortex would be recruited to process object-related acoustic features. To do so, we created a continuum of stimuli such that at one end individual, distinct stimuli corresponding to unique sound sources are heard, while at the other end of the continuum an indistinct noise is heard. This was accomplished by summing together a variable number of separate sound sources: when many such sources are added together, an indistinct mixture is perceived that does not provide clear cues to sound-source identity; when few such sources are added together, the percept is of individual, separate sound sources. We used PET to measure cerebral blood flow as a function of the number of sound sources that were added, and sought out areas that showed significant covariation with this variable. This approach, in which the dimension of interest (distinctiveness of auditory objects) is manipulated in a graded rather than categorical fashion, has some advantages over the commonly used subtraction approach (Friston *et al.*, 1996); and it has proven useful to examine spatial (Zatorre *et al.*, 2002), spectrotemporal (Zatorre and Belin, 2001), and affective (Blood *et al.*, 1999) processes in past studies. The result of this analysis was compatible with the prediction, with a significant response found within the anterior portion of the STS (upper bank), as in Belin and Zatorre (2003), but about 15 mm more posteriorly (Fig. 1). This difference may represent domain specificity for voices, similar to the claim that certain classes of visual objects, such as faces and scenes, engage specialized regions uniquely sensitive to those particular types of stimuli (Kanwisher, 2000).

5. CONCLUSION AND QUESTIONS FOR FUTURE STUDY

Auditory cognitive neuroscience is at an exciting moment in its development. We are now at the point where a variety of complementary data from many labs is becoming available, and a variety of models is available to help us think about how to interpret these data. In the present chapter, we have concentrated on thinking about the idea of two cortical pathways for different types of processing and its ramifications for understanding the functions of these pathways. To summarize the main concepts, we think it is useful to take seriously the idea that evolution has resulted in parallel organization for cortical processing in audition and vision, insofar as both systems need to solve similar problems (e.g. navigating and manipulating the environment, identifying new information, recognizing old information, and so forth). But to do so one needs to consider the type of computations that are involved, and to take into account differences between the two domains. Although we do not have much to go on in terms of computational models so far (a big job for future auditory neuroscientists no doubt), we feel it is best to try to characterize the systems based on the operations that each processing pathway must be involved with. Based on this, we propose that if there are two pathways, the more posterior/dorsal one is involved with processing time-varying information concerning spectral changes—what we call pattern—while the more anterior/ventral one is concerned with time-invariant information concerning features of sounds that characterize individual sources.

As mentioned at the outset, these ideas are meant as much as a heuristic for future research as anything else, in that we hope they will lead to testable hypotheses, and keep auditory neuroscientists gainfully employed for some time to come. In any case, some important points remain to be made concerning the idea of processing streams. First, although we have focused on the idea of two streams because so much anatomical and physiological evidence exists to suggest it, we should not lose track of the distinct possibility that more than two pathways may exist (Kaas *et al.*, 1999). In the posterior stream there may be both a more medial and a more lateral component. In the anterior stream it is of interest to note that the data reviewed here generally fall ventrally, within the STS, leaving open the possibility that the anterior STG participates in additional functions not yet well specified. Functionally speaking the existence of several pathways would not be surprising, considering the many different types of cognitive processes that must be carried out. In this respect, a simple division into what/where, or into pattern/source, is clearly an oversimplification, because auditory cognition is about more than just identifying what an object is and where it is located. There are many subtleties involved at the highest levels of auditory processing, particularly those at which humans excel such as music and speech, which are not yet captured by any of these dichotomies.

This consideration brings us to a second point of interest concerning the processing pathways, which is that so far we do not have much information about how the two (or more) streams may interact with hemispheric specialization. We have purposely avoided this question in the present paper as being beyond our scope, but the reader is referred to other sources for our views on this topic (Zatorre *et al.*, 2002). It is clear, however, that whatever consensus emerges concerning the putative processing streams, they will have to be multiplied by two, so to speak, to account for the fact that streams within each hemisphere are likely to be functionally distinct. Indeed, there is quite a bit of work in the speech domain, which we have not covered here, that relates precisely to this point (e.g. Binder *et al.*, 2000; Narain *et al.*, 2003). It is conceivable that in the human brain (or even in other species) we may need to deal separately with each hemisphere in terms of characterizing the different processing pathways.

A final point to make is that whatever the number and function of pathways one agrees upon, these systems must interact closely. Auditory neuroscience is probably now at the stage where it is still useful to assume a relative independence of the putative streams, so that we can design and test appropriate models. This simplification, which we have tended to adhere to out of necessity, will eventually have to yield to a much more dynamic view in which interactions across streams are considered, as is being done to a greater degree now in vision.

ACKNOWLEDGMENTS

The research described in this paper was supported by funding from the Canadian Institutes of Health Research, and by the Natural Sciences and

Engineering Research Council of Canada. We thank M. Bouffard for technical assistance. For further information on the research cited in this paper, including audio examples, go to our web site: www.zlab.mcgill.ca.

REFERENCES

Alain, C., Arnott, S. R., Hevenor, S., Graham, S., and Grady, C. L., 2001, "What" and "where" in the human auditory system. *Proc Nat Acad Sci USA* **98**: 12301-12306.

Andersen, R. A., 1995, Encoding of intention and spatial location in the posterior parietal cortex. *Cereb. Cortex*, **5**: 457-469.

Baumgart, F., Gaschler-Markefski, B., Woldorff, M., Heinze, H.-J., and Scheich, H., 1999, A movement-sensitive area in auditory cortex. *Nature* **400**: 724-725.

Belin, P., and Zatorre, R. J., 2000, What, where, and how in auditory cortex. *Nature Neurosci.* **3**: 965-966.

Belin, P., and Zatorre, R. J., 2003, Adaptation to speaker's voice in right anterior temporal lobe. *NeuroReport* **14**: 2105-2109.

Belin, P., Zatorre, R. J., and Ahad, P., 2002, Human temporal-lobe response to vocal sounds. *Cog. Brain Res.* **13**: 17-26.

Belin, P., Zatorre, R. J., Lafaille, P., Ahad, P., and Pike, B., 2000, Voice-selective areas in human auditory cortex. *Nature* **403**: 309-312.

Binder, J., Frost, J., Hammeke, T., Bellgowan, P., Springer, J., Kaufman, J., and Possing, J., 2000, Human Temporal Lobe Activation by Speech and Nonspeech Sounds. *Cereb. Cortex* **10**: 512-528.

Blood, A. J., Zatorre, R. J., Bermudez, P., and Evans, A. C., 1999, Emotional responses to pleasant and unpleasant music correlate with activity in paralimbic brain regions. *Nature Neurosci.* **2**: 382-387.

Bushara, K., Weeks, R., Ishii, K., Catalan, M.-J., Tian, B., Rauschecker, J., and Hallett, M., 1999, Modality-specific frontal and parietal areas for auditory and visual spatial localization in humans. *Nature Neurosci.* **2**: 759-766.

Cohen, Y. E., and Wessinger, C. M., 1999, Who Goes There? *Neuron* **24**: 769-771.

Friston, K., Price, C., Fletcher, P., Moore, C., Frackowiak, R., and Dolan, R., 1996, The trouble with cognitive subtraction. *NeuroImage* **4**: 97-104.

Griffiths, T., and Warren, J., 2002, The planum temporale as a computational hub. *Trends Neurosci.* **25**: 348-353.

Griffiths, T. D., Büchel, C., Frackowiak, R. S. J., and Patterson, R. D., 1998, Analysis of temporal structure in sound by the human brain. *Nature Neurosci.* **1**: 422-427.

Griffiths, T. D., Green, G. G. R., Rees, A., and Rees, G., 2000, Human brain areas involved in the analysis of auditory movement. *Hum. Brain Map.* **9**: 72-80.

Grill-Spector, K., and Malach, R., 2001, fMR-adaptation: a tool for studying the functional properties of human cortical neurons. *Acta Psychol.* **107**: 293-321.

Hall, D. A., Johnsrude, I. S., Haggard, M. P., Palmer, A. R., Akeroyd, M. A., and Summerfield, A. Q., 2002, Spectral and Temporal Processing in Human Auditory Cortex. *Cereb. Cortex* **12**: 140-149.

Hart, H. C., Palmer, A. R., and Hall, D. A., 2003, Amplitude and Frequency-modulated Stimuli Activate Common Regions of Human Auditory Cortex. *Cereb. Cortex* **13**: 773-781.

Hyde, K. L., Zatorre, R.J., and Peretz, I., 2003, Evidence for the specific role of the right auditory cortex in fine pitch resolution. *NeuroImage* **19**: S62.

Imaizumi, S., Mori, K., Kiritani, S., Kawashima, R., Sugiura, M., Fukuda, H., Itoh, K., Kato, T., Nakamura, A., Hatano, K., Kojima, S., and Nakamura, K., 1997, Vocal identification of speaker and emotion activates differerent brain regions. *NeuroReport* **8**: 2809-2812.

Kaas, J. H., Hackett, T. A., and Tramo, M. J., 1999, Auditory processing in primate cerebral cortex. *Curr. Opin. Neurbiol.* **9**: 164-170.

Kanwisher, N., 2000, Domain specificity in face perception. *Nature Neurosci.* **3**: 759-763.

Krumhansl, C., and Zatorre, R.J., 2003, Rhythmic and Tonal Violations in Music: An fMRI study. *NeuroImage* **19**: S63.

Maeder, P., Meuli, R., Adriani, M., Bellmann, A., Fornari, E., Thiran, J.-P., Pittet, A., and Clarke, S., 2001, Distinct pathways involved in sound recognition and localization: a human fMRI study. *NeuroImage* **14**: 802-816.

Middlebrooks, J., 2002, Auditory space processing: here, there or everywhere? *Nature Neurosci.* **5**: 824 - 826.

Narain, C., Scott, S. K., Wise, R. J. S., Rosen, S., Leff, A., Iversen, S. D., and Matthews, P. M., 2003, Defining a left-lateralized response specific to intelligible speech using fMRI. *Cereb. Cortex* **13**: 1362-1368.

Patterson, R. D., Uppenkamp, S., Johnsrude, I. S., and Griffiths, T. D., 2002, The processing of temporal pitch and melody information in auditory cortex. *Neuron* **36**: 767-776.

Pavani, F., Macaluso, E., Warren, J., Driver, J., and Griffiths, T., 2002, A common cortical substrate activated by horizontal and vertical sound movement in the human brain. *Curr. Biol.* **12**: 1584-1590.

Pisoni, D., 1997, Some thoughts on "normalization" in speech perception. In *Talker Variability in Speech Processing (*K. Johnson and J. Mullennix, eds.), Academic Press, San Diego, pp. 9-32.

Rauschecker, J. P., 1998, Parallel processing in the auditory cortex of primates. *Audiol. Neurootol.* **3**: 86-103.

Rauschecker, J. P., and Tian, B., 2000, Mechanisms and streams for processing of "what" and "where" in auditory cortex. *Proc. Nat. Acad. Sci. USA* **97**: 11800-11806.

Rauschecker, J. P., Tian, B., Pons, T., and Mishkin, M., 1997, Serial and parallel processing in Rhesus monkey auditory cortex. *J. Comp. Neurol.* **382**: 89-103.

Recanzone, G. H., Guard, D. C., Phan, M. L., and Su, T.-I. K., 2000, Correlation between the activity of single auditory cortical neurons and sound-localization behavior in the Macaque monkey. *J. Neurophysiol.* **83**: 2723-2739.

Romanski, L., Tian, B., Fritz, J., Mishkin, M., Goldman-Rakic, P., and Rauschecker, J.P., 1999, Dual streams of auditory afferents target multiple domains in the primate prefrontal cortex. *Nature Neurosci.* **2**: 1131-1136.

Romanski, L., Tian, B., Fritz, J., Mishkin, M., Goldman-Rakic, P., and Rauschecker, J.P., 2000, Reply to "What', 'where' and 'how' in auditory cortex'. *Nature Neurosci.* **10**: 966.

Thivard, L., Belin, P., Zilbovicius, M., Poline, J., and Samson, Y., 2000, A cortical region sensitive to auditory spectral motion. *Neuroreport* **11**: 2969-2972.

Tian, B., Reser, D., Durham, A., Kustov, A., and Rauschecker, J. P., 2001, Functional Specialization in Rhesus Monkey Auditory Cortex. *Science* **292**: 290-293.

Ungerleider, L., and Haxby, J., 1994, "What" and "where" in the human brain. *Curr. Biol.* **4**: 157-165.

von Kriegstein, K., Eger, E., Kleinschmidt, A., and Giraud, A.-L., 2003, Modulation of neural responses to speech by directing attention to voices or verbal content. *Cog. Brain. Res.* **17**: 48-55.

Warren, J., Zielinski, B., Green, G., Rauschecker, J., and Griffiths, T., 2002, Analysis of sound source motion by the human brain. *Neuron* **34**: 139-148.

Weeks, R., Horwitz, B., Aziz-Sultan, A., Tian, B., Wessinger, C., Cohen, L., Hallett, M., and Rauschecker, J., 2000, A positron emission tomographic study of auditory localization in the congenitally blind. *J. Neurosci.* **20**: 2664-2672.

Zatorre, R. J., and Belin, P., 2001, Spectral and temporal processing in human auditory cortex. *Cereb. Cortex* **11**: 946-953.

Zatorre, R. J., Belin, P., and Penhune, V. B., 2002, Structure and function of auditory cortex: music and speech. *Trends Cog. Sci.* **6**: 37-46.

Zatorre, R. J., Bouffard, M., Ahad, P., and Belin, P., 2002, Where is 'where' in the human auditory cortex? *Nature Neurosci.* **5**: 905 - 909.

Zatorre, R. J., Evans, A. C., Meyer, E., and Gjedde, A., 1992, Lateralization of phonetic and pitch processing in speech perception. *Science* **256**: 846-849.

Zatorre, R. J., Mondor, T. A., and Evans, A. C., 1999, Functional activation of right parietal and frontal cortex during auditory attention to space and frequency. *NeuroImage* **10**: 544-554.

Zatorre, R. J., Bouffard, M., and Belin, P., in press, Sensitivity to auditory object features in human temporal neocortex. *J Neurosci.* **24**: 3637-3642.

Congenital Amusia: Impaired Musical Pitch But Intact Musical Time

KRISTA L. HYDE, and ISABELLE PERETZ
Department of Psychology, University of Montreal, Montreal, Quebec, Canada

1. INTRODUCTION

The existence of a developmental disorder of music, colloquially known as 'tone-deafness', has been entertained for over one century (Grant-Allen, 1878; Geschwind, 1984). However, it is only recently that a large research effort has been made to study the origins of such a disorder, currently termed 'congenital amusia' (Ayotte *et al.*, 2002; Peretz *et al.*, 2002; Hyde and Peretz, 2003; Peretz and Hyde, 2003). Afflicted individuals have a severe, life-long impairment in both the perception and production of music, which cannot be explained by obvious sensory or brain anomalies, low intelligence or lack of environmental stimulation to music. The disorder appears to be limited to the musical domain, and spares language. Congenital amusics process and recognize speech, including speech prosody, common environmental sounds and human voices as well as control subjects (Ayotte *et al.*, 2002).

Collective findings show that congenital amusics have a deficit that affects the processing of musical pitch, while musical time perception seems much less affected (Ayotte *et al.*, 2002). Most notably, afflicted individuals have a severe problem in the detection of anomalous pitches inserted in popular melodies (Ayotte *et al.*, 2002; Kalmus and Fry, 1980). The pitch anomalies consist of tones that do not belong to the key in which the melody has been written and hence sound incongruous in the melodic context, for listeners of Western music. Even young listeners with no formal musical training are quite sensitive to these key violations (Trainor and Trehub, 1994). Apparently, amusic individuals have failed to acquire this implicit knowledge about the pitch structure of the surrounding music.

As mentioned, the deficit seems to be specific to the pitch dimension of music since discrimination of musical rhythms has been shown to be intact in about half of the congenital amusic individuals studied to date (Peretz *et al.*, 2003). However, this relative sparing of time processing has not yet been assessed in music perception. Prior studies have involved a "same-different" discrimination task that is quite demanding in terms of memory (Ayotte *et al.*, 2002). Therefore, the goal of the present study was to assess the selectivity of the musical pitch-based disorder in congenital amusia, by using perceptual tasks that involve the detection of a musical anomaly either in time or in pitch, while maintaining a constant musical context.

Table1: Characteristics of participants, mean percentages of correct responses on subtests of the Montreal Battery of Evaluation of Amusia (MBEA), and significance levels on corresponding *t* tests. Standard deviations are in parentheses.

Characteristics	Amusic	Control	t- test
Age (yrs)	(n=10)	(n=10)	
Gender	58 (6.3)	58 (6.4)	n.s.
Education (yrs)	3 M, 7 F	3 M, 7 F	
	16 (1.5)	16 (1.3)	n.s.
Musical discrimination			
Melodic (MBEA)	61.0 (6.8)	91.4 (9.4)	$P < .001$
Rhythmic (MBEA)	74.3 (13.2)	92.1 (8.2)	$P < .05$

2. METHOD

Participants. The amusic group consisted of 10 adults, 7 who had also participated in previous studies (Ayotte *et al.*, 2002; Hyde and Peretz, 2003). Each amusic individual was matched in age, gender, education, and handedness to a normal control participant with no formal musical training (see Table 1). None of the participants had any previous neurological or psychiatric history.

Upon objective testing with a screening battery of musical tests (Montreal Battery of Evaluation of Amusia - MBEA) (Ayotte *et al.*, 2002; Hyde and Peretz, 2003), all amusic participants scored significantly below matched controls with global composite scores (53.3-72.3%) lying below two standard deviations of the normal mean (89.5% SD 5.6). The musical battery is comprised of six tests, four of which are presented in Table 1. Three melodic tasks assess the ability to discriminate changes in melody (by pitch contour, scale, and interval size), while the rhythmic task tests rhythmic discrimination (by temporal grouping). These tests employ a "same-different" discrimination task, with the same set of novel tonal melodies. Examination of the individual data indicates that each amusic participant was impaired in the melodic tests, while about half of them showed normal performance in the rhythmic test.

Stimuli, apparata and procedure. The musical stimuli were derived from the MBEA and were slightly modified for the present study. There were 14 original melodies, all constructed according to Western tonal music conventions, and in a major mode; they consisted of on average 9.6 (1.4 SD) successive tones, and were computer-generated in a piano timbre at a tempo of 120 beats/min. Each original melody was modified so that the same critical tone was altered either in terms of pitch or time (see Figure 1). The critical tone always fell on the first downbeat in the third bar of the four-bar melody (hence, was metrically stressed) and was 500 ms long. The pitch change consisted of a mistuning by 50 cents (100 cents corresponds to 1 semitone, which is the smallest pitch distance between adjacent keys on a keyboard) of the nearest chromatic tone, hence introducing a "sour" note. The time change consisted of introducing a silence of 5/7 of the beat duration (i.e., 143 ms) directly preceding the critical tone, thereby locally disrupting the meter (i.e., regularity).

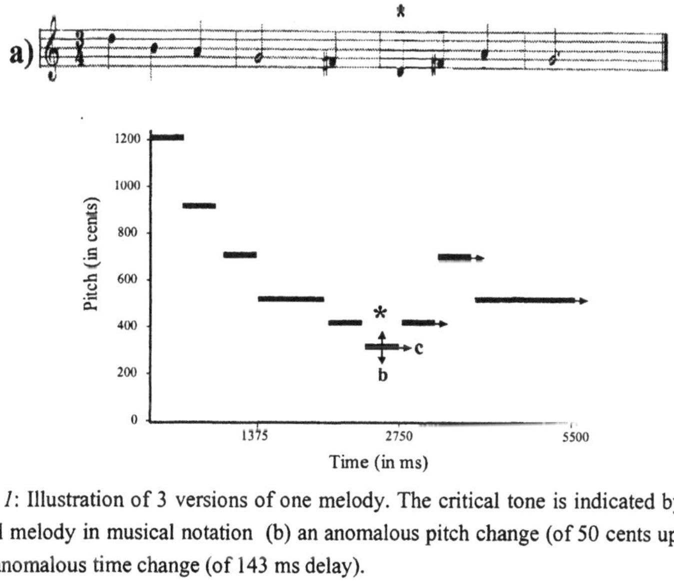

Figure 1: Illustration of 3 versions of one melody. The critical tone is indicated by *. (a) an original melody in musical notation (b) an anomalous pitch change (of 50 cents up or down) (c) an anomalous time change (of 143 ms delay).

In each of two blocks, subjects were presented with 56 melodies (28 original, 14 anomalous pitch and 14 anomalous time melodies) one at a time, in a random order. Since the participants were tested in a 1.5 Tesla magnetic resonance scanner, each melody was presented in silence but was followed by a rapid scanning noise of 2 seconds. Their task was simply to detect whether an incongruity occurred in each melody, by way of a button press: "yes " button whenever they detected a change, and a " no " button when they did not detect a change. Subjects first received 20 practice trials and were provided with feedback after each practice trial. The stimuli were presented bilaterally through headphones (Koss Electronics), at an intensity level of 75 dB SPL.

3. RESULTS

The percentage of hits (correct detection of an incongruity) minus false alarms (detection of an incongruity when there was none) was computed for each participant as a function of the type of change, whether pitch or time. The mean percentages obtained by each group (amusic and control) in each condition (pitch and time) are presented in Figure 2 along with the individual scores represented by dots (the negative values indicate a greater number of false alarms than of hits). As can be seen, the pitch and time conditions were matched in terms of difficulty, with 89.9 and 89.5 % correct in normal controls with $t(18)=.10$, n.s. Yet, amusics failed to detect the pitch changes with scores being close to chance level of 50% ($t(9)=1.979$, n.s.).

The scores were submitted to an ANOVA with Type of change (pitch versus time) as the within-subjects factor and Group (amusic versus control) as the between-subjects factor. A highly significant interaction between these factors was obtained, with $F(1,18)=303.75$, $P<.001$. This effect reflects the fact the amusic group was impaired in the pitch condition only, with $t(18)= 23.93$, $P < .001$, relative to the control group. In contrast, amusics matched control performance in the time condition with $t(18)= .37$, *n.s.*

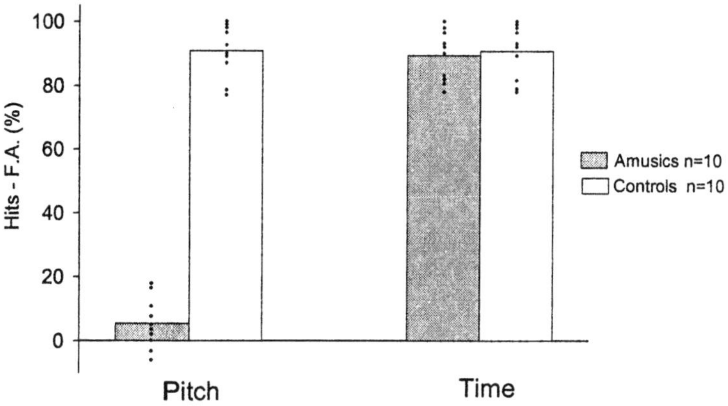

Figure 2. Mean percentage of hits minus false alarms obtained for 10 amusic and 10 control subjects in the anomalous pitch and time detection conditions. Each dot represents an individual score calculated over 28 trials.

4. DISCUSSION

The results confirm that amusic individuals have severe difficulty in detecting an anomalous pitch in conventional melodies (Ayotte *et al.*, 2002; Kalmus and Fry, 1980). There was no overlap between individual amusic and control performance on this task. In contrast, amusics have no problem in detecting a time anomaly in the same musical context. Thus, the poor performance in the pitch condition cannot be attributed to lack of task understanding, nor to any general auditory problems. Rather, the findings provide strong support for the notion that the amusic disorder is essentially a pitch-based defect.

Presently, we can only offer functional explanations, derived from the present behavioural studies. Although we construe that congenital amusia results from a slight disruption in the wiring of the auditory cortex (Peretz and Hyde, 2003), we are presently unable to support this claim since this work is still in progress. In this endeavour, specifying the functional origin of congenital amusia is essential because it narrows down the possible neural loci to a sizeable set of circuitries that can be further inspected for the presence of an anomaly.

ACKNOWLEDGEMENTS

This work was supported by funding from the Canadian Institutes of Health Research.

REFERENCES

Ayotte, J., Peretz I., Hyde K., 2002, Congenital amusia: A group study of adults afflicted with a music-specific disorder. *Brain* **125**: 238-251.

Geschwind, N., 1984, The brain of a learning-disabled individual. *Annals of Dyslexia* 34, 319-327.

Grant-Allen., 1878, Note-deafness. *Mind* **10**: 157-167.

Hyde, K. L., and Peretz, I., 2004, Brains that are out-of-tune but in-time, *Psychol. Sci.* **5**: 356-360.

Kalmus, H., and Fry, D. B., 1980, On tune deafness (dysmelodia) : Frequency, development, genetics and musical background. *Ann. Hum. Genet.* **43**: 369-382.

Peretz, I. *et al.* (2002) Congenital amusia : A disorder of fine-grained pitch discrimination. *Neuron* **33**: 185-191.

Peretz I., and Hyde, K. L. (2003) What is specific to music processing? Insights from congenital amusia. *Trends Cogn. Sci.* **7**: 362-367.

Peretz, I., Champod, A-S., and Hyde, K. L., 2003, Varieties of musical disorders. The Montreal Battery of Evaluation of Amusia. *Ann. N. Y. Acad. Sci.* **999**: 58-75.

Trainor L. J., and Trehub, S. E., 1994, Key membership and implied harmony in Western tonal music: developmental perspectives. *Percept. Psychophys.* **56**: 125-32.

Time-Courses of 40 Hz Steady-State Responses Reveal Temporal Processing in the Central Auditory System

BERNHARD ROSS, and CHRISTO PANTEV
Institute for Biomagnetism and Biosignalanalysis, Münster University Hospital, Kardinal-von-Galen-Ring 10, D-48149 Münster, Germany

1. INTRODUCTION

The increased performance in auditory detection or discrimination with increased stimulus duration suggests temporal integration with time constants of several 100 ms (Gerken *et al.*, 1990; Green *et al.*, 1957; Plomp and Bouman, 1959). On the other hand, just noticeable duration (JND) of gaps in noise were modeled with integration time constants on the order of 3 ms (Buus and Florentine, 1985). Viemeister and Wakefield (1991) resolved the paradox of widely different time constants with the model of "multiple looks". They hypothesized sampling of the acoustical signal at high rate and sequential storage. The temporal integration is realized through combination of these samples by parallel access without the need of a long integration time.

White and Plack (1998) investigated the effect of temporal integration for the detection of the pitch of unresolved harmonics and postulated a reset mechanism for the central pitch processor. The pitch processor uses a long integration time constant for continuous complex tone. However, any discontinuity on the order of 5 ms requires the pitch processor to reset.

In a recent study we observed that the auditory steady-state response (ASSR) developed over the first 200 to 300 ms after the onset of amplitude modulated (AM) tone bursts (Ross *et al.*, 2002) and interpreted this result to represent temporal integration in the auditory system. In the present study we investigated the time-course of ASSR to AM stimuli with a gap and asked, whether the temporal integration continues across the interruption.

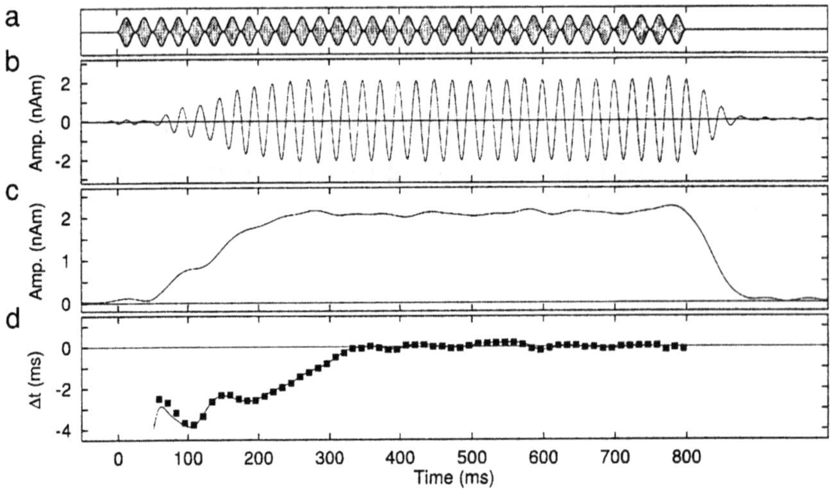

Figure 1. Time-course of the ASSR to AM burst stimulation, **a:** AM burst stimulus of 800 ms duration (500 Hz carrier, 40 Hz modulation frequency), **b:** grand averaged ASSR waveform across nine subjects, **c:** amplitude of the ASSR as a result of Hilbert-transform applied to the ASSR waveform, **d:** peak latency measured as time difference between the ASSR peaks and the maxima of the auditory stimulus. It has been adjusted to zero in the steady-state interval

2. MEG RECORDINGS

Magnetoencephalographic (MEG) recordings were performed on a group of nine healthy young volunteers. The stimuli were amplitude modulated tone bursts (500 Hz carrier, 40 Hz modulation frequency) presented to the right ear at 60 dB sensation level. The 37-channel MEG was recorded from the left hemisphere and combined into a single time-series of the dipole moment after magnetic source localization.

The characteristic time-course of the ASSR is shown in Fig.1 in conjunction with an AM burst stimulus of 800 ms duration. The onset of the ASSR showed a gentle slope and the amplitude reached its final value after about 250 ms (Fig.1c). In the same time interval the peak latency increased continuously. About 100 ms after stimulus onset the peak latency was 4 ms shorter than in the steady-state interval, which is defined by constant amplitude and latency over time and was reached 320 ms after stimulus onset. Finally, the ASSR amplitude decayed quickly within 50 ms after the end of the stimulus burst. .

When the AM stimuli were presented sequentially with a gap-like inter-stimulus interval the succeeding responses did not overlap if the gap was longer than the 50 ms decay time of the leading response. Thus, the time-courses of ASSR to AM bursts separated by 100 and 200 ms, shown in Fig.2, feature a rising slope of more than 200 ms duration and resemble the onset response shown in Fig.1. Some overlap between succeeding responses occurred at gap duration equal to one or two AM periods (25 or 50 ms). However, the amplitude of the ASSR needed still about 200 ms to reach the steady-state value. Similar recovery times of 200 ms became obvious from Fig.2a for gap durations as short as 12, 9, and 6 ms and diminishes at 3 ms.

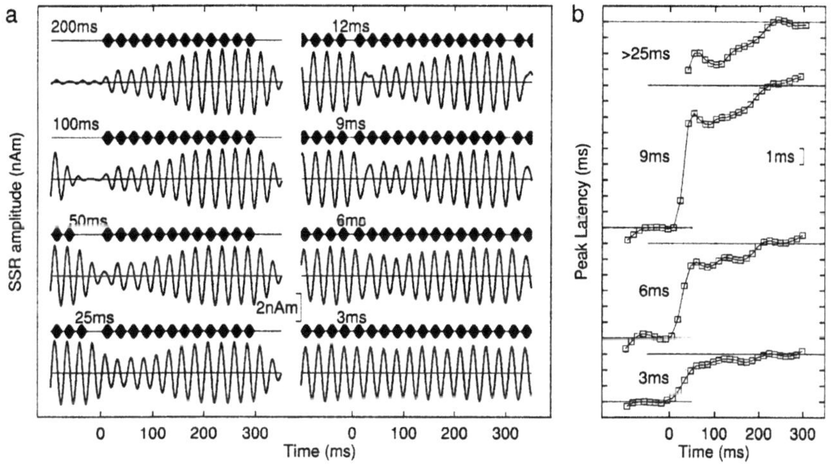

Figure 2. Time-course of the ASSR to AM bursts separated by a brief gap, a: stimulus wave forms and ASSR after omission of multiple periods of the AM stimulus (left) and after short gaps of 3, 6, 9, and 12 ms duration, **b:** time-course of the ASSR latency after a gap of 25 ms or longer duration (top) and after a gap of 9, 6, and 3 ms duration.

The time-courses of the peak latency of ASSR after a gap in the AM stimuli are shown in Fig.2b. At 25 ms gap duration the latency increased almost linearly by 3 ms in the interval between 100 and 250 ms after stimulus onset. A transient response was superimposed to the ASSR in the first 100 ms after stimulus onset, and caused the distortions visible in the latency characteristics. A gap shorter than the AM period is equivalent to a phase step in the stimulus periodicity, which is reflected in the time-course of the response latency. However, after an initial step, the response latency approximated during a 200 ms interval to its final value. The slow recovery of the steady-state phase is visible even after a short gap of 3 ms duration.

3. PSYCHO-ACOUSTICAL TESTS

The JNDs of gaps in the AM sound were obtained with an adaptive two alternative forced choice procedure and are summarized in Tab.1. The stimulus duration was 500 ms with the gap inserted in the middle. The JND was just below 3 ms for the 500/40 Hz AM stimulus, which was used for the MEG investigations. The gap detection performance increased with decreasing carrier frequencies. Thus, the JND at 250 Hz was half the JND at 500 Hz. Longer JNDs were found at higher carrier frequencies.

When investigating the effect of the modulation frequency at a constant 500 Hz carrier the JND of less than 3 ms at 40 Hz was corroborated. The gap detection performance decreased with decreasing modulation frequency. At 40 Hz the JND was about one ninth of the duration of the AM period, at 16 Hz one eighth and one sixth at 56 Hz.

When changing the duration of the leading proportion of the AM stimulus the JND was longest when the gap was inserted after only a single AM period. The performance increases with longer duration until about 250 ms and keeps constant for longer duration of the leading sound.

Table 1. Duration of a just noticeable gap inserted in amplitude-modulated tones as function of the carrier frequency, the modulation frequency and the duration of the leading marker. The error ranges denote the 95% confidence intervals for the group mean.

	Carrier Frequency (Hz)					
	250	350	500	1000	2000	4000
JND (ms)	1.32±0.16	1.43±0.30	2.74±0.10	3.65±0.30	4.09±0.20	4.16±0.25

	Modulation Frequency (Hz)				
	16	24	32	40	56
JND (ms)	7.96±0.66	5.73±0.57	3.43±0.15	2.84±0.11	2.94±0.06

	Duration of the leading AM burst (ms)									
	25	50	75	100	150	200	250	300	400	500
JND (ms)	7.70	6.80	6.15	5.42	3.95	3.28	3.12	3.03	2.90	3.01

4. CONCLUSION

The results of the psycho-acoustical tests support the hypothesis that the 40 Hz ASSR represents central processing of AM sound. The gap duration of 3 ms, which resulted in still detectable perturbations of the ASSR is just above the JND found at 500/40 Hz. Better gap detection performance at low carrier frequencies is consistent with larger 40 Hz ASSR at low carrier frequencies (Ross *et al.*, 2000). The characteristic is in contrast to gap detection in band-pass noise, which is explained mostly by the properties of the auditory periphery and for which larger JND were found, when the centre frequency was smaller. Thus, central processing of gaps in AM sound is concluded. The JNDs were found as almost constant proportion of the AM period, thus, the periodicity of the modulation is assumed being the most important cue used for gap detection. Temporal integration has been reported for central gap detection only. Obviously, the listener benefits from increasing representation of the stimulus periodicity. The time interval for temporal integration resembles the interval of developing ASSR. The time-course of ASSR after a gap in the AM stimulus resembles the time-course of the ASSR after AM burst onset. The ASSR amplitude and phase need 200 to 300 ms to reach the steady-state value. This observation of an "onset-like" response even after a very short gap was interpreted as a reset of the ASSR caused by the gap. Thus, the time-course of ASSR reveals an integrate-and-reset mechanism in central auditory processing, which may solve the apparent contradiction between temporal integration and high temporal acuity.

ACKNOWLEDGEMENTS

This work was supported by grants from the Deutsche Forschungsgemeinschaft (Grand No. Pa392/7-2).

REFERENCES

Buus, S., and Florentine, M., 1985, Gap detection in normal and impaired listeners: The effect of level and frequency. In *Time Resolution in Auditory Systems* (A. Michelsen, ed.), Springer, Berlin, pp. 159-79.

Gerken, G. M., Bhat, V. K., and Hutchison-Clutter, M., 1990, Auditory temporal integration and the power function model. *J. Acoust. Soc. Am.* **88**: 767-78.

Green, D., Birdsall, T., and Tanner, W., 1957, Signal detection as a function of signal intensity and duration. *J. Acoust. Soc. Am.* **86**: 961-70.

Plomp, R., and Bouman, A., 1959, Relation between hearing threshold and duration for tone pulses. *J. Acoust. Soc. Am.* **31**: 749-58.

Ross, B., Borgmann, C., Draganova, R., Roberts, L. E., and Pantev, C., 2000, A high-precision magnetoencephalographic study of human auditory steady- state responses to amplitude-modulated tones. *J. Acoust. Soc. Am.* **108**: 679-91.

Ross, B., Picton, T. W., and Pantev, C., 2002, Temporal integration in the human auditory cortex as represented by the development of the steady-state magnetic field. *Hear. Res.* **165**: 68-84.

Viemeister, N. F., and Wakefield, G. H., 1991, Temporal integration and multiple looks. *J. Acoust. Soc. Am.* **90**: 858-65.

White, L. J., and Plack, C. J., 1998, Temporal processing of the pitch of complex tones. *J. Acoust. Soc. Am.* **103**: 2051-63.

Auditory Cortex Role in Human Directional Hearing

MANON GRUBE[*,#], D. YVES VON CRAMON[*], and RUDOLF RÜBSAMEN[#]
[*]*Dayclinic for Cognitive Neurology, Univ. of Leipzig, Liebigstr. 22a, Leipzig, Germany; and* [#]*Faculty of Biosciences, Pharmacy and Psychology, Univ. of Leipzig, Liebigstr. 18, Leipzig, Germany*

1. INTRODUCTION

Our understanding of human auditory space perception is still limited, with not even a conclusive concept of auditory cortex (AC) role established. In animals, neurons tuned to interaural differences in lateralized signals were reported in the auditory brainstem. In AC, electrophysiology and ablation studies identified contralateral representations. Spatial information is thought to be further processed along a dorsal *'where pathway'*. In man, brainstem-mediated extraction of signal laterality was evidenced as well. Concerning spatial processing in human AC, contralateral representations are still matter of debate. Alternatively, right-hemispheric dominance in spatial processing is proposed to be established at AC level. For further processing, a dorsal *where* stream was postulated irrespectively of interhemispheric asymmetries.

2. ANIMAL MODELS

Electrophysiological animal studies on brainstem processing of signal laterality characterized the medial and lateral superior olive to be sensitive to the most salient cues for low and high-frequency signals, interaural time and intensity differences, respectively (ITDs and IIDs; Irvine, 1992, review); neurons in the inferior colliculi respond to either one or both. AC ablation caused impaired localization in the contralesional hemifield (Jenkins and

Merzenich, 1984; Heffner, 1997); contralaterality dominance was affirmed by cortical units with peak responses for the opposite hemifield. Receptive fields were mostly panoramically tuned, though, with no further topography found (Middlebrooks, 2000).

3. HUMAN SPATIAL HEARING

3.1 Brainstem-Mediated Processing

In humans, brainstem-mediated extraction of signal lateralities on the basis of ITDs and IIDs was approved by EEG investigations of auditory brainstem responses (Pratt and Polyakov, 1996; Riedel and Kollmeier, 2002). The brainstem's critical contribution in signal lateralization was confirmed by reports of specific deficits due to respective lesions (Griffiths *et al.*, 1998; Furst *et al.*, 2001; Bungert, unpublished work). It is not clear yet though, in how far ITDs and IIDs are integrated at brainstem levels.

3.2 Role of Auditory Cortex

It is not known yet, how a unified auditory spatial percept is formed and in how far the cortex relies on brainstem-mediated pre-processing. The AC lies in the medial Heschl's gyrus (HG, Fig. 1), its main input being the acoustic radiation ascending from the thalamus and transcallosal fibres from the contralateral AC (Galaburda and Sanides, 1980).

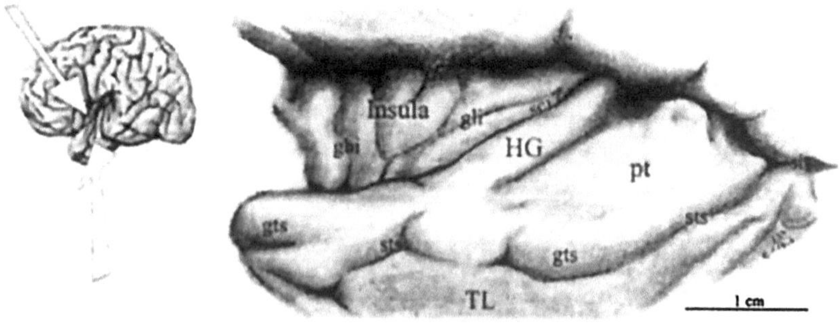

Figure 1. Heschl's gyrus (HG) of the superior temporal lobe (courtesy of M. Schönwiesner).

To date, the majority of investigations on AC relevance in spatial hearing has used headphone tasks. Contralaterality effects observed for monaural stimulus presentation to either ear (Woldorff *et al.*, 1999) were confirmed with contralesional deficits due to unilateral AC damage (Bungert *et al.*,

2004). Further supporting the contralaterality hypothesis adopted from animals, simulated lateralized sound sources used in functional studies induced more prominent activity in the opposite cortical hemisphere, albeit not necessarily in AC itself (Bushara *et al.*, 1999; Maeder *et al.*, 2001). The latter issue can be addressed in patients with acquired brain lesions. As far as headphone-based lateralization is concerned, results depended on the behavioural measure and were in part even contradictory: In some studies, AC damage did interfere with ITD and IID perception (Yamada, 1996; Bisiach *et al.*, 1984). In patients investigated in our laboratory though, interaural discrimination was inconspicuous after AC damage. Another study characterized lesions affecting both right auditory and parietal cortex to impair ITD discrimination (Tanaka *et al.*, 1999). Mapping internalized ITD-induced sound sources to extrapersonal space was reported to be deteriorated after left- and right-hemisphere lesions not necessarily including AC (Clarke *et al.*, 2002). It can not be inferred, though, whether these findings are transferable to free-field situations; to date, there have been only few patient studies on real localization. Early investigations of unilateral damage reported deficits in the contralesional hemifield (Sanchez-Longo and Forster, 1958; Efron *et al.*, 1983), respectively larger impacts of right-hemisphere damage (Klingon and Bontecou, 1966; Altman *et al.* 1979); however, lesions did not necessarily affect AC in most of these studies. Zatorre and Penhune (2001) examined pointing behaviour and same-different discrimination of click pairs presented 30° apart. Both tasks were frequently deteriorated after to right, but rarely after left temporal lesions; deficits were more severe if excisions encroached HG. In order to tackle AC relevance in free-field discrimination, we adaptively measured minimal audible angles (MAAs, 50% correct thresholds) in patients with HG lesion or deafferentation caused by medial temporal artery infarction or cerebral haemorrhage. Low (LF 250-1200 Hz) vs. high frequency (HF 2-8 kHz) noise bursts (250 ms, 37±3 dB SL) differentially addressed ITD- vs. IID-based azimuthal localization. In patients with right HG damage (four lesion cases, one deafferentation case), significantly elevated thresholds frequently occurred for all directions (-90° to +90°; Fig. 2). The group performance significantly differed from the controls ($p<0.05$) for low-frequency stimuli. For high-frequency stimuli, threshold elevations were less pronounced and less frequent ($p<0.05$ in 20%), mostly for pericentral directions. In patients with left HG damage on the contrary (six lesion cases, five deafferentation cases), the majority of MAAs fell below the control group's medians, especially in the left-hemifield ($p<0.05$ for -90°; Fig. 2). Significant elevations were rare while more frequent in the right-hemifield. The effect of lesion side affirmed right AC significance in directional hearing rarely seen in functional studies yet (Kaiser *et al.*, 2000). The fact, that performance was more severely deteriorated for low-frequency noise, supports its high relevance in ITD-based and a less critical role in IID-based processing as was pointed to in headphone experiments (Yamada, 1996). It may furthermore reflect partially independent representations as suggested in EEG studies on ITD vs. IID processing (Schröger, 1996).

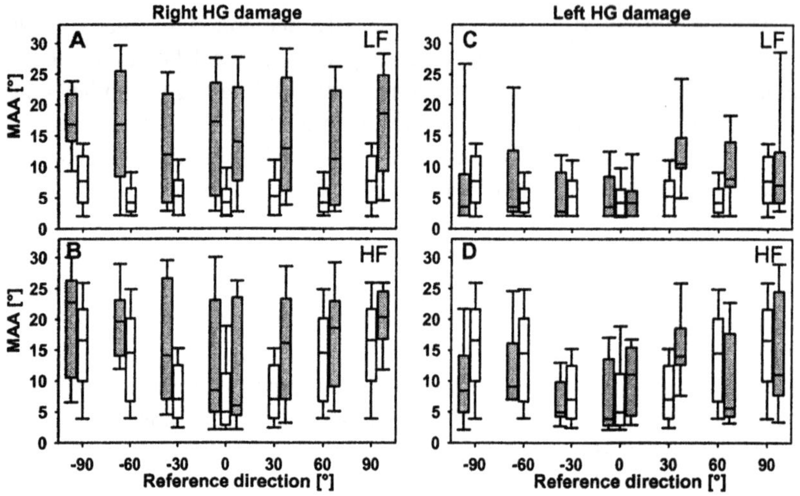

Figure 2. MAAs after right (A, B) vs. left (C, D) HG damage (grey), and in controls (white).

3.3 Higher-Order Parietal Processing

Besides the contralateral dominance hypothesis, a right-hemisphere dominance at parietal lobe levels was postulated on the basis of sound localization deficits in patients with respective lesions (Klingon and Bontecou 1966, Altman *et al.*, 1979). This notion was functionally confirmed in headphone lateralization experiments and a recent free-field PET study (Weeks *et al.*, 1999; Griffiths *et al.*, 2000; Palomäki *et al.*, 2000; Fujiki *et al.*, 2002; Zatorre *et al.*, 2002). In the context of parietal lobe functions, attempts were made to transfer the idea of anatomically distinct, bilaterally existing *what* and *where* pathways (as established in the visual system) to the auditory system in non-human primates and in humans (Tian *et al.*, 2001; Maeder *et al.*, 2001). Supporting this hypothesis, Clarke and co-workers (2002) reported partly dissociated recognition and localization deficits, the latter correlating with temporo-parieto-frontal lesions of either hemisphere.

4. CONCLUSIONS

The results presented here suggest right-hemispheric specialization in spatial processing to be established at AC level already. They further support the notion of right AC being more important in ITD- than IID-mediated lateralization and partially independent representations of the two. In the evolutionary context of the auditory system's bilateral organization, right-hemisphere specialization in spectral and spatial processing may be regarded the pendant to left-hemisphere dominance in temporal processing and speech perception (Zatorre *et al.*, 2003). To what extent this reflects AC processing or asymmetric input to higher-order areas has to be further investigated.

ACKNOWLEDGEMENTS

Our work was supported by the DFG project Cr 43-13. We further thank Katrin Krumbholz for programming help and Dem Doe for statistics advice.

REFERENCES

Altman, J. A., Balonov, L. J., and Deglin, V. L., 1979, Effects of unilateral disorder of the brain hemisphere function in man on directional hearing. *Neuropsychologia* **17**: 295-301.

Bisiach, E., Cornacchia, L., Sterzi, R., and Vallar, G., 1984, Disorders of perceived auditory lateralization after lesions of the right hemisphere. *Brain* **107**: 37-52.

Bungert, P., Biedermann, F., Dörrscheidt, G. J., von Cramon D. Y., and Rübsamen, R., 2004, Psychoacoustic test tools for the detection of deficits in central auditory processing: Normative data. *Zeitschrift für Audiologie,* in press.

Bushara, K. O., Weeks, R. A., Ishii, K., Catalan, M. J., Tian, B., Rauschecker, J. P., and Hallett, M., 1999, Modality-specific frontal and parietal areas for auditory and visual spatial localization in humans. *Nat. Neurosci.* **2**: 759-766.

Clarke, S., Bellmann Thiran, A., Maeder, P., Adriani, M., Vernet, O., Regli, L., Cuisenaire, O., and Thiran, J. P., 2002, What and where in human audition: selective deficits following focal hemispheric lesions. *Exp. Brain Res.* **147**: 8-15.

Efron, R., Crandall, P. H., Koss, B., Divenyi, P. L., and Yund, E. W., 1983, Central auditory processing. III. The "cocktail party" effect and anterior temporal lobectomy. *Brain Lang.* **19**: 254-263.

Fujiki, N., Riederer, K. A., Jousmaki, V., Makela, J. P., and Hari, R., 2002, Human cortical representation of virtual auditory space: differences between sound azimuth and elevation. *Eur. J. Neurosci.* **16**: 2207-2213.

Furst, M., Aharonson, V., Levine, R. A., Fullerton, B. C., Tadmor, R., Pratt, H., Polyakov, A., and Korczyn, A. D., 2000, Sound lateralization and interaural discrimination. Effects of brainstem infarcts and multiple sclerosis lesions. *Hear. Res.* **143**: 29-42.

Galaburda, A., and Sanides, F., 1980, Cytoarchitectonic organization of the human auditory cortex. *J. Comp. Neurol.* **190**: 597-610.

Griffiths, T. D., Elliott, C., Coulthard, A., Cartlidge, N. E., and Green, G. G., 1998, A distinct low-level mechanism for interaural timing analysis in human hearing. *Neuroreport* **9**: 3383-3386.

Griffiths, T. D., Green, G. G., Rees, A., and Rees, G., 2000, Human brain areas involved in the analysis of auditory movement. *Hum. Brain Mapp.* **9**: 72-80.

Heffner, H. E., 1997, The role of macaque auditory cortex in sound localization. *Acta Oto-Laryngol. Suppl.* **532**: 22-27.

Irvine, D., 1992, Physiology of the Auditory Brainstem. In *The Mammalian Auditory Pathway: Neurophysiology* (A. N. Popper and R. R. Fay, eds.), Springer Verlag, New York, pp. 153-231.

Jenkins, W. M., and Merzenich, M. M., 1984, Role of cat primary auditory cortex for sound-localization behavior. *J. Neurophysiol.* **52**: 819-847.

Kaiser, J., Lutzenberger, W., Preissl, H., Ackermann, H., and Birbaumer, N., 2000, Right-hemisphere dominance for the processing of sound-source lateralization. *J. Neurosci.* **20**: 6631-6639.

Klingon, G. H., and Bontecou, D. C., 1966, Localization in auditory space. *Neurology* **16**: 879-886.

Maeder, P. P., Meuli, R. A., Adriani, M., Bellmann, A., Fornari, E., Thiran, J. P., Pittet, A., Clarke, S., and Innocenti, G. M., 2001, Distinct pathways involved in sound recognition and localization: a human fMRI study. *Neuroimage* **14**: 802-816.

Middlebrooks, J. C., 2000, Cortical Representations of Audtory Space. In *The New Cognitive Neurosciences* (M. Gazzaniga, ed.), MIT Press, Cambridge.

Palomäki, K., Alku, P., Makinen, V., May, P., and Tiitinen, H., 2000, Sound localization in the human brain: neuromagnetic observations. *Neuroreport* **11**: 1535-1538.

Pratt, H., and Polyakov, A., 1996, Electrophysiological evidence of a sound localizing binaural subsystem in the human auditory brainstem. *J. Basic Clin. Physiol. Pharmacol.* **7**: 235-244.

Riedel, H., and Kollmeier, B., 2002, Auditory brain stem responses evoked by lateralized clicks: is lateralization extracted in the human brain stem? *Hear. Res.* **163**: 12-26.

Sanchez-Longo, L. P., and Forster, M., 1958, Clinical Significance of Impairment of Sound Localization. *Neurology* **8**: 119-125.

Schröger, E., 1996, Interaural time and level differences: integrated or separated processing? *Hear. Res.* **96**: 191-198.

Tanaka, H., Hachisuka, K., and Ogata, H., 1999, Sound lateralisation in patients with left or right cerebral hemispheric lesions: relation with unilateral visuospatial neglect. *J. Neurol. Neurosurg. Psychiatry* **67**: 481-486.

Tian, B., Reser, D., Durham, A., Kustov, A., and Rauschecker, J. P., 2001, Functional specialization in rhesus monkey auditory cortex. *Science* **292**: 290-293.

Weeks, R. A., Aziz-Sultan, A., Bushara, K. O., Tian, B., Wessinger, C. M., Dang, N., Rauschecker, J. P., and Hallett, M., 1999, A PET study of human auditory spatial processing. *Neurosci. Lett.* **262**: 155-158.

Woldorff, M. G., Tempelmann, C., Fell, J., Tegeler, C., Gaschler-Markefski, B., Hinrichs, H., Heinz, H. J., and Scheich, H., 1999, Lateralized auditory spatial perception and the contralaterality of cortical processing as studied with functional magnetic resonance imaging and magnetoencephalography. *Hum. Brain Mapp.* **7**: 49-66.

Yamada, K., Kaga, K., Uno, A., and Shindo, M., 1996, Sound lateralization in patients with lesions including the auditory cortex: comparison of interaural time difference (ITD) discrimination and interaural intensity difference (IID) discrimination. *Hear. Res.* **101**: 173-180.

Zatorre, R. J., Bouffard, M., Ahad, P., and Belin, P., 2002, Where is 'where' in the human auditory cortex? *Nat. Neurosci.* **5**: 905-909.

Zatorre, R. J., and Penhune, V. B., 2001, Spatial localization after excision of human auditory cortex. *J. Neurosci.* **21**: 6321-6328.

Zatorre, R. J., 2003, Sound analysis in auditory cortex. *Trends Neurosci.* **26**: 229-230.

True Auditory Lateralization Mismatch Responses Can Be Obtained by Changing the Binaural Cues Rather Than by Switching a Monaurally Presented Sound from One Ear to the Other

PEKCAN UNGAN, and SUHA YAGCIOGLU
Department of Biophysics, Hacettepe University, Faculty of Medicine, Ankara, Turkey

1. INTRODUCTION

It has long been known that an occasional deviant sound (an odball) in a sequence of standard sounds evokes in the brain an extra negative wave at a latency between 100-200 ms (Näätänen *et al.*, 1978). A negative difference potential is obtained, therefore, when the response to standards is subtracted from the response to deviants. This difference wave which is called mismatch negativity (MMN) has been related to the mechanisms of passive attention and short-term memory (Näätänen, 1995). There are several studies on the MMN evoked by infrequent changes in the lateralization of a repeated sound (Paavilainen *et al.*, 1989; Schröger and Wolff, 1996; Winkler *et al.*, 1998; Takegata *et al.*, 1999). In these studies lateralization is either changed by altering the interaural disparities dichotically or by switching the sound between two speakers placed symmetrically on two sides. We conducted the present study to see if larger and more robust auditory lateralization MMNs could be recorded by changing occasionally the ear of delivery of monaurally presented repetitive sound pips; in which case recording of this type of MMNs with higher signal-to-noise ratios should have been possible.

2. MATERIALS AND METHODS

2.1 Subjects and Stimulation

Subjects were 5 volunteers (2 females) with no audiological problem. They watched a voiceless film during the recording sessions and ignored the stimuli presented by means of a matched pair of earphones (TDH 39). The experiments were performed in a virtual acoustic environment. That is, the sounds that the ears should receive from the stereo speakers in a room (4m×4m×3m; wooden walls) were computed by means of a room acoustics software (CATT Acoustic v.7.2) with head related transfer function.

Owing to the use of virtual acoustic environment possible errors, which would arise from non-identical dynamic characteristics of the loudspeakers in a real acoustic system, could be avoided. Possible changes in sound pressure level due to head movements were also avoided in this way. The electrical waveform of stimulus to drive the virtual speakers was a 1 kHz sine wave of 100 ms duration, with exponential build-up and decay (τ=6 ms). Probability of the randomly occurring deviants was 20% and the stimuli were presented at 700 ms intervals, as suggested optimal for recording the MMN (Sinkkonen and Tervaniemi, 2000). With each subject two recording sessions were run. In one of them lateralization changes were produced by switching the sound from the right virtual speaker to the left one (BIN) and, in the other, by switching the sound from one the right earphone to the left one (MON). In the latter case monaural stimulation was achieved by cutting off the signal to the earphone contralateral to the active virtual speaker.

2.2 Recording and processing

Potentials were recorded from 19 scalp positions (10/20) and three electrodes attached to the earlobes and nose, Cz serving as reference. From each subject at least 128 single responses to deviants were collected. Record length was 640 ms, including a 100 ms prestimulus period. 0.3-70 Hz (12 dB/octave) pass-band filtered EEG was sampled at 2 ms intervals. Sweeps with EEG voltages exceeding ±70μV in any channel were excluded from averaging. Recorded potentials were converted off-line to nose reference and shifted for zeroing their average prestimulus levels after they were digitally filtered within the pass-band of 1-30 Hz. The difference waveforms were computed by subtracting the response to standards from the response to deviants. The MMN waveforms were computed by subtracting the response to the standard sounds from the response to the rare lateralization deviants.

3. RESULTS AND DISCUSSION

Grand averages of the ERPs obtained in two experiments using either binaural or monaural stimulation are displayed topographically in Fig. 1. There seems to be no apparent lateralization of the responses or fronto-parietal dominance difference in their scalp distribution. It is clearly seen that the amplitudes of the deviant responses to stimuli through the earphones (MON) are at least 4 times larger than those to stimuli through the virtual speakers (BIN), although there is no amplitude difference between the responses to standards in two experiments. Consequently, the deviant-standard difference potential for MON is much larger than that for BIN (Fig. 2). Furthermore, the latency of this difference wave for MON (mean, 102 ms) is significantly shorter than that for BIN (mean, 130 ms).

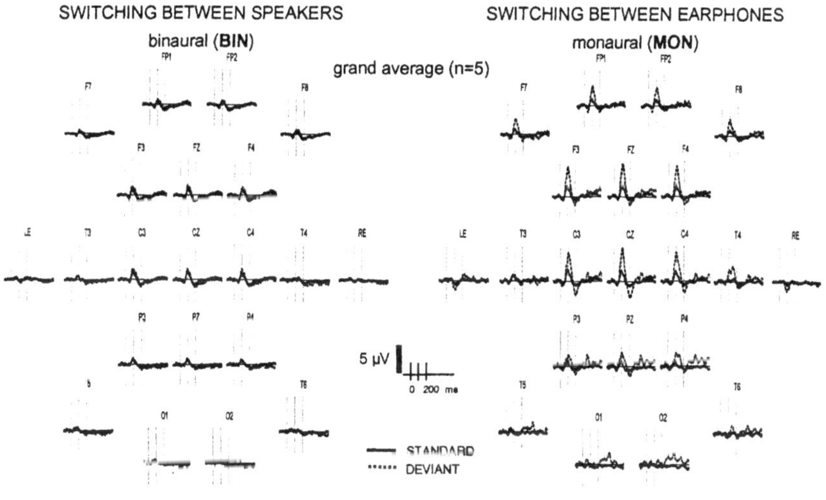

Figure 1. Nose-referenced responses to standard (solid lines) and deviant (dotted lines) stimuli delivered, in two experiments, binaurally and monaurally.

Results of the present study shows that, while typical MMN waves are evoked by infrequent changes in the lateralization of a binaural sound, the negative difference waves obtained by switching a monaural sound between ears can hardly be identified as a MMN. Firstly, its peak latency is exactly the same with that of the wave N1 (ie., 102 ms), contrary to the fact that the MMN should be recorded significantly later than wave N1. Secondly, its magnitude is too large (>4µV) to be a MMN, which has amplitudes typically in the range of 0.5-1µV for comparable stimuluation rates and deviant probabilities. Such a large amplitude cannot also be explained by a slight discriminability advantage that monaural switching might offer. We rather suspect that the difference potentials evoked by monaural lateralization changes are mainly due to refractoriness of the onset-N1 wave (Butler *et al.*, 1969; Jacobsen and Schröger, 2001) and, therefore, they cannot be evaluated as a genuine MMN. This is because the MMN is known as the response of an automatic, pre-attentive, memory-based change detection mechanism (Näätänen and Picton, 1987) and it is presumed that some fresh neurons that have remained silent to the preceding standard stimuli should not be involved in its generation (Jacobsen and Schröger, 2001). This crucial condition, which has also been discussed by several authors (eg., Näätänen *et al.*, 1993; Gomez *et al.*, 1994; Näätänen and Alho, 1995; Schröger, 1997), poses a severe issue in MMN studies and there appears in recent literature some attempts to overcome it (Schröger and Wolff, 1996; Jacobsen *et al.*, 2003).

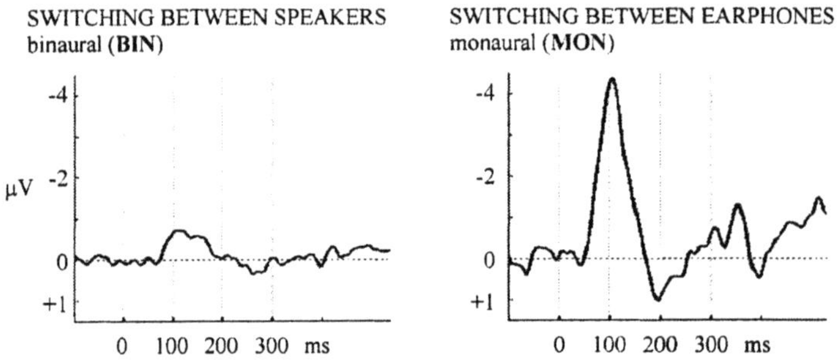

Figure 2. Deviant-standard difference potentials of the responses to auditory lateralization mismatches produced by switching the sound pips between loudspeakers (binaural) and between earphones (monaural). Nose-referenced Fz recordings.

4. CONCLUSION

Although switching a monaurally presented sound from one ear to the other evoked much larger negative difference potentials, we concluded that lateralization change in MMN studies must be made by altering the binaural cues, either by changing interaural time and/or level disparities under dichotic conditions, or by using loudspeakers. This was because our results indicated a significant role of different states of refractoriness for the responses to standard and deviant monaural stimuli, which is a clear violation of conditions for a genuine, memory comparison based MMN.

ACKNOWLEDGEMENTS

This work was supported by Hacettepe University Research Fund and by the Turkish Academy of Sciences.

REFERENCES

Butler, R. A., Spreng, M., and Keidel, W. D., 1969, Stimulus repetition rate factors which influence the auditory potential in man. *Psychophysiology* **5**: 665–672.

Gomes, H., Ritter, W., and Vaughan, H. G., Jr., 1995, The nature of preattentive storage in the auditory system. *J. Cognitive Neurosci.* **7**: 81-94.

Jacobsen, T., and Schröger, E., 2001, Is there pre-attentive memory-based comparison of pitch? *Psychophysiology* **38**: 723–727.

Jacobsen, T., Schröger, E., Horenkamp, T., and Winkler, I., 2003, Mismatch negativity to pitch change: varied stimulus proportions in controlling effects of neural refractoriness on human auditory event related brain potentials. *Neurosci. Let.* **344**: 79–82.

Näätänen, R., 1995, Mismatch negativity: a powerful tool for cognitive neuroscience. *Ear Hear.* **16**: 6-18

Näätänen, R., and Alho, K., 1995, Mismatch negativity-a unique measure of sensory processing in audition. *Intern. J. Neurosci.* **80**: 317-337.

Näätänen, R., Gaillard, A., and Mäntysalo, S., 1978, Early selective attention reinterpreted. *Acta Psychol.* **42**: 313–329.

Näätänen, R., Jiang, D., Lavikainen, J., Reinikainen, K., and Paavilainen, P., 1993, Event-related potentials reveal a memory trace for temporal features. *Neuroreport* **5**: 310-312.

Näätänen, R., and Picton, T. W., 1987, The N1 wave of the human electric and magnetic response to sound: a review and an analysis of component structure. *Psychophysiology* **24**: 375–425.

Paavilainen, P., Karlsson, M.-L., Reinikainen, K. and Näätänen, R., 1989, Mismatch negativity to change in spatial location of an auditory stimulus. *Electroenceph. Clin. Neurophysiol.* **73**: 129–141.

Schröger, E., and Wolff, C., 1996, Mismatch response of the brain to changes in sound location. *Neuroreport* **7**: 3005-3008.

Schröger, E., 1997, On the detection of auditory deviations: A pre-attentive activation model. *Psychophysiology* **34**: 245-257.

Sinkkonen, J., and Tervaniemi, M., 2000, Towards optimal recording and analysis of the mismatch negativity. *Audiol. Neurootol.* **5**: 235-246.

Takegata, R., Paavilainen, P., Näätänen, R., and Winkler, I., 1999, Independent processing of changes in auditory single features and feature conjunctions in humans as indexed by the mismatch negativity. *Neurosci. Let.* **266**: 109-112

Winkler, I., Tervaniemi, M., Schröger, E., Wolff, C., and Näätänen, R., 1998, Preattentive processing of auditory spatial information in humans. *Neurosci. Let.* **242**: 49-52.

Perception of the Direction of Frequency Sweeps in Moving Ripple Noise Stimuli

SUSAN L. DENHAM
Plymouth Institute of Theoretical and Computational Neuroscience, University of Plymouth, Drake Circus, Plymouth UK

1. INTRODUCTION

Frequency and amplitude modulations are prominent features of animal vocalisations and human speech, and play an important role in the communication of information (Hauser, 1996). In these sounds it is important to distinguish between spectrotemporal properties of the envelope and those of the fine structure. The envelope is controlled by changes in the shape of the vocal tract and the resulting spectrotemporal patterns constitute the 'information bearing elements' of speech. For this reason, the acoustic models used as inputs to automatic speech recognition systems are normally derived from smoothed spectra, for example cepstral coefficients. On the other hand, the spectrotemporal fine structure is controlled by means of tension in the vocal chords, and can be independently varied to convey prosodic information. However, given the importance of harmonicity as a grouping cue (Darwin & Carlyon, 1995), fine structure may have more to do with auditory scene analysis, than with precise information transmission.

Most experiments on the perception of frequency modulations and frequency sweeps have used pure tone stimuli (Pardo and Sams, 1993; Moore and Sek, 1998; Sek and Moore, 1999; Brechman *et al.*, 2001; Gordon and Poeppel, 2001). However, since the envelope and fine structure have the same pattern, it is not clear whether subjects are responding to the FM sweep pattern of the envelope, the fine structure or both.

Moving ripple noises (MRN) are broadband complex sounds with sinusoidal spectral profiles that drift along the frequency axis. They have been used in many physiological experiments since they can be parametrically varied to produce spectrotemporal envelope patterns that approximate those of natural sounds. The response properties of the cells can then be represented in the form of spectro-temporal response fields (Schreiner and Calhoun, 1994; Shamma *et al.*, 1995; Kowalski *et al.*, 1996; Miller *et al.*, 2002). MRNs are analogous to drifting gratings in vision, and consist of random phase component sinusoids, distributed to give equal energy per octave. The sinusoidal spectral profile, s(t,x), is defined by:

$$s(t,x) = 1 + \Delta\mathrm{A} \cdot \sin(2\pi\omega t + 2\pi\Omega x + \phi)$$
$$where: \mathrm{x} = \log_2(f / f_0)$$

Ω is the ripple density in cycles/octave, ω is the ripple velocity in cycles/second or Hz, ϕ is the ripple phase, and ΔA the modulation depth. MRNs evoke a vigorous phase-locked response in auditory cortex, often tuned to particular values of Ω and ω.

The aim of these experiments was to investigate human perception of MRN stimuli. We wanted to characterise the perception of movement in the spectral envelope and establish under what circumstances subjects could distinguish upward from downward moving spectral envelope patterns in the absence of any changes in the frequency of the carriers.

2. EXPERIMENT 1

In the first experiment the discrimination of upward and downward moving ripples was investigated for a range of ripple densities and velocities. The task was to make an up/down judgement for MRNs in which the ripple density and velocity were varied. The aim was to determine the relationship between these parameters and the difficulty of the up/down decision.

Subjects heard MRN stimuli with bandwidth 6 octaves centred on 1000 Hz, and duration 250 ms. Six ripple densities, (0.5, 1, 2, 4, 8, 16) cyc/oct, and fourteen ripple velocities, +/- (64, 16, 4, 2, 1, 0.5, 0.1) Hz, were used, giving 84 cases in all. Each was presented five times with the presentation order randomised.

The results are plotted in Figure 1. In summary, subjects were only able to perform the task reliably for ripple densities less than about 5 cyc/oct. There was some variability across subjects, and for one the upper limiting ripple density was nearer 2 cyc/oct. The influence of ripple density on performance is consistent with what is known about the spectral resolution of the auditory periphery (Glasberg and Moore, 1990), but is somewhat lower than the upper limit for spectral ripple detection of about 8 cyc/oct found by Supin *et al.* (1998). Performance was impaired at both low and high ripple velocities, with reliable discrimination only possible between +/- 0.5 and 16 Hz. The results for upward and downward moving ripples were approximately the same, and no advantage was observed for upward moving ripples. This is not consistent with some experiments which have shown a clear advantage for upward sweeps, for example (Gordon and Poeppel, 2001).

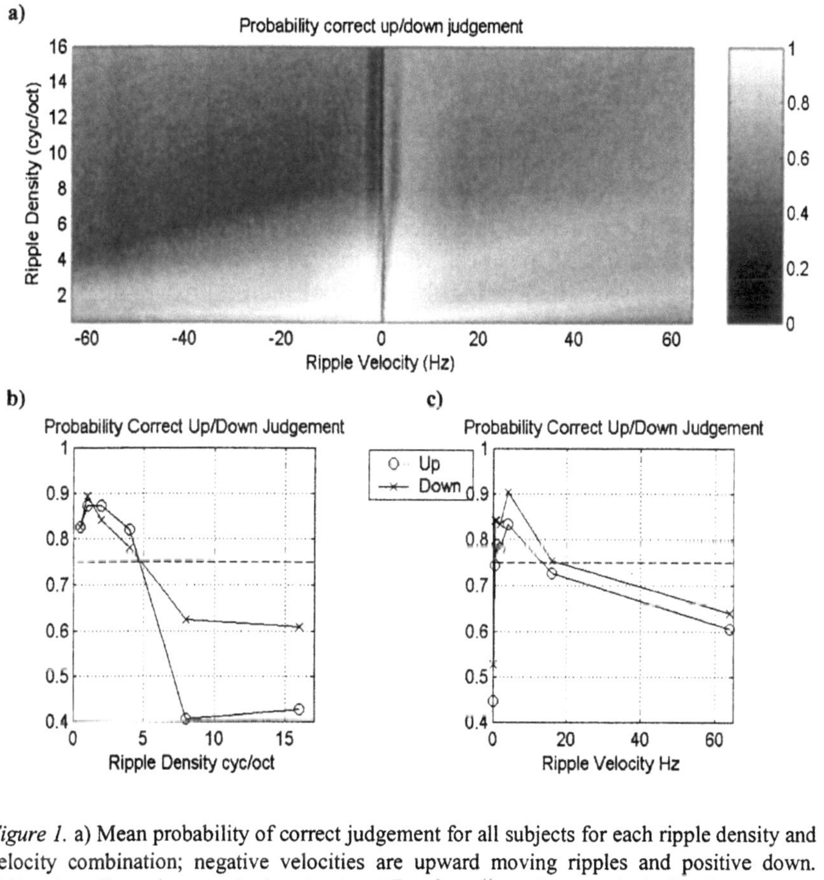

Figure 1. a) Mean probability of correct judgement for all subjects for each ripple density and velocity combination; negative velocities are upward moving ripples and positive down. b) Results collapsed across ripple velocity. c) Results collapsed across ripple density.

One possible source of confusion with other frequency sweep experiments is the way in which the stimuli are parameterised. Most experiments have used FM tone sweeps where the rate of sweep is measured in octaves/second. Ripple velocity in MRN stimuli is actually a within-channel measure of the number of peaks per second. The rate at which the ripples move up or down depends on the relationship between the ripple velocity and the density, and is given by: ω / Ω octaves/second. If the results are recalculated as a function of FM sweep rate, it is found that reliable judgements are possible for both upward and downward sweeps at rates between +/- 0.5 and 30 oct/sec, which is consistent with the range for FM sweep detection found by Gordon and Poeppel (2001).

3. EXPERIMENT 2

In MRNs there are multiple peaks moving in parallel in log frequency space. We were interested to discover whether the presence of multiple peaks was beneficial or not. In the second experiment the task was to make an up/down judgement for a single moving peak in a wide band noise. Once again the carrier frequencies remained constant. The spectral profile simply had a single Gaussian-shaped peak which moved upwards or downwards at a fixed rate in oct/sec. The starting frequency of the peak was randomised to reduce the cue offered by the position of the peak at the beginning and end. The duration for all stimuli was 250 ms, so very rapidly moving peaks did not occupy the entire duration of the stimulus.

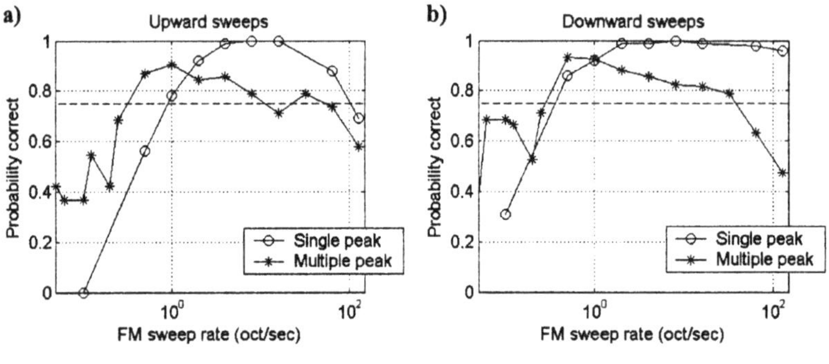

Figure 2. Comparison between responses to wideband noise stimuli containing multiple moving peaks and single moving peaks, for a) upward sweeps and b) downward sweeps. In both cases discrimination is better for low FM sweep rates if there are multiple spectral peaks, but the reverse is true for rates greater than about 1 octave/second.

4. CONCLUSION

Subjects can differentiate between upward and downward moving spectral envelope patterns within a circumscribed range of spectral profiles and drift rates. The range for accurate performance appears to be consistent with spectral resolution in the periphery and the receptive field characteristics found in auditory cortex. The presence of multiple peaks in the spectral profile helps discrimination at low sweep rates, while at higher rates discrimination is better if only one moving peak is present. In conclusion, these experiments have shown that subjects are sensitive to the spectrotemporal envelope of stimuli in the absence of meaningful fine structure, within the range of spectral peak densities and sweep rates characteristic of speech.

ACKNOWLEDGEMENTS

Supported by EU Open FET IST-2001-38099

REFERENCES

Brechman, A., Baumgart, F., and Scheich, H., 2001, Sound-level-dependent representation of frequency modulations in human auditory cortex: A low-noise fMRI study. *J. Neurophysiol.* **87**: 423-433.

Darwin, C.J., and Carlyon, R. P., 1995, Auditory grouping. In *Hearing*, (B. C. J. Moore, ed.), Academic Press.

Glasberg, B. R., and Moore, B. C. J., 1990, Derivation of auditory filter shapes from notched noise data. *Hear. Res.* **47**: 103-138.

Gordon, M., and Poeppel, D., 2001, Inequality in identification of direction of frequency change (up vs. down) for rapid frequency modulated sweeps. *ARLO* **3**: 29-34.7.

Hauser, M. D., 1996, *The Evolution of Communication*, MIT Press.

Kowalski, N., Depireux, D., and Shamma, S., 1996, Analysis of dynamic spectra in ferret primary auditory cortex, *J.Neurophysiol.* **76**: 3503-3523.

Miller, L. M., Escabi, M. A., Read, H. L., and Schreiner, C. E., 2002, Spectrotemporal receptive fields in the lemniscal auditory thalamus and cortex. *J. Neurophysiol.* **87**: 516-527.

Moore, B. C. J., and Sek, A., 1998, Discrimination of frequency glides with superimposed random glides in level. *J. Acoust. Soc. Am.* **104**: 411-421.

Pardo, P. J., and Sams, M., 1993, Human auditory cortex responses to rising versus falling glides. *Neuroscience Letters,* **159**: 43-45.

Schreiner, C. E., and Calhoun, B. M., 1994, Spectral envelope coding in cat primary auditory cortex: distribution of distributed excitation. *Aud. Neurosci.* **1**: 39-61.

Sek, A., and Moore, B. C. J., 1999, Discrimination of frequency steps linked by glides of various durations. *J. Acoust. Soc. Am.* **106**: 351-359.

Shamma, S. A., Versnel, H., and Kowalski, N., 1995, Ripple analysis in the ferret primary auditory cortex. *J. Aud. Neurosci.* **1**: 233-254.

Supin, A. Y., Popov, V. V., Milekhina, O. N., and Tarakanov, M. B., 1998, Ripple density resolution for various rippled-noise patterns. *J. Acoust. Soc. Am.* **103**: 2042-2050.

Rippled-Spectrum Resolution as a Measure of Frequency Resolving Power of Hearing

ALEXANDER Ya. SUPIN
Institute of Ecology and Evolution, Russian Academy of Sciences. 33 Leninsky Prosp., 119071 Moscow, Russia

1. INTRODUCTION

Frequency resolving power (FRP) is an important characteristic of the auditory system determining the ability to discriminate complex acoustical signals. Many functional and plastic changes in the auditory system influence FRP, therefore FRP measurements are important or necessary to monitor many manifestations of the auditory plasticity.

There are a number of methods for measurement of the auditory frequency tuning. Most of them are based on the masking technique with various masker types: tonal (frequency-tuning curves, rev. Zwicker, 1974), band-noise (critical bands, rev. Zwicker, 1982), notch-noise (Patterson, 1976; Patterson *et al.*, 1982) or comb-filtered noise (Houtgast, 1974, 1977; Pick, 1980). All these methods are based on the masking effect and provide indirect estimates of FRP.

Elaboration of reliable methods for direct frequency resolution measurements still remains a challenge. We explored this problem using the rippled noise as a representative of complex-spectrum sounds. Rippled noise spectrum contains periodically alternating peaks and valleys (Fig. 1). Similarly to many natural sounds, rippled noise has a complex spectrum pattern, but it may be quantitatively described by the ripple density (the number of ripples per a certain frequency band) and depth. The highest ripple density which is still resolvable may be taken as a measure of the frequency resolving power (FRP) of hearing.

To find out which rippled spectrum patterns are resolvable and which are not, we used a ripple phase reversal test (Supin *et al.*, 1994). Rippled noise of a certain ripple density was presented to a listener. At a certain instant it was replaced by the noise of the same intensity and ripple density and depth but of the opposite position of spectral peaks and valleys (compare solid- and thin-line spectra in Fig. 1). During this phase reversal, the listener may detect a change in the noise timbre, but only if the listener is capable to discriminate the rippled spectrum pattern (suppose, spectra of Fig. 1A). If the ripples are spaced too densely to be discriminated (suppose, spectra Fig. 1B), the switch cannot be detected because the noise before and after the switch is quite the same in all respects except for the peak and valley positions. Thus the highest ripple density at which the phase reversal is detectable can be regarded as a FRP measure.

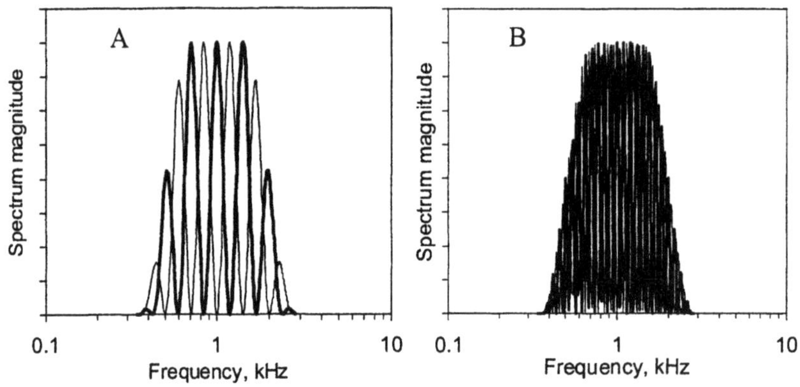

Figure 1. Examples of rippled noise spectra. A – lower ripple density, B – higher ripple density. Solid and thin lines present spectra with opposite peak-valley positions which replace one another at the phase-reversal test. Ripples are frequency-proportional (look as equally spaced on the log frequency scale).

2. METHODS

Six normal subjects aged 23 to 53 years were tested.

Sound probes were bursts of digitally generated narrow-band rippled noise of central frequency varying in half-octave steps from 0.5 to 11 kHz. Relative ripple density (ratio of the ripple central frequency to ripple frequency spacing) varied from 4 to 64 (dimensionless units) in approximately quarter-octave steps. Test noise band was restricted by an envelope which, being presented in the spectrum-power domain, was one octave (oct)-wide cycle of a cosine function of frequency logarithm. Sounds were presented diotically over earphones.

A three-interval two-alternative forced-choice procedure was used. Each trial consisted of three 4-sec noise bursts: either the first and third bursts contained the phase-reversal stimulus and the second did not contain, or the second burst contained the stimulus and the first and third did not. The task of the listener was to report either any modification of the noise appeared in the first and third bursts or in the second one. Each measurement run consisted of a total of 100 trials: five ripple densities around the anticipated resolution limit, with each ripple density randomly presented 20 times (the constant-stimuli method). The ripple density providing 75% of correct responses was adopted as the ripple-density resolution limit.

3. RESULTS AND DISCUSSION

Figure 2 combines the results obtained from all the listeners (Supin *et al.*, 1998).. The results are presented in the two measures: as absolute measure of the ripple density, i.e. the number of ripples per 1 kHz (Fig. 2A) and as relative measure, i.e. the ratio of the ripple center frequency to ripple frequency spacing (Fig.2B). At low frequencies, below 0.5 kHz, the absolute FRP approximated a level of about 15 kHz^{-1}. Accordingly, the relative FRP measure in this range was near proportional to the central probe frequency. At higher frequencies, above 1 kHz, the relative measure increased slowly, from about 12 relative units at 1 kHz to 17 units at 8 kHz.

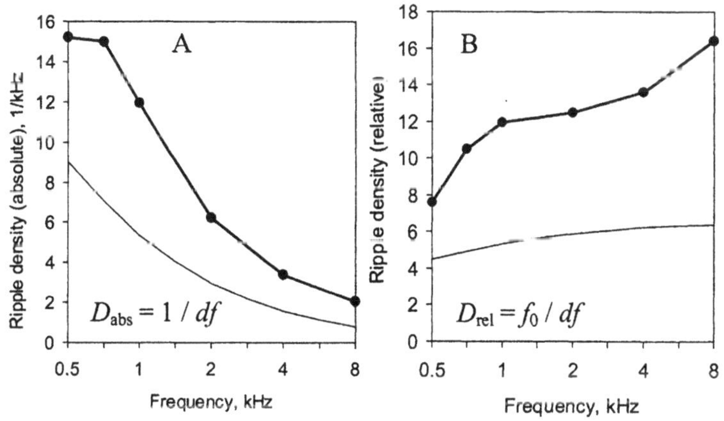

Figure 2. FRP of normal listeners as a function of center frequency of the rippled-noise band (inter-individual means). A – presented in absolute units (kHz^{-1}), B – presented in relative units. Solid line – experimental data, thin line – calculated for a bank of frequency-tuned filters with a quality according to equation of Glasberg and Moore (1990).

It is of interest to compare these data with predictions derived from the auditory filter quality. For a linear system, the transfer of any spectrum pattern through a bank of frequency filters of known quality may be easily computed. For rippled spectrum such computation resulted in FRP prediction (thin lines in Fig. 2) about twice lower than that found in direct measurements (solid line). This comparison shows that *the ability of the auditory system to discriminate complex spectrum patterns can not be predicted in a simple way basing on frequency tuned filter characteristics.*

Another case of disagreement between predictions from the auditory filter quality and the real FRP was found for sounds of varying intensity. It is known that increasing the sound level results in lowering the filter quality (Patterson *et al.*, 1982; Glasberg *et al.*, 1984; Rosen *et al.*, 1998). Lower filter quality predicts worse spectrum discrimination at high sound levels.

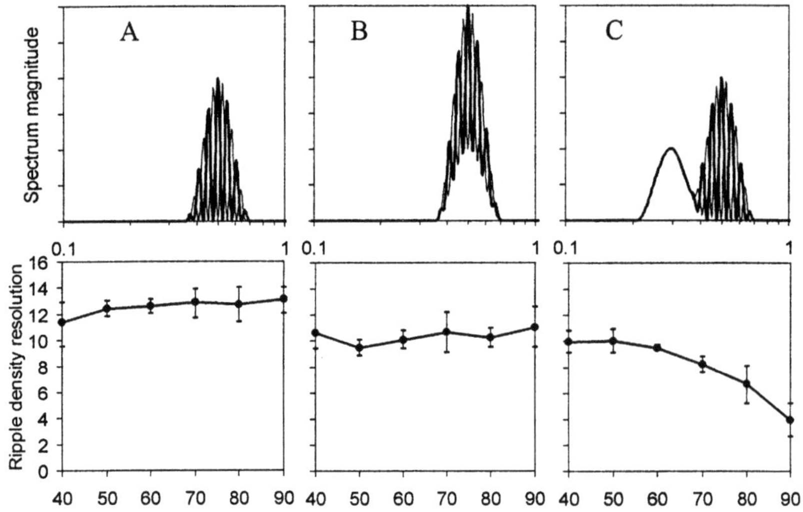

Figure 3. FRP dependence on sound level and the presence of additional noise. Upper panels – stimulus spectra (abscissa – frequency in kHz, ordinate – spectrum power in relative units), lower panels – ripple-density resolution (ordinate, relative ripple density) dependence on intensity (abscissa, dB SPL). A – no additional noise, B – additional noise band coincides with the probe band, C – additional noise band below the probe band. Solid and thin lines – spectra with alternative peak-valley position replacing one another in the phase-reversal test; note that the lines coincide in the low-frequency noise band (no change in this band during the phase reversal in the probe band).

Contrary to this prediction, direct measurements showed no FRP decrease with sound level increase if the signal contained only the test rippled spectrum and no additional noise (Fig. 3A). In a similar fashion independence of sound level was observed with an additional noise band overlapping the rippled probe spectrum (Fig. 3B): FRP values were a little less than without noise (as a result of decreased ripple depth), but almost constant throughout all the tested intensity range. However, in a quite different manner FRP was influenced by the sound level when an additional noise band was below the rippled probe band on the frequency scale (Fig. 3C). In this case, FRP dramatically decreased with sound level increase. At high sound intensities (90 dB SPL in Fig. 3B) FRP was several times less than without additional noise. Thus at high intensities the low-frequency additional noise much more effectively decreases FRP than the noise, coinciding with the probe band (Supin *et al.*, 2001) – the effect which is not predicted from frequency-tuned filter properties either.

These data demonstrate again that the real ability of the auditory system to discriminate complex spectrum patterns of sound stimuli in many cases can not be predicted by linear computation based on the known parameters of the auditory frequency-tuned filters and requires direct measurements. The ripple density resolution limit may be a reliable estimate of this ability providing a direct and quantitative FRP measure.

ACKNOWLEDGEMENTS

The study was supported by the Russian Foundation for Basic Research, Grant 02-04-48188 and the Russian Ministry of Science and Technology, Grant NSh.2152.2003.4.

REFERENCES

Glasberg, B. R., and Moore, B. C. J., 1990, Derivation of auditory filter shapes from notched-noise data. *Hear. Res.* **47**: 103-138.

Glasberg, B. R., Moore, B. C. J., Patterson, R. D., and Nimmo-Smith, I., 1984, Dynamic range and asymmetry of the auditory filter. *J. Acoust. Soc. Am.* **76**: 419-427.

Houtgast, T., 1974, Masking patterns and lateral inhibition. In *Facts and Models in Hearing* (E. Zwicker and E. Terhardt, eds.), Springer, Berlin, pp. 258-265.

Houtgast, T., 1977, Auditory-filter characteristics derived from direct-masking and pulsation-threshold data with a rippled-noise masker. *J. Acoust. Soc. Am.* **62**: 409-415.

Patterson, R. D., 1976, Auditory filter shapes derived with noise stimuli. *J. Acoust. Soc. Am.* **59**: 640-654.

Patterson, R. D., Nimmo-Smith, I., Weber, D. L., and Milory, R., 1982, The deterioration of hearing with age: Frequency selectivity, the critical ratio, the audiogram, and speech threshold. *J. Acoust. Soc. Am.* **72**: 1788-1803.

Pick, G. F., 1980, Level dependence of psychophysical frequency resolution and auditory filter shape. *J. Acoust. Soc. Am.* **68**: 1085-1095.

Rosen, S., Baker, R. J., Darling, A., 1998, Auditory filter nonlinearity at 2 kHz in normal hearing listeners. *J. Acoust. Soc. Am.* **103**: 2539-2550.

Supin, A. Ya., Popov, V. V., Milekhina, O. N., and Tarakanov, M. B., 1994, Frequency resolving power measured by rippled noise. *Hear. Res.* **78**: 31-40.

Supin, A. Ya., Popov, V. V., Milekhina, O. N., and Tarakanov, M. B., 1998, Frequency resolving power at various rippled-noise patterns. *J. Acoust. Soc. Am.* **103**: 2042-2050.

Supin, A. Ya., Popov, V. V., Milekhina, O. N., and Tarakanov, M. B., 2001, The effect of masking noise on rippled-spectrum resolution. *Hear. Res.* **151**: 157-166.

Zwicker, E., 1974, On a psychoacoustical equivalent of tuning curves. In *Facts and Models in Hearing* (E. Zwicker and E. Terhardt, eds.), Springer, Berlin, pp. 132-141.

Zwicker, E., 1982, *Psychoacustic*. Springer, Berlin.

Low Frequency Repetitive Transcranial Magnetic Stimulation (rTMS) in Brain Hyperexcitability Disorders Like Tinnitus and Auditory Hallucinations

BERTHOLD LANGGUTH*, PETER EICHHAMMER*, JOERG MARIENHAGEN**, TOBIAS KLEINJUNG#; PHILIPP SAND*, and GOERAN HAJAK*

** Department of Psychiatry and Psychotherapy; **Department of Nuclear Medicine; and #Department of Otorhinolaryngology and Audiology, University of Regensburg, Germany*

1. INTRODUCTION

Chronic tinnitus, the perception of sound or noise in the absence of acoustical stimulation, is a debilitating condition affecting some 0.5-1% of the population of the western world. At the present time there is no satisfactory treatment and no objective method for diagnosis and evaluation of treatment effects (Dobie, 1999). Like auditory hallucinations, tinnitus is considered as an auditory phantom perception, related to plastic alterations in the auditory cortex (Mühlnickel, 1998). Electrophysiologic (Kaltenbach, 2000) and functional neuroimaging studies (Arnold, 1996; Mirz, 2000) demonstrated, that chronic tinnitus is associated with excessive spontaneous activity in the central auditory system.

Repetitive transcranial magnetic stimulation (rTMS) offers a noninvasive method for altering excitability of the brain (George, 1996). Involving the production of intermittent magnetic fields in the range of 1,5 – 2 T (tesla), rTMS is powerful enough to cause neuronal depolarisation in the cortex of humans, taking advantage of the fact that magnetic fields pass largely undistorted through the scalp and scull. Electrophysiological as well as functional imaging studies have demonstrated that repetitive transcranial magnetic stimulation induces alterations of excitability in distinct cortical areas that outlast the stimulation period. Low frequency rTMS (<= 1 Hz) not only reduces cortical activity in the direct stimulated area, but can also alter cortical excitability in functionally connected areas (Gilio, 2003). It was suggested that low frequency rTMS produces effects paralleling those of long-term depression (LTD), induced by low-frequency direct electrical stimulation (Post, 1997). In rodents, changes of neuronal activity of the auditory cortex induced by one session of rTMS lasted up to 24 hours (Wang, 1996). In schizophrenic patients 1 Hz rTMS of the temporoparietal cortex, a brain region critically involved in speech perception, resulted in a reduction of auditory hallucinations (Hoffman, 2000). In this context low frequency rTMS was proposed as an efficient method for the treatment of brain hyperexcitability disorders (Hoffman and Cavus, 2002). For this reason we investigated low-frequency rTMS for the treatment of tinnitus complaints.

2. TMS FOR THE TREATMENT OF TINNITUS

Within a pilot study (Eichhammer, 2003) patients with chronic tinnitus underwent a [^{18}F]deoxyglucose- positron emission tomography (PET) and magnetic resonance imaging (MRI) measurement. Fusioning of the individual PET scan with the structural MRI-scan (T1, MPRAGE) allowed to exactly identify the area of increased metabolic activity in the auditory cortex, which was selected as the target point for rTMS. In this context, a neuronavigational system adapted for TMS positioning (Eichhammer, 2003) enabled to monitor the exact position of the figure 8-shaped magnetic coil in relation to the target area. rTMS (110% motor threshold; 1 Hz; 2000 stimuli/ day over 5 days) was performed using a placebo controlled cross-over design. Patients were blind regarding the stimulus condition. For the sham stimulation a specific sham-coil system was used. Treatment outcome was assessed with a specific tinnitus questionnaire (Goebel, 1994). In all investigated tinnitus patients an increased activity within the auditory cortex could be detected. After 5 days of verum rTMS a remarkable improvement of the tinnitus score was found in the majority of the treated patients. The other patients showed minor treatment effects or no effects at all. During sham treatment only a slight and transient reduction of tinnitus was observed. In most treatment responders the reduction of tinnitus sensation outlasted the treatment for several months.

2.1 Treatment Effects on Cortical Excitability

Arising from these results was the question whether the subjective effects of reduced tinnitus sensation were paralleled by stimulation induced changes in cortical excitability.

Until now there is no noninvasive method available to directly measure the excitability of the auditory cortex in humans. Dense anatomical and functional connections between the auditory cortex and the motor cortex (Graziano, 1999; Cacace, 2003) and the fact that rTMS of cortical areas may affect functionally connected brain areas, provide the framework for our hypothesis that rTMS effects within the primary auditory cortex (PAC) are reflected by changes in the excitability of the motor cortex. The cortical excitability of the left motor hand area was serially measured with paired-pulse TMS, a paradigm that allows to detect the functional activity of inhibitory and excitatory circuits within the motor cortex (Kujirai, 1993; Ziemann, 2000). After five days of active rTMS paired pulse TMS revealed increased intracortical facilitation, that lasted for several weeks and paralleled the subjective effects on tinnitus sensation (Langguth, 2003).

3. DOES TMS INDUCE NEUROPLASTICITY ?

Our finding of increased metabolic activity in the PAC in tinnitus patients supports the notion of a hyperexcitability of the auditory cortex in these patients and justifies the classification of tinnitus as brain hyperexcitability disorder. Low frequency rTMS applied to the area of increased activity resulted in subjective tinnitus improvement. The heterogeneous response to rTMS therapy found in our tinnitus patients is in line with other clinical rTMS studies demonstrating a high variability in the rTMS response (Hoffman, 2002). Our finding of a correlation between stimulation induced reduction of tinnitus complaints and changes in cortical excitability links the heterogeneous clinical effects with the previously described high interindividual variability of the modulatory effects of rTMS on cortical excitability (Maeda, 2000).

As genetic makeup and behavioral state affect the induction of long-term depression in animals (Manahan-Vaughan, 1999, Manahan-Vaughan, 2000, Xu, 1997) these factors have to be considered to understand individual differences in the reagibility of neuronal networks on rTMS, as has been suggested by Hoffman and Cavus (Hoffman, 2002).

The increased intracortical facilitation as well as the clinical effect of reduced tinnitus occure more pronounced six days after termination of rTMS and not immediately after treatment. These delayed effects could reflect the temporal dynamics of neuroplastic changes: It was suggested that an increase of intracortical facilitation can reflect the induction of synaptic plasticity (Ziemann 1998). Low-frequency rTMS was shown to induce synaptic plasticity in presence of local disinhibition (Ziemann, 2001). The hyperexcitability of the PAC in tinnitus patients could reflect such a local disinhibition phenomenon. This would go in line with the hypothesis that low frequency rTMS selectively depotentiates enhanced synaptic weights associated with pathological states (Hoffman, 2002).

The alterations of cortical excitability last for several weeks after rTMS treatment. This parallels results of low-frequency electrical stimulation in animal experiments, where daily stimulation for 10 days induced long-term depression, lasting for at least two weeks (Froc, 2000).

4. CONCLUSION

Our findings support the hypothesis that tinnitus - like auditory hallucinations - is associated with hyperexcitability in the auditory cortex. The several weeks-long lasting effects assessed with objective (cortical excitability) and subjective (tinnitus sensation) measures demonstrate the potential of rTMS to alter cortical excitability and to improve the symptomatology of neuropsychiatric syndroms classified as hyperexcitability disorders.

ACKNOWLEDGEMENTS

We would like to thank Helene Niebling, Sandra Pflügl and Marc Zowe for technical assistance with transcranial magnetic stimulation.

REFERENCES

Arnold, W., Bartenstein, P., Oestreicher, E., Romer, W., and Schwaiger, M., 1996, Focal metabolic activation in the predominant left auditory cortex in patients suffering from tinnitus: a PET study with [18F]deoxyglucose. *ORL J. Otorhinolaryngol. Relat. Spe.* **58**: 195-199.

Cacace, A. T., 2003, Expanding the biological basis of tinnitus: crossmodal origins and the role of neuroplasticity. *Hear. Res.* **175**: 112-132.

Dobie, R. A., 1999, A review of randomized clinical trials in tinnitus. *Laryngoscope* **109**:1202-1211.

Eichhammer, P., Langguth, B., Marienhagen, J., Kleinjung, T., and Hajak, G., 2003, Neuronavigated Repetitive Transcranial Magnetic Stimulation in Patients with Tinnitus. *Biol. Psychiatry* **54**: 862-865.

Froc, D. J., Chapman, C. A., Trepel, C., and Racine, R. J., 2000, Long-term depression and depotentiation in the sensorimotor cortex of the freely moving rat. *J. Neurosci.* **20**: 438-445.

George, M. S., Wassermann, E. M., and Post, R. M., 1996, Transcranial magnetic stimulation: a neuropsychiatric tool for the 21st century. *J. Neuropsychiatry. Clin. Neurosci.* **8**: 373-382.

Gilio, F., Rizzo, V., Siebner, H. R., and Rothwell, J. C., 2003, Effects on the right motor hand-area excitability produced by low-frequency rTMS over human contralateral homologous cortex. *J. Physiol.* **551**: 563-573.

Goebel, G., and Hiller, W., 1994, The tinnitus questionnaire. A standard instrument for grading the degree of tinnitus. Results of a multicenter study with the tinnitus questionnaire. *HNO* **42**: 166-172.

Graziano, M. S., Reiss, L. A., and Gross, C. G., 1999, A neuronal representation of the location of nearby sounds. *Nature* **397**: 428-430.

Hoffman, R. E., Boutros, N. N., Hu, S., Berman, R. M., Krystal, J. H., and Charney, D. S., 2000, Transcranial magnetic stimulation and auditory hallucinations in schizophrenia. *Lancet* **355**: 1073-1075.

Hoffman, R. E., and Cavus, I., 2002, Slow transcranial magnetic stimulation, long-term depotentiation, and brain hyperexcitability disorders. *Am. J. Psychiatry* **159**: 1093-1102.

Kaltenbach, J. A., 2000, Neurophysiologic mechanisms of tinnitus. *J. Am. Acad. Audiol.* **11**: 125-137.

Kujirai, T., Caramia, M. D., Rothwell, J. C., Day, B. L., Thompson, P. D., Ferbert, A., Wroe, S., Asselman, P., and Marsden, C. D., 1993, Corticocortical inhibition in human motor cortex. *J Physiol.* **471**: 501-519.

Langguth, B., Eichhammer, P., Wiegand, R., Marienhegen, J., Maenner, P., Jacob, P. and Hajak, G., 2003, Neuronavigated rTMS in a patient with chronic tinnitus. Effects of 4 weeks treatment. *Neuroreport* **14**: 977-980.

Maeda, F., Keenan, J. P., Tormos, J. M., Topka, H., and Pascual-Leone, A., 2000, Interindividual variability of the modulatory effects of repetitive transcranial magnetic stimulation on cortical excitability. *Exp. Brain Res.* **133**: 425-430.

Manahan-Vaughan, D., 2000, Long-term depression in freely moving rats is dependent upon strain variation, induction protocol and behavioral state. *Cereb. Cortex* **10**: 482-487.

Manahan-Vaughan, D., and Braunewell, K. H., 1999, Novelty acquisition is associated with induction of hippocampal long-term depression. *Proc. Natl. Acad. Sci. U. S. A.* **96**: 8739-8744.

Mirz, F., Gjedde, A., Ishizu, K., and Pedersen, C. B., 2000, Cortical networks subserving the perception of tinnitus--a PET study. *Acta Otolaryngol. Suppl.* **543**: 241-243.

Muhlnickel, W., Elbert, T., Taub, E., and Flor, H., 1998, Reorganization of auditory cortex in tinnitus. *Proc. Natl. Acad. Sci. U. S. A.* **95**: 10340-10343.

Post, R. M., Kimbrell, T. A., Frye, M., George, M., McCann, U., Little, J., Dunn, R., Li, H., and Weiss, S. R. B., 1997, Implications of kindling and quenching for the possible frequency dependence of rTMS. *CNS Spectrums* **2**: 54-60.

Wang, H., Wang, X., and Scheich, H., 1996, LTD and LTP induced by transcranial magnetic stimulation in auditory cortex. *Neuroreport* **7**: 521-525.

Xu, L., Anwyl, R., and Rowan, M. J., 1997, Behavioural stress facilitates the induction of long-term depression in the hippocampus. *Nature* **387**: 497-500.

Ziemann, U., Corwell, B., and Cohen, L. G., 1998, Modulation of plasticity in human motor cortex after forearm ischemic nerve block. *J. Neurosci.* **18**: 1115-1123.

Ziemann, U., Hallett, M., 2000, Basic neurophysiological studies with rTMS. In *Transcranial magnetic stimulation in psychiatry* (M. S. George and R. H. Belmaker, eds.), American Psychiatry Press, Washington, DC, pp.45-98.

Ziemann, U., Muellbacher, W., Hallett, M. and Cohen, L. G., 2001, Modulation of practice-dependent plasticity in human motor cortex. *Brain* **124**: 1171-1181.

Effects of Long Term Unilateral Hearing Loss on the Lateralization of fMRI Measured Activation in Human Auditory Cortex

DAVID R. MOORE[*], JOSEPH T. DEVLIN[**], JOSEPHINE RALEY[**], ELIZABETH TUNBRIDGE[**], KATHERINE LANARY[**], ANNA FLOYER-LEA[**], CHARVY NARAIN[**], IAN COHEN[**], PETER JEZZARD[**], MARTIN J. BURTON[#], and PAUL M. MATTHEWS[**]
[*]MRC Institute of Hearing Research, University Park, Nottingham NG7 2RD, U. K; [**]Centre for Functional Magnetic Resonance Imaging of the Brain, Department of Clinical Neurology, University of Oxford, John Radcliffe Hospital, Headley Way, Headington, Oxford OX3 9DU, U. K.;and [#]Department of Otolaryngology, Nuffield Department of Surgery, University of Oxford, Radcliffe Infirmary, Oxford OX2 6HE, U. K

1. INTRODUCTION

Profound deafness greatly reduces or abolishes neural activity in the auditory nerve (Tucci *et al.*, 1987). This leads to a variety of well-documented cellular and molecular effects in the cochlear nucleus on the side or sides of the deafness (Rubel *et al.*, 2003) that are highly dependent on the age at which the deafness occurs (Tierney *et al.*, 1997). In the auditory brainstem, midbrain and primary auditory cortex (PAC) of non-human mammals, a number of changes in neural projection pathways have been described following unilateral deafening (Moore and King, 2003). These seem less dependent on age at the time of deafening. Physiologically, neurons in the midbrain (Kitzes and Semple, 1985; McAlpine *et al.*, 1997) and PAC (Reale *et al.*, 1987; Popelar *et al.*, 1994) have been found to increase their response to acoustic stimulation of the ipsilateral, undeafened ear following unilateral deafening (Figs. 1A,C). This enhanced physiological response, thought to be a form of neural unmasking, occurs very rapidly following deafening in adults (Mossop *et al.*, 2000) as well as in infants.

These animal studies, recently reviewed in detail (Parks *et al.*, 2003), suggest that a variety of neural mechanisms underlie the effects of deafness on the brain. In humans, much less is known because of the obvious inability to access cellular events *in vivo*. However, the issues raised have clinical and, in particular, rehabilitative significance. One such issue is cochlear implantation. If central neurons die or rewire following deafness, what effect will electrical stimulation, or its withdrawal, have on these processes? It is possible that the substantially better outcomes reported from cochlear implantation performed soon after deafness, or in very young children, could result in part from the prevention of maladaptive brain plasticity.

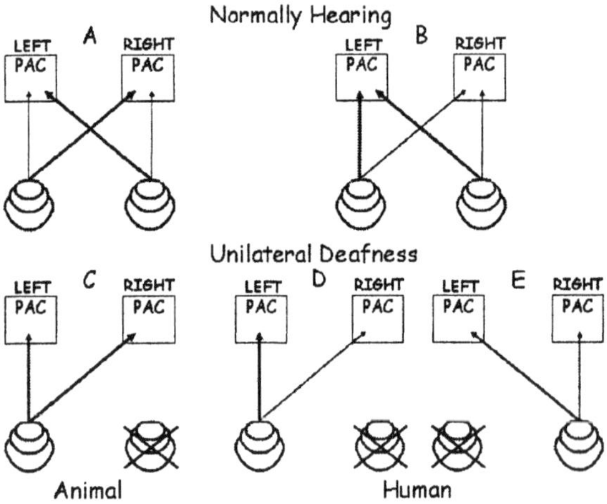

Figure 1. Activation of PAC by stimulation of each ear in animals (A, C) and humans (B, D, E). In each figure, the thickness of the arrow connecting the ear to the PAC represents the strength of cortical activation produced by unilateral tone stimulation of the ear. A. In animals (non-human mammals), stimulation of either ear produces strong activation of the contralateral PAC but, in humans (B), stimulation of either ear produces stronger activation of the left PAC. C. Following unilateral deafness in animals, activation of the two PACs becomes more balanced, without any apparent loss of contralateral PAC activation. D. In humans, right ear deafness doesn't affect the laterality of brain activation produced by left ear stimulation, but left ear deafness (E) results in a significant increase in relative activation of the right PAC by right ear stimulation.

Recently, there has been some neuroimaging research on the effects of unilateral deafness on human brain activation. Using fMRI, Scheffler *et al.* (1998) found that 5 participants with unilateral deafness had a shift in the balance of sound-induced activation between the two cerebral hemispheres, relative to control, normally hearing participants. In the controls, stimulation of either ear was found to activate the contralateral hemisphere more than the ipsilateral hemisphere. In the deaf participants, the inter-hemisphere activation ratio shifted significantly, so that the hemisphere ipsilateral to the unimpaired ear was relatively more activated. Subsequently, Bilecen *et al.* (2000) reported on the time course of this laterality shift in one of the 5 participants studied by Scheffler *et al.* who was imaged for the first time soon after a sudden onset deafness in adulthood. The laterality shift in this participant was found to occur gradually over a period of at least one year.

In other work, Vasama and colleagues (1995, 1997, 1998) used auditory evoked magnetic fields to show, first, that extra cortical areas may contribute to a reduced response laterality following unilateral deafness in humans and, second, that a unilateral conductive hearing loss can influence brain activity evoked by stimulation of the normally-hearing ear. Ponton *et al.* (2001), using long latency event related potentials, found a gradual balancing in the laterality of cortical activation that occurred over at least 2 years following late-onset, unilateral deafness. Khosla and colleagues (2003) found reduced laterality only in those participants for whom the unilateral deafness was in the left ear (Figs. 1 D,E).

We initiated the fMRI study described here to explore further the cerebral localization of changes in brain activation following either sensorineural or conductive unilateral hearing loss. However, our initial control studies (Devlin *et al.*, 2003) produced a surprising result (Fig. 1B). We found that the left PAC was more strongly activated by unilateral tone stimulation whichever ear was stimulated. While a 'laterality index' (ratio of contralateral to ipsilateral activation) was much larger for right than for left ear stimulation, stimulation of the left ear consistently produced stronger activation in the left than in the right PAC. In non-primary areas, no significant difference was found between the pattern of activation produced by stimulation of either ear. In light of this result and the report of Khosla *et al.* (2003), we were intrigued to discover what activation would be observed in participants with hearing loss in either the right or left ear.

2. METHODS

2.1 Participants

Eleven patients (5F, 6M) with long term unilateral hearing loss participated in this study (see Table 1). Seven had a sensorineural loss and four a conductive loss, as determined by standard air and bone stimulation procedures (Katz, 2001). In addition, twelve adults (4F, 8M, 20-30 years old) with normal hearing (pure tone average ≤ 20 dB HL) served as the control population (Devlin *et al.*, 2003). Each patient had a history of at least eight years of hearing loss and all but one was right handed. No participant (control or patient) had a personal or family history of epilepsy or any other neurological condition and each gave informed consent after the experimental methodology was explained.

Gender	Age	Affected side	Age of diagnosis	Severity (threshold - PTA)	Etiology
Sensorineural patients					
F	19	L	10	Profound (>95 dB)	Possibly rubella
M	19	L	11	Profound (>95 dB)	Possibly viral
M	22	L	2	Profound (≥90 dB)	Possibly mumps
F	19	R	3	Profound (>95 dB)	Possibly mumps
F	26	R	3	Mod.-Severe (65-75 dB)	Possibly mumps
M	19	R	7	Profound (>95 dB)	Unknown
M	20	R	8	Profound (>95 dB)	Possibly congenital
Conductive patients					
F	50	L	30	Moderate (58 dB)	Otosclerosis
F	41	L	39	Mild (36 dB)	Otosclerosis
M	18	R	Birth	Mod.-Severe (60 dB)	Atresia
M	23	R	12	Moderate (46 dB)	Perforation of ear drum

Table 1.Patient details. For the patients with sensorineural loss, etiology was based on parental report.

2.2 Scanning Procedure

Details of scanning, analysis and auditory cortex definition were as reported by Devlin *et al.* (2003). Participants were asked to discriminate between high (4000Hz) or low (250Hz) frequency tones, amplitude modulated at 5 Hz, by pressing one of two buttons as quickly as possible after the tone's onset. An equal number of silent trials was randomly interleaved. The purpose of the task was to force subjects to attend to the auditory tones throughout the scan. Stimuli were generated and presented monaurally using the MRC Institute of Hearing Research MR sound system (Palmer *et al.*, 1998) at 90dB SPL. Controls participated in two scanning runs (one ear per run) and patients participated in one run during which their normally hearing ear was tested.

Sparse sampling was used to collect the functional data. This technique allows one to image auditory stimulus evoked BOLD responses unaffected by the sound of the scanner (Hall *et al.*, 1999). A trial began with 6s of stimulation (i.e. a high or low frequency tone or a silent trial), during which the gradient coils were turned off, followed immediately by 3s of whole head EPI acquisition. Inter-trial intervals varied between 7-14s in order to minimize anticipation of the upcoming trial. Scans were carried out using a Varian-Siemens 3T scanner and a Magnex head-dedicated gradient insert coil in conjunction with a birdcage head radio frequency coil tuned to 127.4MHz. Functional imaging consisted of 21 T2*-weighted echo-planar image (EPI) slices with a 4 × 4 × 5mm resolution. A high resolution T1-weighted scan was acquired for registering the functional data to an individual's anatomy.

2.3 Image Analysis and Regional Definition

Functional images were realigned to correct for small head movements, registered to the participant's structural scan and to the MNI 152-mean brain, and smoothed with a Gaussian filter. FSL software (http://www.fmrib.ox.ac.uk/fsl) was used to compute individual subject analyses using the general linear model. Tone trials (high and low) were modelled as a single factor of interest and the estimated motion parameters for each subject were included as covariates of no interest to reduce spurious activations due to head motion, thereby increasing statistical sensitivity. Random effects group analyses identified significantly activated brain regions. A cluster-based significance test was used with a height threshold of $Z>3.5$ and $P<0.05$, corrected for multiple comparisons (Friston *et al.*, 1994).

To evaluate hemispheric asymmetries within auditory areas, we computed laterality indices (LI) based on two separate measures of activity in the selected regions-of-interest for each participant: LI = (Contralateral BOLD signal – Ipsilateral BOLD signal) x 100 /(Contralateral BOLD signal + Ipsilateral BOLD signal).

LI values ranged from +100, indicating completely contralateral activation, to 100, indicating completely ipsilateral activation. Equal activations contralateral and ipsilateral to the stimulated ear produced a value of zero. BOLD signal was measured as the mean percent BOLD signal change within the region-of-interest. This measure is not sensitive to volume differences between the hemispheres nor is it effected by any arbitrary statistical thresholds (unlike the number of "active" voxels). Separate analyses were conducted for the primary and non-primary auditory cortices, identified according to anatomical and functional criteria

PAC was defined from the structural scan (following Rademacher *et al.*, (1993) and Patterson *et al.*, (2002)) as the first transverse temporal gyrus (Heschl's gyrus), labelled in both hemispheres of each participant using 3D visualisation software (http://www.psychology.nottingham.ac.uk/staff/cr1/mricro.html). The location of non-primary auditory cortex was based on the functional activation maps. Voxels activated (Z>3.5) by tones relative to silence were masked to include only those within the auditory regions of the supratemporal plane and insular cortex. For each participant, Heschl's gyrus was removed from the non-primary regions-of-interest to exclude the PAC.

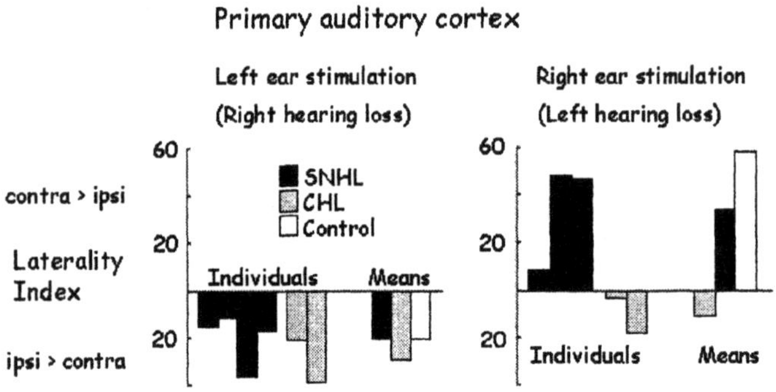

Figure 2. Laterality of PAC activation in patients with long term hearing loss. Left ear stimulation produced results comparable with Controls (see also Fig. 1B), but right ear stimulation produced, in 3/5 patients, a shift towards laterally balanced activation.

3. RESULTS

The laterality indices obtained from the hearing impaired patients are shown in Figs. 2 and 3. In the PAC, left ear stimulation produced, in all patients with right ear hearing loss, stronger activation of the left than of the right cortex. This confirmed the surprising result found in the control participants (Figs. 1B, 2; Devlin *et al.*, 2003), but showed that long-standing right ear loss did not lead to a change in the laterality of activation in PAC. Right ear stimulation in patients with left ear hearing loss yielded other unexpected results. Two of the 3 patients with sensorineural loss produced clear, contralateral LIs that were similar to those of the control participants. However, the third sensorineural loss patient had an LI that was lower than all but one of the controls. Two further patients with a conductive loss in the left ear had greater activation in the right, ipsilateral PAC (Fig. 2).

The small number of patients belonging to these groups prevented full statistical comparison, and the results must be considered preliminary. However, it is noteworthy that, of the 12 controls and 5 patients tested with right ear stimulation in this series to date, the 2 patients with left ear conductive loss were the only ones who produced an ipsilaterally biased LI. When all the patients with left ear hearing loss were grouped, the LI was significantly less ($t(15) = 2.71$, $p = 0.016$, two-tailed) than that of the controls. The PAC results are summarised in Figs. 1 D,E.

LI results from non-primary cortex (Fig. 3) were all close to zero, in line with control data (Devlin *et al.*, 2003).

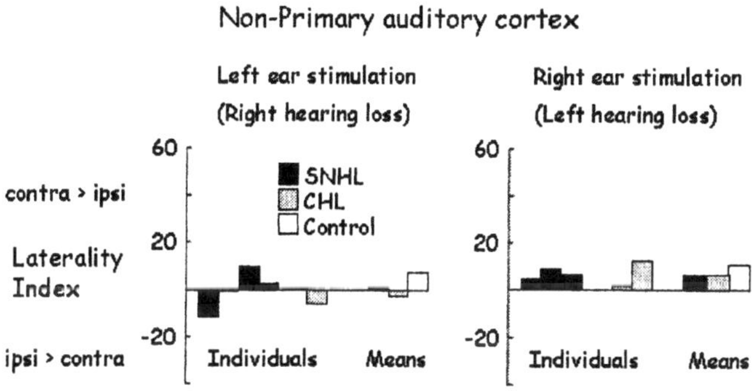

Figure 3. Laterality of non-primary auditory cortex activation in patients with long term hearing loss. Stimulation of either ear balanced activation in the right and left cortex in all groups.

4. DISCUSSION

Two major results were obtained in this study of the effects of stimulation of the normally hearing ear on auditory cortex activation in patients with unilateral hearing loss. First, stimulation of the left ear produced stronger activation of the left than of the right PAC. This surprising result was, nevertheless, consistent with previously reported data from normally hearing listeners (Devlin *et al.*, 2003). Second, patients stimulated in the right ear had, on average, a reduced level of left cortical activation dominance compared with normally hearing listeners. This result confirmed, with fMRI, that recently reported by Khosla and colleagues (2003), who used ERPs. However, there was complexity in our data from patients with left hearing loss that would not have been predicted from Khosla's study. We found that three sensorineural loss patients (as studied by Khosla) had variable left cortex dominance, whereas both of two conductive loss patients had reduced left dominance.

The finding, confirmed here, that human PAC left lateralization effects appear to be stronger than the contralateral dominance of pure-tone, auditory evoked activity seen in other species leads to several further questions. One is the specificity of the left lateralization effect to the particular experimental conditions used in this and our previous (Devlin *et al.*, 2003) study. Because this is an important and novel result, it is clearly desirable to perform further studies that vary the complexity both of the listening task and of the acoustic stimulus. A further question is whether non-human primates show a left lateralization effect under test conditions like those we used. The answer to this also requires further, functional study, but it is noteworthy that no structural asymmetry between the left and right superior temporal plane of old world monkeys has been found, in contrast to a significant asymmetry in great apes (Gannon *et al.*, 1998) and the major structural and functional asymmetry in humans. If monkeys do not have a functional asymmetry, their particular usefulness as animal models for auditory cortical function (Poremba *et al.*, 2003) is questionable. Having demonstrated the left lateralization effect in the PAC of humans, a further issue is whether it is restricted to the PAC, or is seen further up and down the auditory system. In our previous study, we failed to find lateralisation in the medial geniculate nucleus and in both that and the present study no significant lateralisation was found in non-primary cortex. Nevertheless, these results require verification, possibly using, in the case of the MGN, a scanner with a higher field strength and, for other cortical areas, stimuli of graded complexity.

As mentioned above, the left ear loss (right ear stimulated) results must be treated with particular caution because of the small number of patients tested. But they do raise intriguing possibilities. Regarding the sensorineural loss cases, it could be that the individual history of the hearing loss, or some aspect of its aetiology unknown to us in the cases reported here, may have contributed to the variable results. For the conductive loss patients, a change in the laterality of activation of the PAC produced by what is, in effect, a change in auditory experience (Moore and King, 2003) would have several important implications. As outlined in the Introduction, it is generally assumed that changes in auditory brain structure and function produced by 'total' deafening are due to a cessation of activity in the auditory nerve. Conductive hearing loss does not abolish auditory nerve activity, just reduces it somewhat. To respond to this reduction, a central neuron would therefore have somehow to compare the new level with the previous level of input. This could be done either on an absolute scale, integrating across time, or on a relative scale, by comparing with another input, such as the signal from the intact ear. The cellular mechanisms mediating such a comparison would be likely to differ from those thought responsible for mediating 'deafferentation' responses, currently considered to be intracellular signalling paths connected with metabotropic glutamate receptors (Zirpel *et al.*, 2000). Whatever the mechanisms, a large change in cortical activation, as seen in human neuroimaging, resulting from a conductive loss implies a dynamic and influential effect of auditory experience on the PAC. Recent studies in animals of the influence of auditory experience on cortical functional organization (e.g. Beitel *et al.*, 2003; Weinberger, 2003) support this view.

The asymmetry between the effect of right and left ear hearing losses is intriguing. Any change in cortical activation resulting from a right ear loss was either too small to detect using our methods or was matched between the two hemispheres. In either case, the left cortical dominance persisted. In contrast, a left ear loss, while presenting a variable response, nevertheless produced a significant shift in activation away from the dominant left hemisphere. While this overall pattern of results resembles that reported by Khosla and colleagues (2003), the details are rather different. In normally hearing participants, Khosla found contralateral activation that was both larger in amplitude and peaked earlier in time than ipsilateral activation, regardless of the ear that was stimulated. For their deaf participants, left ear (but not right ear) hearing loss produced more symmetrical responses. However, because a lateral asymmetry was not found in the controls, Khosla suggested the deafness-induced asymmetry may be due to sub-cortical plasticity. Wherever the origin of this effect, one plausible explanation of the asymmetry may be derived from a consideration of the relative strengths of the ipsilateral and contralateral pathways in normally hearing individuals. In animal studies (e.g. McAlpine *et al.*, 1997; Mossop *et al.*, 2000), the effect of deafening on activation ipsilateral to the intact ear is much greater than that contralateral to the intact ear (Fig. 1C). This may be at least partly because stimulation of the ipsilateral pathway normally produces only a low level of activation. Thus, a given level of increased activation would, as a relative measure, be more apparent ipsilateral to the intact ear. In humans, the ipsilateral path from the left is normally stronger than the contralateral path (Fig. 1B). The failure to find any effect of right ear hearing loss may therefore be due to a relatively small effect of the deafening on the already strong ipsilateral path. On the other hand, right ear hearing loss would have a relatively greater effect on the ipsilateral path from that ear, which is normally weak.

If it turned out that the principle driver of these changes in activation of the higher levels of the auditory system is a reduced influence of ascending inhibition from the side of the hearing loss, as the animal data suggest, an important further question is whether the distribution of excitatory and inhibitory influences on human auditory cortical neurons is also asymmetric between the left and right sides. Studies of clinical (Hartvig Jensen *et al.*, 1989) and experimental (Gustafsen and Hamill, 1995) unilateral hearing loss in humans suggest that right ear loss is more detrimental to verbal and localization performance, respectively, than left ear loss. Left/right differences in auditory brain function may thus be rather extensive in humans and should certainly be controlled for in future animal research.

ACKNOWLEDGEMENTS

This research was supported by Programme Grants from the Medical Research Council (MRC), an MRC ROPA award, and Wellcome Trust and EPSRC studentships.

REFERENCES

Beitel, R. E., Schreiner, C. E., Cheung, S. W., Wang, X. and Merzenich, M. M., 2003, Reward-dependent plasticity in the primary auditory cortex of adult monkeys trained to discriminate temporally modulated signals. *Proc. Natl. Acad. Sci. U. S. A.* **100**: 11070-5.

Bilecen, D., Seifritz, E., Radu, E. W., Schmid, N., Wetzel, S., Probst, R., and Scheffler, K., 2000, Cortical reorganization after acute unilateral hearing loss traced by fMRI. *Neurology* **54**: 765-7.

Devlin, J. T., Raley, J., Tunbridge, E., Lanary, K., Floyer-Lea, A., Narain, C., Cohen, I., Behrens, T., Jezzard, P., Matthews, P. M., and Moore, D. R., 2003, Functional asymmetry for auditory processing in human primary auditory cortex. *J. Neurosci.* **23**: 11516-22.

Friston, K. J., Worsley, K. J., Frackowiak, R. S. J., Mazziotta, J. C., and Evans, A. C., 1994, Assessing the significance of focal activations using their spatial extent. *Hum. Brain Map.* **1**: 214-20.

Gannon, P. J., Holloway, R. L., Broadfield, D. C., and Braun, A. R., 1998, Asymmetry of chimpanzee planum temporale: humanlike pattern of Wernicke's brain language area homolog. *Science* **279**: 220-2.

Gustafson, T. J., and Hamill, T. A., 1995, Differences in localization ability in cases of right versus left unilateral simulated conductive hearing loss. *J. Am. Acad. Audiol.* **6**: 124-8.

Hall, D. A., Haggard, M. P., Akeroyd, M. A., Palmer, A. R., Summerfield, A. Q., Elliott, M. R., Gurney, E. M. and Bowtell, R. W., 1999, "Sparse" temporal sampling in auditory fMRI. *Hum. Brain Mapp.* **7**: 213-23.

Hartvig Jensen, J., Angaard Johansen, P., and Borre, S., 1989, Unilateral sensorineural hearing loss in children and auditory performance with respect to left/right ear differences. *Brit. J. Audiol.* **23**: 207-13.

Khosla, D., Ponton, C. W., Eggermont, J. J., Kwong, B., Don, M., and Vasama, J. P., 2003, Differential ear effects of profound unilateral deafness on the adult human central auditory system. *J. Assoc. Res. Otolaryngol.* **4**: 235-49.

Kitzes, L. M. and Semple, M. N., 1985, Single-unit responses in the inferior colliculus: effects of neonatal unilateral cochlear ablation. *J. Neurophysiol.* **53**: 1483-500.

McAlpine, D., Martin, R. L., Mossop, J. E. and Moore, D. R., 1997, Response properties of neurons in the inferior colliculus of the monaurally deafened ferret to acoustic stimulation of the intact ear. *J. Neurophysiol.* **78**: 767-79.

Moore, D. R., King, A. J., 2003, Plasticity of binaural systems. In *Development and Plasticity of the Auditory System* (T. N. Parks, E. W. Rubel, A. N. Popper, and R. R. Fay, eds.), Springer, New York, in the press.

Mossop, J. E., Wilson, M. J., Caspary, D. M., and Moore, D. R., 2000, Down-regulation of inhibition following unilateral deafening. *Hear. Res.* **147**: 183-7.

Palmer, A. R., Bullock, D. C., and Chambers, J. D., 1998, A high-output, high quality sound system for use in auditory fMRI. *NeuroImage* **7**: S359.

Patterson, R. D., Uppenkamp, S., Johnsrude, I. S., and Griffiths, T. D., 2002, The processing of temporal pitch and melody information in auditory cortex. *Neuron* **36**: 767-76.

Ponton, C. W., Vasama, J. P., Tremblay, K., Khosla, D., Kwong, B. and Don, M., 2001, Plasticity in the adult human central auditory system: evidence from late-onset profound unilateral deafness. *Hear. Res.* **154**: 32-44.

Popelar, J., Erre, J. P., Aran, J. M., and Cazals, Y., 1994, Plastic changes in ipsi-contralateral differences of auditory cortex and inferior colliculus evoked potentials after injury to one ear in the adult guinea pig. *Hear. Res.* **72**: 125-34.

Poremba, A., Saunders, R. C., Crane, A. M., Cook, M., Sokoloff, L., and Mishkin, M., 2003, Functional mapping of the primate auditory system. *Science* **299**: 568-72.

Rademacher, J., Caviness, V. S., Jr., Steinmetz, H., and Galaburda, A. M., 1993, Topographical variation of the human primary cortices: implications for neuroimaging, brain mapping, and neurobiology. *Cereb. Cortex* **3**: 313-29.

Reale, R. A., Brugge, J. F., and Chan, J.C., 1987, Maps of auditory cortex in cats reared after unilateral cochlear ablation in the neonatal period. *Brain Res.* **431**: 281-90.

Rubel, E. W., Parks, T. N., Zirpel, L., 2003, Constructing, innervating and maintaining the developing cochlear nucleus. In *Development and Plasticity of the Auditory System* (T. N. Parks, E. W. Rubel, A. N. Popper, and R. R. Fay, eds.), Springer, New York, in the press.

Scheffler, K., Bilecen, D., Schmid, N., Tschopp, K., and Seelig, J., 1998, Auditory cortical responses in hearing subjects and unilateral deaf patients as detected by functional magnetic resonance imaging. *Cereb. Cortex* **8**: 156-63.

Tierney, T. S., Russell, F. A., and Moore, D. R., 1997, Susceptibility of developing cochlear nucleus neurons to deafferentation-induced death abruptly ends just before the onset of hearing. *J. Comp. Neurol.* **378**: 295-306.

Tucci, D. L., Born, D. E., and Rubel, E. W., 1987, Changes in spontaneous activity and CNS morphology associated with conductive and sensorineural hearing loss in chickens. *Ann. Otol. Rhinol. Laryngol.* **96**: 343-50.

Vasama, J. P., and Makela, J. P., 1995, Auditory pathway plasticity in adult humans after unilateral idiopathic sudden sensorineural hearing loss. *Hear. Res.* **87**: 132-40.

Vasama, J. P., and Makela, J. P., 1997, Auditory cortical responses in humans with profound unilateral sensorineural hearing loss from early childhood. *Hear. Res.* **104**: 183-90.

Vasama, J. P., Makela, J. P., and Ramsay, H.A., 1998, Modification of auditory pathway functions in patients with hearing improvement after middle ear surgery. *Otolaryngol. Head Neck Surg.* **119**: 125-30.

Weinberger, N. M., 2003, The nucleus basalis and memory codes: Auditory cortical plasticity and the induction of specific, associative behavioral memory. *Neurobiol. Learn. Mem.* **80**: 268-84.

Zirpel, L., Janowiak, M. A., Veltri, C. A., and Parks, T.N., 2000, AMPA receptor-mediated, calcium-dependent CREB phosphorylation in a subpopulation of auditory neurons surviving activity deprivation. *J. Neurosci.* **20**: 6267-75.

Arguments in Favor of Auditory Reorganization in Human Subjects with Cochlear Damage

HUNG THAI-VAN*,#, DAMIEN GABRIEL#, and LIONEL COLLET*,#
*Unité CNRS UMR 5020 Laboratoire "Neurosciences et Systèmes Sensoriels", CNRS GDR 2213 "Prothèses auditives", Université Claude Bernard Lyon I, Lyon, France; and #Service d'Audiologie et d'Explorations Orofaciales, Hôpital Edouard Herriot, Lyon, France

1. INTRODUCTION

During the last decades, numerous studies have shown that organization of sensory maps is not fixed after development, but retains a large degree of plasticity, even in adulthood. Dramatic alterations in sensory maps at the cortical and subcortical levels have been reported after peripheral lesions affecting restricted portions of sensory receptor surfaces: rather than becoming silent, neurones deprived of peripheral input become responsive to stimuli applied to peri-lesion regions of the receptor surface, to which they formerly did not respond. Following upper-limb amputation in monkeys, neurones in the somatosensory cortex that initially responded specifically to the severed limb could later be activated by stimuli applied to the face (Jones and Pons, 1998). Because the face is represented adjacent to the upper arm in the somatosensory cortex, this observation can be explained by re-routing of neural inputs between neighbouring cortical regions. By analogy, in the auditory modality, cochlear lesions affecting a restricted portion of the cochlea have been shown to affect dramatically the tonotopic functional organization of the primary auditory cortex: as a general rule, auditory neurones that are deprived of direct cochlear inputs due to the lesion all become responsive to cochlear sites at which significant input is still present (Robertson and Irvine, 1989). Moreover, when hearing thresholds for a given frequency region have become abnormally elevated due to a restricted cochlear lesion, neurones with initial characteristic frequencies falling in such a region will develop low-threshold responses to frequencies whose cochlear place was at the edge of the lesion (lesion-edge or cutoff frequencies). These receptive field modifications may lead to an over-representation of the lesion-edge frequencies (Rajan *et al.*, 1993; Schwaber *et al.*, 1993; Irvine and Rajan, 1995). In turn, the observation that injury-induced plasticity results in reorganization of the primary auditory cortex raises the hypothesis that cochlear lesions might induce frequency-specific enhancements in auditory discrimination performance. In this paper, we summarize the results of research on perceptual consequences of restricted cochlear lesions and present arguments in favour of central auditory reorganization in subjects with cochlear damage.

2. LOCAL ENHANCEMENT IN AUDITORY FREQUENCY DISCRIMINATION AROUND THE HEARING LOSS CUTOFF

McDermott *et al.* (1998) first reported that local improvements in frequency discrimination performance could occur around the audiogram cutoff frequency (Fc) in human subjects with steeply sloping high-frequency hearing loss. The general objective of our first study (Thai-Van *et al.*, 2002) was to confirm this effect and to investigate further its relationships with various audiometric variables: the slope, the extent, and the pattern of the hearing loss. Specifically, we assessed whether the DLF enhancement effect could occur in subjects with more moderately sloping hearing loss and/or other patterns of loss. Frequency difference limens (DLFs) were measured in 20 subjects (8 female, 12 male, median age = 55.5 years) with high-frequency hearing loss. A minimum of 12 frequencies were tested at intervals of 1/8 octave over a range of one octave and a half around the Fc. Corresponding to the audiogram edge-frequency, the Fc was defined as the highest test frequency, at the beginning of the slope, with a hearing threshold of no more than 5 dB 'HL' above that of the best hearing frequency. Customized software randomised the level of the test tones over a +/-3 dB range around a nominal level, following an equal-loudness contour curve measured at 1/2-octave intervals. This methodological precaution was taken to prevent subjects from basing their judgements on frequency-related changes in loudness rather than on changes in pitch. Results demonstrated that DLFs were significantly smaller in a 1/4-octave wide frequency band centred on Fc than in the other bands. Moreover, the average DLF value measured in this band was found to be negatively correlated with the hearing-loss slope. No statistical link was found with the other audiometric parameters considered: either the extent and maximum amount of hearing loss, or the Fc value. The 20 subjects were split into three groups according to the slope of their hearing loss relative to Fc (steep:> 25 dB/½ octave; medium: between 12 and 25 dB/½ octave; and shallow: < 12 dB/½ octave). Local improvements in DLFs near Fc were seen in the steep- and medium-slope groups, and statistically confirmed in the steep-slope group. Fig. 1 shows a typical DLF improvement at Fc in a case of steeply sloping high-frequency hearing loss.

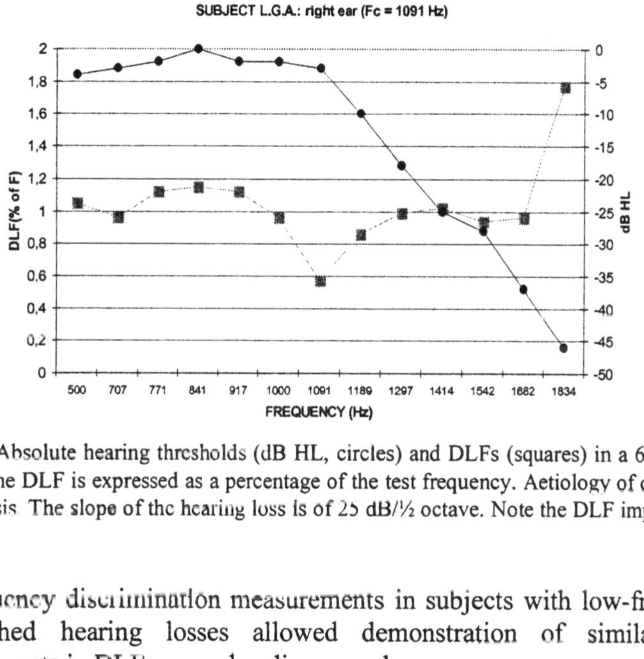

Figure 1. Absolute hearing thresholds (dB HL, circles) and DLFs (squares) in a 66 year-old woman. The DLF is expressed as a percentage of the test frequency. Aetiology of deafness is presbyacusis. The slope of the hearing loss is of 25 dB/½ octave. Note the DLF improvement at Fc.

Frequency discrimination measurements in subjects with low-frequency or notched hearing losses allowed demonstration of similar local improvements in DLFs around audiogram edges.

3. DOES LOCAL ENHANCEMENT IN FREQUENCY DISCRIMINATION REFLECT CENTRAL PLASTICITY?

The above results were consistent with an interpretation in terms of central mechanisms, i.e; injury-induced central reorganization. That is, a lesion-induced increase in the number of neurones responding to a narrow range of frequencies near the hearing loss cutoff may explain the local enhancement in discrimination performance that we observed. However, alternative interpretations in terms of peripheral mechanisms had not been completely eliminated. When acoustic stimulus are delivered at a constant physical level, changes in loudness are likely to occur with changes in frequency in subjects whose hearing thresholds and/or loudness function slopes differ across frequencies. In our first study, like in the study of McDermott *et al.* (1998), such systematic loudness changes were weakened by the use of randomised level stimulation together with stimulus levels falling along an equal-loudness contour in each subject. Nevertheless, because the contours were determined only at 1/2-octave intervals, they may not have sufficiently taken into account the rapid changes in loudness with frequency that occur around the audiogram edge frequency. Thus, it might be that loudness cues were still present in frequency regions where the slope of the audiogram changed rapidly. The fact that local improvements in DLFs were observed in subjects with steeply-sloping audiograms but not in those with shallow-sloping audiograms is consistent with this interpretation. Another "peripheral interpretation" for local improvements in DLFs around the Fc involves spontaneous otoacoustic emissions (SOAEs). SOAEs, which are thought to be generated by the spontaneous activity of the cochlear outer hair cells (Kemp, 1978), may interact with external tones of neighbouring frequencies and provide additional cues during the DLF task. It was recently shown that, in normal-hearing subjects, DLFs can be locally enhanced in the vicinity of SOAEs (Norena *et al.*, 2002).

In a second study (Thai-Van *et al.*, 2003), we assessed the possible role of these two peripheral mechanisms (loudness cues and SOAEs) in a group of subjects with high-frequency hearing loss. First, we tested whether the DLF enhancement effect was still observed under conditions where these subjects could not rely on loudness cues to perform the DLF task. To this aim, the nominal level of each stimulus was adjusted so that it fell on an equal-loudness contour measured at very fine (1/8 octave) frequency intervals, and the level of each stimulus was roved over a large range (12 dB). Under these conditions, the DLF enhancement effect was still observed in all subjects, demonstrating that this effect cannot be explained simply by the use of loudness cues. Second, we tested the subjects for SOAEs, to assess whether the DLF enhancement could be explained by the presence of such emissions in the vicinity of Fc. None of the subjects exhibited SOAEs. Finally, we tested whether these subjects had cochlear dead regions, i.e., regions lacking functional inner hair cells and/or auditory nerve fibres. Using a refined version of a non-invasive clinical test for the identification of dead regions (Moore *et al.*, 2000), we assessed the presence of such regions in fine frequency steps (1/4 octave) up to very high frequencies. All of the subjects had cochlear dead regions. The first two findings strengthen the hypothesis that DLF enhancement is due to injury-induced central reorganization in the auditory system. The last one is consistent with physiological data in animals, which suggest that total suppression of neural input from certain cochlear regions is a necessary condition for the occurrence of injury-induced auditory reorganization.

4. REVERSAL OF DLFS CHANGES IN COCHLEAR-DAMAGED SUBJECTS USING HEARING AIDS: AN ARGUMENT FOR A SECONDARY PLASTICITY

In cochlear-damaged subjects fitted by monaural or binaural amplification, the reintroduction of auditory input to CNS by hearing aids is thought to induce reversal of injury-induced plasticity (Willott, 1996). That is, the sensory maps that were initially altered by partial deafferentation may be normalized after amplification, suggesting a secondary plasticity. If we assume that deterioration of the frequency encoding properties of auditory neurones relates to deafferentation, then reintroduction of auditory input should restore these properties.

The purpose of our third experiment was to monitor the frequency discrimination performances before and after amplification in cochlear-damaged subjects. Nine subjects with steeply sloping high-frequency hearing loss were screened for DLFs before fitting with a hearing aid, then 1 month and 3 months after.

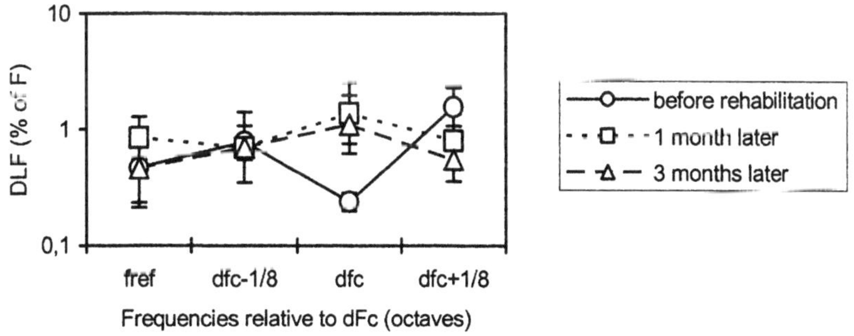

Figure 2: Changes in DLFs across frequencies after reintroduction of auditory input by amplification (fref = frequency 1 octave below Fc; dfc – frequency with the best DLF before amplification).

At 1 and 3 months post-amplification, a significant deterioration in discrimination performance was demonstrated at the frequency which initially exhibited the best DLF. These results clearly suggest that perceptual consequences of auditory deprivation could be reversed. A possible underlying explanation would be a return of tonopic maps in the primary of auditory cortex to normal status. The time course of this secondary plasticity could be quite fast since changes in DLFs were observed as soon as 1 month post-amplification.

5. CONCLUSION

Overall results confirm the improvement in frequency discrimination performance around the cutoff frequency in subjects with steeply sloping hearing loss. This effect does not appear to depend on the pattern of the hearing loss, as it can be observed in subjects with low-frequency, as well as notched hearing losses. Our findings, pointing to the steepness of the hearing loss and the presence of dead cochlear regions as the most important factors of the local DLF enhancement effect, are consistent with an interpretation in terms of injury-induced central auditory reorganization. The reversal of this plasticity after amplification may open some interesting therapeutic horizons. In particular, the benefit of hearing amplification might be conditioned by the occurrence of a secondary plasticity. This issue needs to be investigated by further research.

ACKNOWLEDGEMENTS

We are grateful to Christophe Micheyl and Brian C.J. Moore for their very fruitful help and involvement in this research. We also wish to thank Dexter Irvine and Hugh McDermott for constructive comments on our preliminary work and Arnaud Norena and Evelyne Veuillet for their helpful contributions. This research was supported by grants from CNRS, CCA Groupe, Entendre, Oticon, Phonak, and Siemens Audiologie.

REFERENCES

Irvine, D. R. F., and Rajan, R., 1998, Plasticity in the mature auditory system. In *Advances in Hearing Research* (G. A. Manley, G. M. Klump, C. Köppl, C. Fastl, and H. Oeckinghaus, eds.), World Scientific, Singapore, pp. 3-23.

Jones, E. G., and Pons, T. P., 1998, Thalamic and brainstem contributions to large-scale plasticity of primate somatosensory cortex. *Science* **282**: 1121-5.

Kemp, D. T., 1978, Stimulated acoustic emissions from within the human auditory system. *J. Acoust. Soc. Am.* **64**: 1386-91.

McDermott, H. J., Lech, M., Kornblum, M. S., and Irvine, D. R. F., 1998, Loudness perception and frequency discrimination in subjects with steeply sloping hearing loss: possible correlates of neural plasticity. *J. Acoust. Soc. Am.* **104**: 2314-25.

Moore, B. C. J., Huss, M., Vickers, D. A., Glasberg, B. R., and Alcantara, J. I., 2000, A test for the diagnosis of dead regions in the cochlea. *Br. J. Audiol.* **34**: 205-24.

Norena, A., Micheyl, C., Durrant, J. D., Chery-Croze, S., and Collet, L., 2002, Perceptual correlates of neural plasticity related to spontaneous otoacoustic emissions? *Hear. Res.* **171**: 66.

Rajan, R., Irvine, D. R. F., Wise, L. Z., and Heil, P., 1993, Effect of unilateral partial cochlear lesions in adult cats on the representation of lesioned and unlesioned cochleas in primary auditory cortex. *J. Comp. Neurol.* **338**: 17-49.

Robertson, D., and Irvine, D. R. F., 1989, Plasticity of frequency organization in auditory cortex of guinea pigs with partial unilateral deafness. *J. Comp. Neurol.* **282**: 456-71.

Schwaber, M. K., Garraghty, P. E., and Kaas, J. H., 1993, Neuroplasticity of the adult primate auditory cortex following cochlear hearing loss. *Am. J. Otol.* **14**: 252-8.

Thai-Van, H., Micheyl, C., Norena, A., and Collet, L., 2002, Local improvement in auditory frequency discrimination is associated with hearing-loss slope in subjects with cochlear damage. *Brain* **125**: 524-37.

Thai-Van, H., Micheyl, C., Moore, B. C. J., and Collet, L., 2003, Enhanced frequency discrimination near the hearing loss cutoff: A consequence of central auditory plasticity induced by cochlear damage? *Brain* **126**: 2235-45.

Willott, J. F., 1996, Physiological plasticity in the auditory system and its possible relevance to hearing aid use, deprivation effects, and acclimatization. *Ear Hear.* **17**: 66S-77S.

Central Auditory Processing and Language Learning Impairments: Implications for Neuroplasticity Research

PAULA TALLAL[*], and R. HOLLY FITCH[#]
[*]*Center for Molecular & Behavioral Neuroscience, Rutgers, The State University of New Jersey, 197 University Avenue, Newark, New Jersey, 07102, USA; and* [#]*Department of Psychology, Division of Behavioral Neuroscience, University of Connecticut, Box 4154, Storrs, CT, 06269, USA*

1. INTRODUCTON

Developmental disorders of language (including both oral and written language) are diagnosed in about 20% of preschool and school age children (Beitchman *et al.*, 1986). Such problems are often associated with peripheral hearing impairments, paralysis of the speech musculature, and/or higher order cognitive disorders. For example, language impairment is commonly seen as a feature of general mental retardation, as well as pervasive developmental disorders such as infantile autism. Language problems tend to be co-morbid with childhood disabilities such as attention deficit disorders (ADD/ADHD), social and emotional problems, and to be associated with neurological anomalies such as intra-cranial bleeds of prematurity or seizures. While some children with language problems also demonstrate impaired development of speech articulation, others do not (for review see Leonard, 1998).

Amongst the 20% of children with identifiable language related problems, a subset of children whose disability cannot be attributed to any of the associated causes listed above can be identified. These children appear to be healthy and of normal intelligence, but fail to progress through the normal language milestones at or near the expected age. Children with language delays who meet these exclusionary criteria are considered to have specific language impairment (SLI). Approximately 7-8% of five year olds are diagnosed with SLI in the USA (Tomblin *et al.*, 1997). Impairments seen in this population may manifest as a disability in expressive language (with near normal receptive comprehension), or as a disability in both receptive and expressive language (see DSM-IV, 1994).

Family aggregation studies have shown that about 40% of children with SLI have a primary family member with some form of language disorder, leading to the search for one or more genes linked to the disorder (Bartlett *et al.*, 2002; Fisher *et al.*, 2002; for review see Flax *et al.*, 2003). In addition, a high proportion of individuals with SLI also go on to develop reading problems (e.g., dyslexia), and the co-occurrence of spoken and written language problems in many individuals has contributed to a combined classification of language learning impairment (LLI; Catts, 1993; Flax *et al.*, 2003; Habib, 2000; McArthur *et al.*, 2000).

2. WHAT CAUSES LLI?

It is obvious from observing deaf children that dysfunction in the sensory registration of acoustic input produces an associated deficit in receptive and expressive aural language skills. What may be less obvious is that deafness also negatively impacts the acquisition of reading, even in individuals who are fluent in a visual sign language (e.g., see Rapin, 1978). This link between the auditory system and reading seems to occur via the phonological system. Specifically, in order to learn to perceive and produce aural language, the acoustic waveform of speech must be "chunked" into individual phonemes (the smallest unit of sound that can change the meaning of a word), and these individual phonemes must acquire neural representation. In order to learn to decode print (and become a proficient reader), an individual must become aware that words can be broken down into smaller sounds, and that these sounds (phonemes) are associated with letters (Vellutino and Scanlon, 1987; Shaywitz *et al.*, 1998; see Snow *et al.*, 1998 for review).

Given the primary role that phonological processing plays in aural and written language, considerable research has focused on the neural mechanisms sub-serving phonology. Indeed, a prominent debate in the field of language and reading research concerns whether phonological processing deficits (which have been shown to predict and characterize both aural and written language impairments) are specific to linguistic systems, or rather, reflect constraints in more basic information processing systems (for differing views see Ramus *et al.*, 2003; Tallal and Benasich, 2002).

Perspectives favoring a linguistic explanation for developmental language disorders parallel a more general view that speech is processed in a manner completely different from non-speech acoustic information, and that speech and non-speech processing are subserved by separate neural pathways (Chomsky, 1968; Fodor, 1983; Liberman *et al.*, 1967; Pinker, 1997). Evidence supporting a speech-specific neural sub-system derives from studies of categorical perception (the phenomenon wherein speech sounds are discriminated only when they are identified as being linguistically different), as well as studies of dichotic listening and acquired brain lesions in adults (which show that specific areas of the brain - generally in the left hemisphere - are specialized for speech, language and reading; Broadbent, 1958; Kimura, 1967; Liberman *et al.*, 1967). Evidence also draws from more recent neuroimaging studies, for example showing activation in left hemisphere regions using speech stimuli (e.g., Shaywitz *et al.*, 1995) versus homologous right hemisphere regions for non-linguistic acoustic discriminations (e.g., Zatorre *et al.*, 1992).

In order to experimentally test the notion that specific brain regions are truly specialized for speech processing, however, comparison "non-speech" control stimuli must be acoustically identical to the speech stimuli in every way except linguistic meaning (see Breier *et al.*, 2001 for discussion). When controls of this type are employed, highly overlapping neural activation patterns are seen for speech and comparable non-speech analogues (Belin *et al.*, 1998; Binder *et al.*, 2000; Fiez *et al.*, 1995; Poldrack *et al.*, 1998; Schwartz and Tallal, 1980; Scott *et al.*, 2000; Temple *et al.*, 2000). Such data suggest that spectro-temporal content may be a critical parameter modulating the processing of speech within the left hemisphere. This in turn supports a view that neural mechanisms sub-serving speech perception and complex acoustic processing share at least some common neural pathways, even at the level of cortex. This "bottom-up" view of speech processing is also supported by a range of electrophysiological and neuroimaging research (e.g., see Giraud and Price, 2001; Kaas and Hackett, 1999, 2000; Rauschecker and Tian, 2000; Romanski *et al.*, 1999; Scott and Johnsrude, 2003; Scott *et al.*, 2000; Steinschneider *et al.*, 1999; Wise *et al.*, 2001; Wise, 2003; Zattore, 2003).

Along similar lines, considerable evidence has accumulated to support the notion that developmental language learning impairments stem (at least in part) from non- speech-specific deficits in processing complex acoustic signals (Farmer and Klein, 1995; Hari and Kiesla, 1996; Hari and Renvall, 2001; Kraus *et al.*, 1996; McAnally and Stein, 1996, 1997; McCrosky and Kidder, 1980; Neville *et al.*, 1993; Reed, 1989; Robin *et al.*, 1989; Wright, *et al.*, 1997; Witton *et al.*, 1998; see Tallal and Benasich, 2002 for review). While some researchers dispute a causal role for non-linguistic acoustic deficits (arguing that language disabilities reflect a processing deficit specific to linguistic information, e.g., see Mody, Studdert-Kennedy, and Brady, 1997), other researchers question whether non-linguistic acoustic processing deficits (even if they do exist in this population) are either necessary or sufficient to cause language disability (Bishop *et al.*, 1999; Ramus, 2003).

Arguments against an acoustic deficit hypothesis based on null findings must be regarded very carefully, however, given that failure to replicate auditory processing deficits in LLI populations can be attributed to a number of methodological issues. Most notably, differences across studies in subject selection (such as degree of impairment or subject age), as well as differences in the difficulty of a psychophysical task (producing ceiling or floor effects), may be expected to obscure meaningful interpretation of findings. As one example, a frequently cited argument against the acoustic deficit hypothesis is made by Mody, Studdert-Kennedy and Brady (1997), based on null results from a study that suffers from a number of significant methodological problems (for detailed critique see Denenberg, 1999, 2001). Denenberg points out, for example, that the majority of subjects included as "poor readers" in this study actually performed well within the normal range for their age on standardized reading tests, and were only considered "poor readers" in comparison to subjects selected for the control group (who performed several years above their grade level on these same tests). Moreover, the studies that these authors claim to have failed to replicate used speech contrasts carefully selected to differ only along a single distinctive feature (primarily a single temporal cue, such as VOT, or a temporospectral cue, such as place of articulation), whereas the speech sound contrasts that Mody *et al.* employed differed along multiple acoustic features (thus the stimuli used in the Mody *et al.* studies could not appropriately test the hypothesis of interest). Yet another problem in interpreting data across studies designed to assess the causal relationship between auditory processing deficits and LLI is that many current studies tend to focus on dyslexic adults (who have had many years to compensate for their early impairments), while the original studies demonstrating significant auditory processing deficits in individuals with LLI were conducted on young children. It is not possible to dissociate the initial underlying cause(s) of developmental disabilities from compensatory effects and strategies that develop over many years when studying older individuals.

In sum, despite some findings to the contrary, the vast majority of research performed on young children who meet a clinical diagnosis of LLI (at least one standard deviation from the mean of normal) clearly supports the notion that deficits in processing rapidly changing acoustic information play a key deleterious role in the emergence of language deficits. In an extensive review of this literature, Leonard (1998) asserts, "the conclusion that children with SLI have difficulty processing brief or rapidly presented stimuli seems indisputable. These findings are so consistent and demonstrable across task and stimulus variations that it is difficult to imagine that they are not an important piece of the SLI puzzle" (p. 145).

In sum, issues regarding the neurobiology of auditory processing deficits - including whether they are specific to the auditory modality, and whether they are causal to or merely co-occurring with linguistic deficits – continue to catalyze experimental and theoretical research in this area. The growing body of associated research has led to several related hypotheses (see Habib, 2000 for review). Examples include the rate processing constraint hypothesis (see Fitch, Miller and Tallal, 1997 and Tallal *et al.*, 1998 for review); the magnocellular deficit hypothesis (see Stein and Talcott, 1999 and Stein, 2001 for review); the cerebellar deficit hypothesis (see Nicolson et. al., 2000 and 2001, for review); the double deficit hypothesis (see Wolf, 1986 for review); and the attentional dwell time hypothesis (see Hari, et. al., 1999 and Hari and Renvall, 2001 for review). All of these hypotheses have in common a constraint in the speed of information processing and/or production that is postulated to disrupt essential components of language learning, beginning with the acquisition of phonological representations. Thus, if we are interested in ultimately understanding the neurobiological basis of language development and disorders, it is important to assess the integrity of the component acoustic processes critical to analyzing the complex acoustic waveform of speech.

3. AUDITORY PROCESSING DEFICITS AND LLI

Tallal and Piercy (1973a, b) showed that young children diagnosed with LLI have particular difficulty both discriminating and sequencing two 75 millisecond (ms) steady state complex tones differing in fundamental frequency, but only when they were presented rapidly in succession. Figure 1 shows that whereas age- matched controls were able to reach 75% correct performance at inter-stimulus-intervals (ISIs) of 8 ms or longer, no LLI subject reached a criteria of 75% correct at an ISI of 150 ms or shorter.

Tallal and Piercy (1973b) dubbed this deficit (for processing acoustic changes in the tens of millisecond time range) a "temporal processing deficit." They suggested that the magnitude of the deficit would be sufficient to disrupt phoneme perception, as well as subsequent aspects of language and reading that depend on discreet phonemic representations. Specifically, they hypothesized that children with LLI might be expected to be impaired in discriminating speech sounds (such as consonant-vowel (CV) syllables like /ba/ vs /da/), which incorporate rapidly changing acoustic spectra.

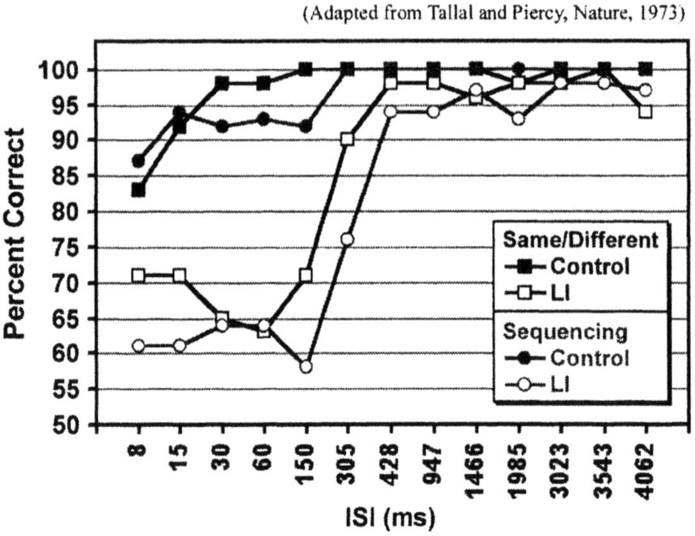

Figure 1. Children with LLI (n=12) and age matched controls (n=12) were presented with 2-tone sequences separated by an inter stimulus interval (ISI) of varying duration. In the Same/Different task, subjects were trained to press one button on a two-button response panel if the two were the same in frequency, and the other button if they were different. In the Sequencing task, subjects were trained to press one button for the higher frequency tone and the other for the lower tone. Next they were presented with 2-tone sequences of these tones and required to press the buttons in the order of stimulus presentation. Percent correct is plotted as a function of ISI; duration of tones = 75 ms. Tone 1 fundamental frequency = 100 Hz; Tone 2 = 305 Hz.

Conversely, children with LLI would not be expected to exhibit difficulties with steady-state stimuli, such as vowels. To test this hypothesis, speech stimuli were computer synthesized (providing explicit control of the temporal/spectral components within the acoustic waveform). Results demonstrated that children with LLI were unimpaired in their ability to discriminate between steady state vowels, but were significantly impaired in their ability to discriminate CV syllables incorporating 40 ms duration formant transitions (Tallal and Piercy, 1974)

To determine whether these results were tied to the duration of the formant transitions per se (as opposed to linguistic properties of vowels versus consonant-vowel syllables), an additional set of stimuli were computer synthesized. The same CV syllables (/ba/ vs /da/) were used, but in this study the duration of the formant transitions were extended from 40 to

80 ms. Results showed that the same children who were unable to discriminate between CV syllables with 40 ms duration formant transitions were able to discriminate the same CV syllables when presented with extended duration (80 ms) transitions (Tallal and Piercy, 1975).

This was the first report suggesting that speech discrimination may be improved through computer manipulation of temporo-spectral stimulus parameters, and led eventually to the development of effective remediation strategies that utilize acoustically modified speech to "re-train" individuals with speech, language and reading deficits (Tallal *et al.*, 1996; Merzenich *et al.*, 1996; see discussion below).

The degree of deficit in auditory temporo-spectral processing further appears to be correlated with higher-level aspects of linguistic processing - well beyond the phonemic level. For example, a significant correlation between degree of rapid auditory processing impairment and overall degree of receptive language impairment in children with LLI (derived from a comprehensive battery of standardized language tests) has been reported (Tallal, Stark and Mellitz, 1985). Interestingly, a significant correlation has also been found between the degree of rapid auditory processing impairment, and phonological decoding (as measured by the ability to read nonsense words) in children with dyslexia (Tallal, 1980; Talcott *et al.*, 2000).

4. PROCESSING MECHANISMS UNDERLYING LLI: A DEVELOPMENTAL CASCADE

Why should the ability to process brief, rapidly successive or rapidly changing acoustic cues play such a critical role in language development? Each language has its own set of phonemes that must be learned in infancy through experience-dependent (environmental) exposure, and mapped (represented) as distinct neural firing patterns in the auditory cortex. Thus the complex acoustic waveform of speech must activate neurons specialized to respond to distinct acoustic features (e.g., see Recanzone et. al., 1993). As early as 1949, Hebb proposed a neural mechanism wherein learning-induced "selection" of behaviorally relevant information (that which excites neurons nearly simultaneously in time) would "bind" together to form perceptually relevant representations. Thus repeated exposure to consistent sensory events (such as repeating components of speech) would increase the statistical probability that a particular neural firing pattern ensemble (cell assembly) would come to be distinctly associated with that sensory event.

It would be easy to understand how phonemes might come to be represented in the brain of an infant if each individual phoneme either occurred in isolation, or was represented by an invariant acoustic pattern within the ongoing speech waveform. However, this is not the case. No distinct acoustic boundaries separate phonemes within syllables or words. Rather, phonemes occur within an ongoing acoustic stream, and differ acoustically (sometimes dramatically so, due to co-articulation) within different contexts. Thus, one might hypothesize that in order for an infant to learn to represent the acoustic waveform of speech, the brain must segment the acoustic stream into "chunks" of time, and then form neural representations based on the consistency and frequency of the occurrence of associated neural firing patterns occurring within these "chunks".

Importantly, consistencies within the ongoing speech waveform can be found in various duration "chunks." For example, chunking over a tens of milliseconds time window would in theory allow for the fine-grain analysis required to represent individual phonemes. Chunking over a longer time window (hundreds of milliseconds) might result in building representations consistent with syllables or words (but at the expense of a much greater memory demand to store the higher number of individual syllables, and also with potentially degraded coding of subtle yet distinct within-syllable features).

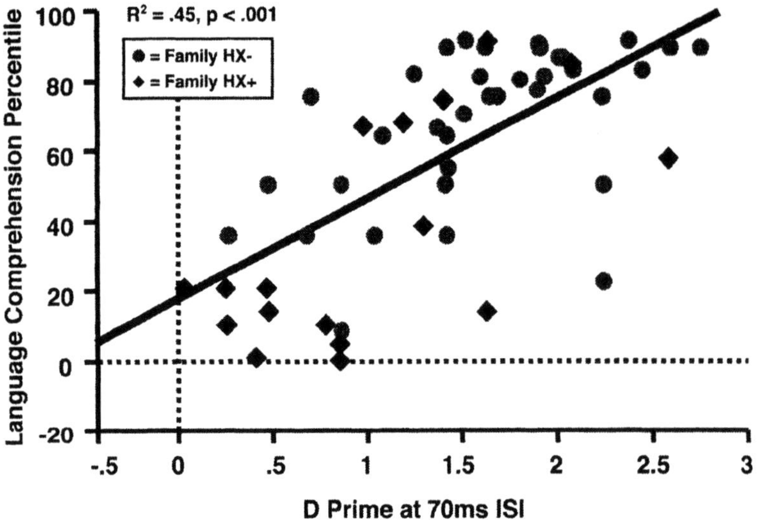

(Adapted from Benasich & Leevers, Progess in Infancy Research, 2002).

Figure 2. A bivariate plot showing the relationship between rapid auditory processing thresholds at 6 months (as measured by d' at 70ms ISI), and language comprehension percentile at 24 months (standardized Preschool Language Scale). Data shown for 33 control infants (HX-), and 17 infants with a family history of LLI (HX+). ISI=interstimulus interval

Tallal and colleagues have recently hypothesized that children with LLI may in fact chunk acoustic information in the hundreds of millisecond time window, and that this may contribute to their considerable phonological difficulties (Tallal, 2003). This hypothesis was tested indirectly by assessing the acoustic processing abilities of young infants beginning at 6 months of age, and then tracking their language development longitudinally through 36 months of age (Benasich and Tallal, 2002). Results from this study showed significantly higher auditory processing thresholds for infants born into families with a history of LLI (HX+) as compared to controls with a negative family history (HX-; Benasich and Leevers, 2002; Benasich and Tallal, 1996). In fact, assessment of infant acoustic processing abilities and emerging language through 36 months of age revealed that individual auditory processing thresholds were the single best predictor of language outcome as measured by the verbal scale of the Stanford- Binet intelligence test, as well as a standardized preschool language comprehension test. Interestingly, this proved to be the case for infants born into both HX+ as well as HX- families (see Figure 2). Specifically, auditory processing thresholds established in infancy not only quite accurately identified children who subsequently were "delayed" in language development, but also predicted language outcome for children developing language within the normal range. Importantly, none of the infant variables were capable of discriminating between children on the non-verbal portions of the Standford-Binet, demonstrating the specificity of the relationship between early auditory processing thresholds and subsequent individual differences in language development (as opposed to general intelligence; Benasich and Tallal, 2002).

Benasich's longitudinal research provides the first data obtained prospectively, in infancy, to demonstrate that: 1) individual perceptual processing thresholds for brief, rapidly presented auditory cues can be individually determined in the first year of life; 2) individual differences in auditory processing thresholds can predict subsequent language outcomes in toddlers; and 3) auditory processing skills in the first year of life are not only significantly related to subsequent language achievement in "at-risk" children, but are remarkably accurate in predicting linguistic outcomes at the age of 24 and 36 months across the full range of individual differences. These findings also are consistent with retrospective evidence that early gap detection thresholds predict language outcome measures in normal children (Trehub and Henderson, 1996), and are consistent with evidence that auditory event related potentials (ERPs) of infants at risk for language problems differ from controls, and also predict later language outcomes (Leppänen and Lyytinen, 1997; Leppänen *et al.*, 1999; Molfese and Molfese, 1997; Pihko *et al.*, 1999). These findings in infants also address the controversial issue of whether non-linguistic auditory processing deficits play a role in the eitiology of LLI by providing the first evidence that auditory temporal processing differences in infancy can be found pre-verbally, prior to an age at which language deficits might be expected to arise. While causality cannot be definitively established, these early indices are certainly predictive of later phonological and language problems.

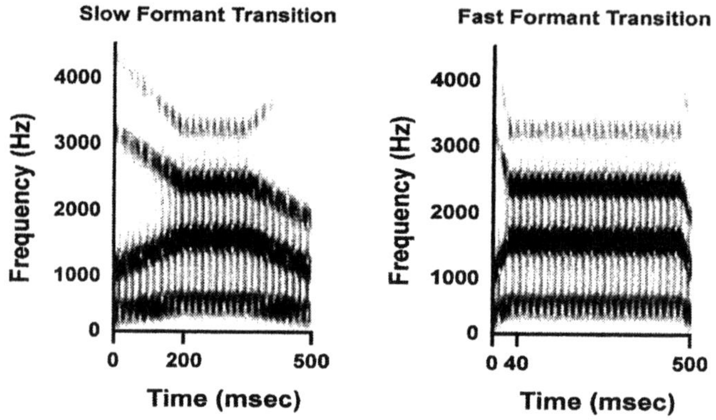

(Adapted from Temple et al., PNAS, 2000)

Figure 3. Non-speech analogues. Spectrograms of stimuli are shown with frequency (Hz) on the vertical axis and time (ms) on the horizontal axis. The spectro-temporal structure of the non-speech analogues with fast formant transitions was similar to that of consonant-vowel-consonant speech syllables (i.e., rapid acoustic changes occurring over 40 ms and surrounding a 520 ms steady-state period). In the slow formant transition stimuli, the duration of the acoustic transitions were extended to 200 msec.

5. NEUROIMAGING STUDIES OF LLI

Advances in in vivo non-invasive neuroimaging technologies have led to significant progress in understanding the neurobiological substrates of language development and disorders. In 1991, Jernigan and colleagues reported the results of the first structural magnetic resonance imaging (MRI) study of children with LLI. The authors found a significant reduction in gray matter volume in subcortical structures (including striatum and thalamus) in children with LLI as compared to age- matched controls, as well as bilateral reduction in cortical structures known to subserve language. Aberrant patterns of cerebral lateralization were also demonstrated both in prefrontal and parietal regions. These results have been confirmed and extended in subsequent MRI studies of individuals with language problems, including a lack of the normal left greater than right pattern in the planum temporale (located in Broca's area), and aberrant asymmetry in the parietal and frontal regions (Cowell *et al.*, 1995; Larsen *et al.*, 1990; Leonard *et al.*, 1993). Decreased physiological activation in these same regions has been reported for children with LLI using evoked potential recording (Neville *et al.*, 1993). Functional neuroimaging techniques (PET, fMRI) have also been used to assess adult dyslexics, and have consistently shown aberrant metabolic activity in both frontal and temporo-parietal language areas in LLI as compared to controls during a variety of phonological processing tasks (Hagman *et al.*, 1992; Paulesu *et al.*, 1996; Rumsey *et al.*, 1992, 1997; Shaywitz *et al.*, 1998).

Recently, Temple *et al.* (2000) used fMRI to examine the neural response to rapidly changing non-linguistic auditory stimuli in dyslexic and normal reading adults. Stimuli were comprised of computer synthesized non-speech (originally designed by Belin *et al.*, 1998) which incorporated rapid (40 ms) or slow (200 ms) acoustic transitions at the beginning and end of a steady state component. These stimuli were designed to mimic the spectro-temporal acoustic changes that characterize CVC syllables, but to have no linguistic relevance (see Figure 3). Subjects performed a pitch discrimination task while alternating between blocks of stimuli incorporating rapid or slow formant transitions. Results demonstrated that the largest area of activation to the rapidly (as compared to the slowly changing stimuli) for normal readers was in the left prefrontal region (between the middle and superior frontal gyri in Brodmann area 46 /10/9). However, the dyslexic readers revealed no left frontal activation to the rapidly (as compared to slowly) changing stimuli (see Figure 4; Temple *et al.*, 2000).

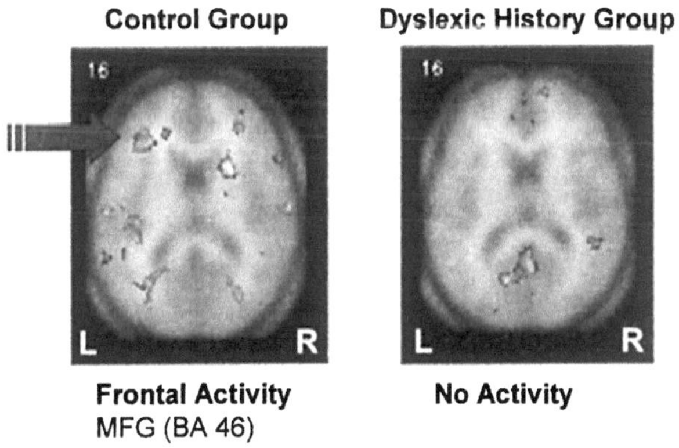

Figure 4. fMRI response to rapid auditory stimuli in normal (N=10) and dyslexic (N=8) readers. Normal readers show a significant frontal activity difference between rapid and slow non-speech analogues. Dyslexic readers show (on the smoothed averaged anatomies of dyslexic readers) no differential left frontal response to rapid vs slow stimuli. Adapted from Temple *et al.*, PNAS, 2000.

Functional neuroimaging studies of dyslexic adults and children have also revealed disruption in the neural responses to transient phonological auditory stimuli. Several studies have consistently found that both dyslexic adults and children failed to show activation in posterior language brain regions while performing phonological processing tasks, but did show activation (in some cases increased or more diffuse) in left frontal language areas. This atypical neural processing in both frontal and posterior left hemisphere areas of the language network has led to the hypothesis that developmental dyslexia could be described as a "disconnection" syndrome (Paulesu *et al.*, 1996). Such a hypothesis has been supported by PET studies (Paulesu *et al.*1996; Rumsey *et al.*, 1992) showing that dyslexic adults exhibit a decreased correlation in functional activity between temporal parietal and frontal brain areas.

6. NEUROMIGRATIONAL ANOMALIES

One possible etiology for anomalous neurocircuitry associated with developmental language learning impairments involves neuromigrational anomalies (Barth, 1987). Neuromigrational anomalies occur when control mechanisms regulating the final positioning of newly generated migrating neural cells are disrupted. Galaburda and Kemper (1979), for example, reported a post mortem cellular case study of the brain of an adult dyslexic male (who by history had been of normal intelligence, but had experienced delayed speech development and difficulties reading and spelling in elementary school). Galaburda and Kemper reported the presence of cortical polymicrogyria in the left temporal lobe, and cortical dysplasias throughout the left hemisphere. Later, Galaburda *et al.* (1985) performed post-mortem analyses on additional brains of dyslexics and reported numerous cellular anomalies, including neuronal ectopias and cortical dysplasias, primarily in left perisylvian/temporal (language) regions. The authors ascribed these cellular anomalies to focal disruption of neocortical neuronal migration, which probably arose during the prenatal period (see Galaburda, 1994, for review).

More recently, the same brains that had shown focal cortical disruption were further analyzed and found also to show disruption at the thalamic level. In comparison to controls, dyslexic brains had significantly smaller magnocellular (large) LGN cells (28% smaller in surface area), but no differences in parvocellular (small) LGN neurons (Livingstone *et al.*, 1991). Interestingly, evidence of thalamic anomalies has also been seen in the auditory (medial geniculate) nucleus (MGN) of the thalamus. Specifically, dyslexics exhibit a significant shift toward more small and fewer large cells in the left MGN as compared to controls (Galaburda *et al.*, 1994). Though the functional relevance of these findings remains unclear, they have been viewed in light of concurrent evidence that language-disabled subjects exhibit deficits in processing transient auditory information.

7. ANIMAL STUDIES OF NEUROMIGRATIONAL ANOMALIES AND AUDITORY PROCESSING

Following the discovery of cellular developmental anomalies in the brains of dyslexics, rodents were used to model these anomalies. Humphreys and colleagues (1991) reported that focal freezing lesions on the skull of postnatal Day 1 (P1) rats produced neocortical microgyria with striking histological similarity to those seen in dyslexic humans (see also Dvorák and Feit, 1977; Dvorák *et al.*, 1978; Rosen *et al.*, 1992). Further, male rats with neonatally induced microgyria also exhibited morphological changes in the medial geniculate nucleus (MGN) of the thalamus—specifically, a significant shift towards fewer large cells and a greater number of small cells (Herman *et al.*, 1997) - much like that seen in the MGN of human dyslexic brains (Galaburda *et al.*, 1994). Somewhat surprisingly, this effect was seen regardless of focal lesion location in cortex.

Thus a rodent model of at least one form of neuropathology associated with human dyslexia was developed. But how could this model provide the jump to neurobehavioral mechanisms underlying language disabilities? While rodents clearly do not learn to talk or read, evidence from many other areas of cognitive neuroscience research shows that animal models can be extremely informative in understanding the neurobiological basis for lower level processes that sub-serve higher cortical functions. Furthermore, there is considerable empirical evidence to suggest that the neural systems involved in the basic acoustic analysis of the complex waveform of speech operate similarly to those involved in comparable acoustic analysis of non-speech signals. Therefore, studying these auditory systems in animals should provided a good model for investigating questions that are central to understanding the neurobiological basis of central auditory processing disorders that have been shown to characterize LLI in humans. Accordingly, Fitch *et al.* (1994) adapted an operant conditioning paradigm to test auditory temporal perception in sham and microgyric male rats. They used two-tone stimuli similar to those found to detect auditory processing disorders in children with LLI, and manipulated the total stimulus duration (the duration of the two tones plus the duration of the ISI). Total stimulus duration was shortened across days of testing, so that subjects were tested with auditory sequences with progressively shorter total stimulus duration. Results showed that there was no significant difference between the performance of the sham operated compared to microgyric male rats at the longest stimulus duration; however, the microgyric male rats (particularly those with bilateral lesions) failed to discriminate between tone sequences at the short stimulus durations (see Figure 5; see also Clark *et al.*, 2000b; Fitch *et al.*, 1997). As can be seen by comparing results shown in Figure 5 to those in Figure 1, the pattern of performance seen in microgyric rats was strikingly similar to that seen in children with LLI (Tallal and Piercy, 1973a, b), and provided the first evidence of a link between a known neuropathological correlate of dyslexia and a known behavioral deficit associated with LLI.

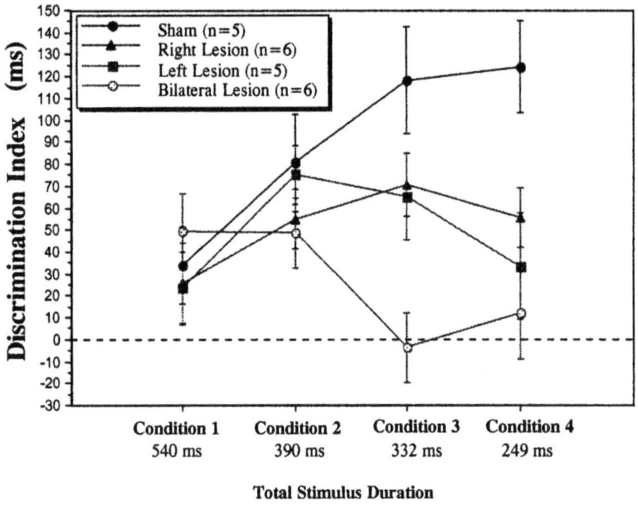

Figure 5. Discrimination indices for sham and microgyric rats tested in a two-tone sequence target identification task, as a function of treatment group and total stimulus duration. For Discrimination Indices, zero represents chance, while positive values reflect discrimination of the target. From Fitch et al., 1994.

More recently, an acoustic startle reduction (reflex modification) paradigm was used to assess discrimination in microgyric male rats. The startle reduction paradigm utilizes loud bursts of white noise, presented at random intervals unpredictable to the subject, and which, when presented, cause large-amplitude motor (startle) responses to occur. However, when the noise burst is preceded by benign stimulus or "cue," a significant reduction in the startle response can be measured (see Hoffman and Ison, 1980; Ison and Hoffman, 1983, for reviews). Stimulus detection assessments (including gap detection thresholds assessed via presentation of silent gaps of varying duration embedded in low-level background white noise; see Ison, 1982;

Ison and Pinckney, 1983; Ison *et al.*, 1991; Leitner *et al.*, 1993) can thus be employed without training. Clark *et al.* (2000a) adapted such a paradigm to the presentation of a repeating background of two-tone sequences, paired with the presentation of a reverse sequence (oddball) immediately before the noise burst on cued trials. (This format derives from the oddball stimulus presentation design used in electrophysiological research, e.g., Kraus *et al.*, 1994). Microgyric and sham male rats were presented with this paradigm with two-tone sequences of varying duration. The results (Figure 6) showed that both the sham and microgyric rats exhibited a highly significant reduction in startle response on cued (oddball) trials for "long" duration two-tone sequences, and thus both the sham and microgyric rats were able to detect the oddball (deviant) two-tone sequence as being different from the background two-tone sequence at these presentation rates. However, a significant difference was found between the sham and microgyric rats for more rapidly presented stimuli (total durations less than 89 ms); microgyric males showed significantly less response attenuation for these "short" duration stimuli (Clark *et al.*, 2000a).

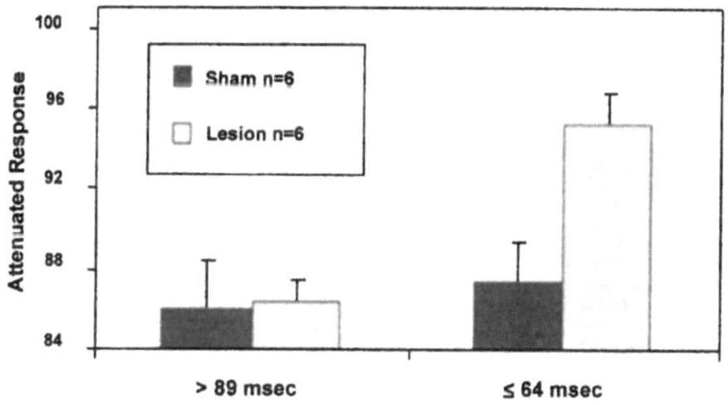

Figure 6. Attenuated startle response as a function of stimulus duration, for sham & microgyric male rats (from Clark *et al.*, 2000a). Responses at/near 100% indicate lack of stimulus discrimination.

Since identical acoustic stimuli were used in both the operant target ID auditory discrimination task and the startle reduction paradigm, a comparison could be made between the stimulus properties that elicited processing deficits in microgyric animals as a function of these different test paradigms. This was an important comparison, because the operant task is a fairly demanding task that requires subjects to learn a target, listen to a stimulus, make a comparison and then determine if that stimulus is the subject's target, whereas the startle reduction paradigm requires only the passive recognition of a stimulus change (and is thus considerably less demanding). Interestingly, while two-tone sequences of 249 ms duration have been found to elicit highly significant processing deficits in microgyric animals in the operant paradigm, the startle reduction paradigm (in which the same tone sequences were presented but in a passive oddball detection format) showed that only tone sequences shorter than 89 ms total duration would elicit response differences between control and microgyric animals. This suggests in turn that cognitive demand or "load" may interact with basic auditory processing deficits to produce behavioral impairment.

This insight from animal studies may help explain some of the discrepancies in the clinical (LLI) literature pertaining to central auditory processing disorders. Even when tasks are identical, the cognitive "load" may be different for young children than it is for older children or adults. While most studies using two-alternative forced choice paradigms to test auditory processing abilities in children with LLI have reported significant deficits in processing rapidly presented tone sequences, some studies using less cognitively demanding psychophysical methods (especially in adult subjects) have failed to find group differences. These discrepant results may reflect differential cognitive load across ages and tasks, and hence differential power to elicit deficits across studies.

8. ANIMAL MODELS OF SENSORY TRAINING AND PLASTICITY

Animal models have clearly been useful for investigating hypotheses pertaining to the neurobiological basis of auditory processing deficits. Animal studies have also proved invaluable in studying neuroplasticity in the auditory system. For example, studies have demonstrated that sensory experience can alter physiological indices of sensory processing in the brain, as well as behavioral performances. Specific to our interest in central auditory processing disorders, studies in non-human primates have shown that the acoustic frequency maps in temporal cortex can be altered by experience with particular frequencies (Recanzone, Schreiner and Merzenich, 1993; Kilgard and Merzenich, 1998). Such effects can be seen even in adult animals, suggesting a previously unimagined level of plasticity in the adult cortex with respect to central auditory processing (see Merzenich and Jenkins, 1995; Merzenich *et al.*, 1999 for reviews). Similar effects have been seen at the behavioral and neurophysiological level in humans (Karni and Sagi, 1991; Karni *et al.*, 1995), suggesting that similar organizational changes might be induced (with the right experience) in the brains of individuals with LLI.

9. REMEDIATION STUDIES

It has been convincingly demonstrated that phonological systems are developed in infancy through experience-dependent exposure to the native language (Kuhl *et al.*, 1992). Neurophysiologists have shown that within each sensory modality, the features that represent the physical world are mapped in the CNS in a highly organized fashion. It was previously thought that these neural maps had to be established during critical windows in development. However, recent research has challenged this view by demonstrating that changes in sensory thresholds, as well as neurophysiological sensory maps, can be driven by experience dependent learning in adult human and non-human primates (Buonomano and Merzenich, 1998; Karni *et al.*, 1995; Karni and Sagi, 1991; Kilgard and Merzenich, 1998; Merzenich and Jenkins, 1995). It remains unclear whether there are differences in learning that occurs early in life during "critical periods" as compared to learning that occurs latter in life. Nonetheless, it is hypothesized that neuroplasticity throughout the lifespan is driven by Hebbian mechanisms. These mechanisms operate to increase the connection strengths between "nearly simultaneous" firing of neurons within cortical networks to create cortical cell assemblies (Hebb, 1949) or "neural groups" (Finkel and Edelman, 1985). These "neural groups" might be strengthened progressively through explicit neuroplasticity based training, resulting in increasingly stronger positive coupling, with the neuronal cell assemblies responding increasingly more synchronously in time (see Merzenich *et al.*, 1999).

Tallal, Merzenich and colleagues (Tallal *et al.*, 1996; Merzenich *et al.*, 1996) hypothesized that Hebbian learning principles, as shown to drive neuroplasticity in monkey studies, might be adapted to treat the auditory and phonological processing constraints in children with LLI. Specifically, it was hypothesized that if language learning problems are characterized by a basic processing constraint in the way in which incoming sensory information is segmented (or "chunked" in time) and represented, and also are affected by the frequency and obligatory nature of phonological and morphological structures within a language (see Leonard, 1998), then these factors should drive the design of remediation strategies. To investigate this hypothesis, a hierarchy of computer-based neural training exercises was developed. The goal of the training was to: 1) drive neural processing of rapidly successive acoustic stimuli to faster and faster rates; and 2) improve speech perception, phonological analysis, phonological awareness, and language and reading comprehension by providing intensive training exercises within obligatory linguistic contexts (phonological, morphological, semantic and syntactic).

A unique aspect of this training was the development of a computer algorithm that acoustically modified speech signals. The algorithm temporally extended and amplitude-enhanced the brief, rapidly successive intra-syllabic acoustic cues that changed within a 3-30 Hz window within ongoing speech (see Nagarajan *et al.*, 1998 for a detailed description of the speech modification algorithm). Exercises were developed in the form of computer "games" that were programmed to be individually adaptive. The goal was to find for each participant a level of acoustic and linguistic functioning that elicited a high rate of accuracy. Once established, the exercises were programmed to change adaptively, moving from easier to progressively harder stimuli, based on each individual's trial-by-trial performance. As training progressed the goal was to drive the performance of each individual from a reliance on acoustically modified speech towards age-appropriate linguistic competence with natural speech. Non-linguistic acoustic training also was explicitly designed to drive temporal integration thresholds for rapidly successive frequency modulated sweeps to within the range needed to process the rapidly changing formant transitions in natural speech.

The first laboratory study to test the efficacy of this dual approach remediation strategy demonstrated dramatic success (Merzenich *et al.*, 1996; Tallal *et al.*, 1996, 1998). To explicitly assess the effects of the acoustic modification algorithm, separately from other aspects of the neuroplasticity-based experimental training, a study was conducted using a randomized, control study design. Two groups of children with LLI (matched on non-verbal intelligence and degree of receptive language impairment) participated. The experimental group received the training as described above, while the control group received essentially identical language training, but using natural, unmodified speech. In addition, instead of the non-linguistic acoustic training, the control group played computer "games" for equivalent periods of time, but these games were visual rather than auditory, and were not temporally adaptive. Both groups attended training together and received the same amount of training, reinforcement, and rewards for compliance throughout the study.

Results shown in Figure 7 demonstrate that, post- training, both the experimental and control group improved in speech discrimination, language processing and grammatical understanding. However, the group that received language training with the acoustically modified speech (as well as explicit training aimed at decreasing temporal integration thresholds for rapidly successive non-linguistic acoustic stimuli) showed significantly greater improvements in language than the comparison group. Furthermore, the temporal integration thresholds of the experimental group were dramatically decreased post- training. Importantly, there was a significant correlation between this change in temporal integration threshold and language improvement in the experimental group (Merzenich *et al.*, 1996; Tallal *et al.*, 1996).

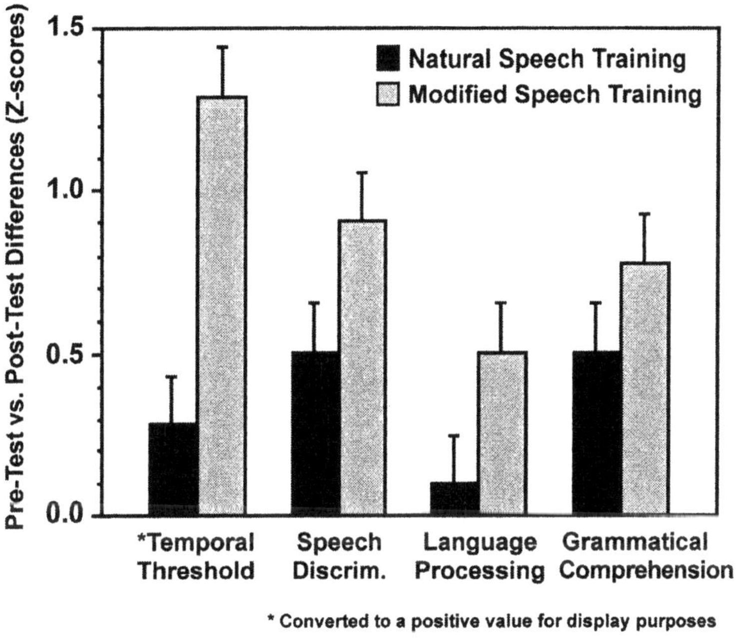

(Adapted from Tallal et al., Science, 1996)

Figure 7. Post-training minus pre-training difference (Z-scores) are shown for subjects with language impairment (LLI) who received training either with natural speech or acoustically modified speech. Raw scores on standardized test measures were converted to Z-scores on the basis of the pre-training performance of all subjects for each individual test. Mean and standard error values for each standardized test measure demonstrated that significantly greater improvement was achieved by the LLI children who received training with the acoustically modified speech (grey bars) as compared to those who received the same training with natural speech (black bars).

These results demonstrate that basic temporal thresholds are remarkably plastic, well beyond early critical periods of development. They also demonstrate that phonological processing, as well as higher level aspects of linguistic processing, can be driven towards rapid improvement through the use of neuroplasticity-based training aimed at ameliorating low level acoustic (temporo-spectral) processing constraints. These data support the hypothesis that when successive inputs are delivered to the same cortical processing channels at progressively higher speeds, they drive stronger positive coupling of smaller "chunked" cooperative "cell assemblies." This stronger coupling within a shorter (more fine-grained) time windows leads to increased precision in acoustic analyses. Precision within this tens of ms time window would appear to be critical for building up sharp (categorical) phonological representations, which in turn represent the building blocks of both aural language and reading.

The experiments described above were adapted from physiological studies with non-human primates demonstrating that alterations in sensory cortical maps could be induced through behavioral training (Recanzone, Schreiner and Merzenich, 1993; Kilgard and Merzenich, 1998). Given the magnitude of positive behavioral changes seen for children following the training described above, it was of considerable interest to determine if training-induced changes in metabolic activity also could be observed in human subjects with LLI (as measured pre- and post- training) using fMRI.

Initially, Temple *et al.* (2000) used fMRI to confirm previous reports of functional cortical disruption in dyslexics, and specifically reported that the neural response to rapidly changing non-speech analogues were disrupted in dyslexics. The authors subsequently asked whether this aberrant metabolic activation pattern could be "normalized" after training aimed at increasing the rate of temporo-spectral integration? To address this question, the commercially available set of neuroplasticity-based training exercises based on the research published in Merzenich *et al.*, 1996 and Tallal *et al.*, 1996 (Fast ForWord) was used. Three adult dyslexic subjects (two of whom showed significantly elevated temporal integration thresholds, and one who showed a threshold within the normal range) were scanned before and after training with Fast ForWord. Recall that this training program was explicitly designed to improve rapid successive processing of non-speech as well as speech stimuli. However, the training did not include exposure to the non-speech analogs or tasks used in the fMRI experiment itself. Results comparing pre- and post-training metabolic activity showed that the two subjects characterized by rapid auditory processing and auditory language comprehension deficits before training exhibited: 1) anomalous activation in left prefrontal cortex to rapid stimuli prior to training; and 2) significantly increased activity in left prefrontal cortex after training. The individual with normal acoustic thresholds before training did not show increased activity in left frontal cortex, and also did not show behavioral improvements. These preliminary data were among the first results to show both behavioral and neurophysiological effects of neural plasticity-based training in individuals with language disability.

As a follow-up to this pilot study, Temple *et al.* (2003) further assessed twenty children with dyslexia and twelve normal reading controls who received behavioral testing and fMRI scans both before and after Fast

ForWord training. The results replicated previous findings from dyslexic adults, in that an anomalous activation in left prefrontal cortex to rapid stimuli prior to training was observed, and significantly increased activity in this region occurred post training.

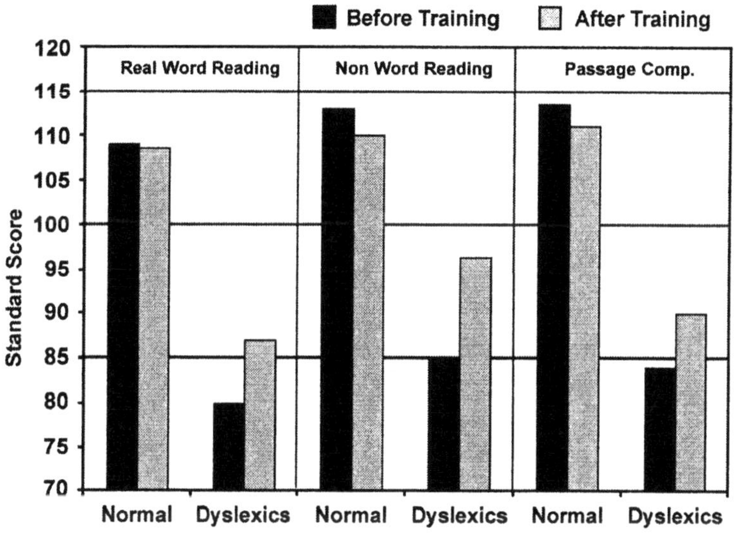

(Adapted from Temple et al., PNAS, 2003)

Figure 8. Standard scores on the three subtests of the Woodcock-Johnson Reading Mastery Test-Revised for normal (N=12) as compared to dyslexic readers (N=20) before (black bars) and after (grey bars) training with Fast ForWord. Mean=100; SD=15; <85 =below average.

An additional goal of this same study was to determine whether the characteristic neural signature showing decreased activation in left temporo parietal language regions, reported consistently in fMRI studies with adult dyslexics using phonological tasks, could be replicated in dyslexic children. If it were, it was of interest to determine whether metabolic activity in this region could be "normalized" by Fast ForWord training. To address these questions, the same twenty dyslexic and 12 normal reading children performed a letter-rhyme (phonological) task in the scanner. Prior to training, the normal reading children showed activity in both left frontal (inferior frontal, BA 44-6) and left temporal parietal (middle temporal and angular gyrus, BA 39) brain regions during the rhyming task. The dyslexic children, in a manner similar to what has previously been reported for dyslexic adults, showed more diffuse left frontal activity, and no significant left temporal parietal response during this phonological task. Next, the 20 dyslexic children completed 6 to 8 weeks of Fast ForWord training. Post-training behavioral results showed that the dyslexic children exhibited significantly improved performance on standardized reading measures assessing single word reading, non-word reading, and passage comprehension (see Figure 8), as well as substantial "normalization" of brain activation during these verbal tasks. Specifically, post-training fMRI results showed that the dyslexic children exhibited significantly increased left temporal parietal activation during the phonological processing (letter rhyming) task (whereas before training, the children with dyslexia had shown a complete failure to activate the left temporal parietal region), and a less diffuse area of activation in the left prefrontal cortex (see Figure 9; Temple *et al.*, 2003). Finally, significant correlations between behavioral improvement on language tasks and metabolic activation changes in dyslexics were found. Similar effects in control children, who had received two scans but did not receive intervening remediation, were not seen.

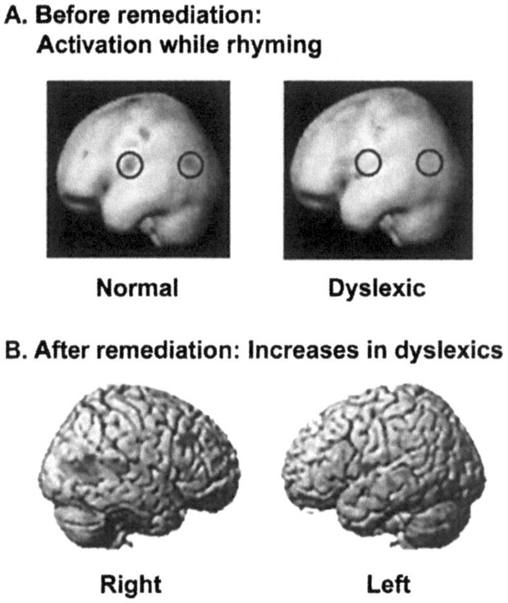

(Adapted from Temple et al., PNAS, 2003)

Figure 9. A) Left hemisphere activations of normal readers and dyslexic readers during rhyming (as compared to matching) letters. B) Areas showing increased activation during phonological processing in dyslexic readers after remediation with Fast ForWord. Black circles highlight left temporo-parietal and frontal regions which are disrupted in dyslexic children and significantly affected by remediation.

10. CONCLUSIONS AND FUTURE DIRECTIONS

In this review we focus on rapid auditory processing deficits often seen in individuals with language learning impairments. This line of research has proven useful in the study of neural substrates underlying language learning systems. However, it is important to emphasize that we do not suggest that this is the only potential model of LLI worthy of study. While temporo-spectral aspects of auditory processing clearly play an important role in the development of phonological systems that form the building blocks for both aural and written language, this does not imply that other acoustic, non-acoustic, or linguistic-specific systems are not also important contributors to language development and disorders. We focus here specifically on reviewing the human and animal literature pertaining to rapid auditory processing disorders because there is a broad body of data to bring to bear on questions of particular interest to systems and cognitive neuroscientists that have been - and more importantly, may become - increasingly amenable to research on the neurobiological substrates of language learning.

It is important to point out that there are likely many different potential "causes" for language learning deficits, and indeed, many different types of LLI. Moreover, language learning clearly is experience-dependent, and above all, developmental. The considerable confusion, and indeed controversy, that has characterized the theoretical assumptions driving this field of research may result, to a large extent, from a failure to study the neural basis of language systems within a developmental framework (for an extensive discussion of this important issue see Thomas and Karmiloff-Smith, 2003). Historically, our knowledge pertaining to the neural basis of language has derived primarily from studies of adult patients who have lost specific functions after sustaining discrete brain damage. Such observations from adult patients (i.e., apparently encapsulated, modular, language-specific systems) largely fail to match the effects of similar brain injuries sustained early in life. Moreover, patterns of impairment, aberrant correlations between variables, and even anatomical or morphological structural anomalies observed in young children with particular types of developmental language impairments may or may not match those observed in older children or adults (who have had a lifetime to develop compensatory strategies, and their resultant neuroplastic effects on brain physiology and anatomy).

At the same time, animal models have proven invaluable in studying the neurobiology of sensory, perceptual and motor systems and also the investigation of neurobiological systems sub-serving learning and memory. But until recently, no animal model has seemed appropriate for doing comparable studies of the neurobiological systems sub-serving speech, language and reading. While non-invasive in vivo neuroimaging technologies hold the potential to significantly increase our ability to explore the neural substrates underlying speech, language and reading in humans, they remain insensitive to the study of finer grain neurobiological circuitry. Thus evidence pin-pointing specific deficits in processing fine-grain acoustic information (in the tens of millisecond time range) seen in many individuals with LLI provides one impetus for the development of relevant animal models. These emergent animal models now hold the potential not only to increase our understanding of the neural basis of central auditory processing systems, but also their potential relationship to systems sub-serving the complex acoustic analysis of speech.

While it is clearly important to understand the neurobiological basis of LLI from a theoretical perspective, such studies also have the potential to significantly improve clinical and educational services for affected individuals. Specifically, advances in neuroplasticity research in the auditory modality has opened the window for developing neuroscience-based methods for remediating the acoustic, phonological language and reading deficits of individuals with LLI. Recent studies have shown the efficacy of these new approaches, both for improving the educational outcomes of affected individuals, as well as for "normalizing" metabolic activity in brain regions specialized for language and reading.

REFERENCES

Barth, P. G., 1987, Disorders of neuronal migration. *Can. J. Neurol. Sci.* **14**: 1–16.

Bartlett, C. W., Flax, J. F., Logue, M. W., Vieland, V. J., Bassett, A. S., Tallal, P., and Brzustowicz, L. M., 2002, A major susceptibility locus for specific language impairment is located on 13q21. *Am. J. Hum. Genet.* **71**: 45-55.

Beitchman, J. H., Nair, R., Clegg, M., Ferguson, B., and Patel, P. G., 1986, Prevalence of psychiatric disorders in children with speech and language disorders. *J. Am. Acad.Child Adolesc. Psychiatry* **25**: 528-535.

Belin, P., Zilbovicius, M., Crozier, S., Thivard, L., Fontaine, A., Masure, M. C., and Samson,Y., 1998, Lateralization of speech and auditory temporal processing. *J. Cogn. Neurosci.* **10**: 536-540.

Benasich, A. A., and Tallal, P., 1996. Auditory temporal processing thresholds, habituation and recognition memory over the first year. *Infant Behav. Dev.* **19**: 339-357.

Benasich, A. A., and Tallal, P., 2002, Infant discrimination of rapid auditory cues predicts later language impairment. *Behav. Brain Res.* **136**: 31-49.

Benasich, A. A., and Leevers, H.J., 2002, Processing of rapidly presented auditory cues in infancy: Implications for later language development. In *Progress in Infancy Research* (J. Fagana, and H. Haynes, eds), Lawrence Erlbaum Associates, Mahwah, NJ, Vol.3, pp.245-288.

Binder, J. R., Frost, J. A., Hammeke, T. A., Bellgowan, P. S. F., Springer, J. A., Kaufman, J. N., and Possing, E.T., 2000, Human temporal lobe activation by speech and non-speech sounds. *Cereb. Cortex* **10**: 512 – 528.

Bishop, D. V. M., Carlyon, R. P., Deeks, J. M., and Bishop, S. J., 1999, Auditory temporal processing impairment: Neither necessary nor sufficient for causing language impairment in children. *J. Speech Hear. Res.* **42**: 1295-1310.

Breier, J. I., Gray, L., Fletcher, J. M., Diehl, R. L., Klaas, P., Foorman, B. R., and Molis, M. R., 2001, Perception of voice and tone onset time continua in children with dyslexia with and without Attention Deficit/Hyperactivity Disorder. *J. Exp. Child Psychol.* **80**: 245 – 270.

Broadbent, D. E., 1958, *Perception and Communication.* Pergamon Press, New York.

Buonomano, D. V., and Merzenich, M. M., 1998, Cortical plasticity: from synapses to maps. *Annu. Rev. Neurosci.* **21**: 149-186.

Catts, H. W., 1993, The relationship between speech-language impairments and reading disabilities. *J. Speech Hear. Res.* **36**: 948-958.

Chomsky, N., 1968, *Language and Mind.* Harcourt, Brace & World, New York.

Clark, M., Rosen, G., Tallal, P., and Fitch, R. H., 2000a, Impaired processing of complex auditory stimuli in rats with induced cerebrocortical microgyria. *J. Cogn. Neurosci.* **12**: 828–839.

Clark, M., Rosen, G., Tallal, P., and Fitch, R. H., 2000b, Impaired two-tone processing at rapid rates in male rats with induced microgyria. *Brain Res.* **871**: 94–97.

Cowell, P., Jernigan, T., Denenberg, V., and Tallal, P., 1995, Language and learning impairment and prenatal risk: An MRI study of the corpus callosum and cerebral volume. *J. Med. Speech Lang. Pathol.* **3**: 1-13.

Denenberg, V. H., 1999, A critique of Mody, Studdert-Kennedy, and Brady's "Speech perception deficits in poor readers: Auditory processing or phonological coding?" *J. Learn. Disabil.* **32**: 379 – 383.

Denenberg, V. H., 2001, More power to them – Statistically, that is: A commentary on Studdert-Kennedy, Mody and Brady's criticism of a critique. *J. Learn. Disabil.* **34**: 299 – 301.

Diagnostic and Statistical Manual of Mental Disorders-Fourth Edition, 1994, American Psychiatric Association, Washington, D.C.

Dvorák, K., and Feit, J., 1977, Migration of neuroblasts through partial necrosis of the cerebral cortex in newborn rats—contribution to the problems of morphological development and developmental period of cerebral microgyria. *Acta Neuropathol. (Berl)* **38**: 203–212.

Dvorák, K., Feit, J. and Juránková, Z., 1978, Experimentally induced focal microgyria and status verrucosus deformis in rats — Pathogenesis and interrelation histological and autoradiographical study. *Acta Neuropathol. (Berl)* **44**: 121–129.

Farmer, M. E., and Klein, R. M., 1995. The evidence for a temporal processing deficit linked to dyslexia: A review. *Psychon. Bull. Rev.* **2**: 460-493.

Fiez, J. A., Raichle, M. E., Miezin, F. M., Petersen, S. E., Tallal, P., and Katz, W. F., 1995, Pet studies of auditory and phonological processing: effects of stimulus characteristics and task demands. *J. Cogn. Neurosci.* **7**: 357-375.

Finkel L.H., and Edelman, G. M., 1985, Interaction of synaptic modification rules within populations of neurons. *Proc. Natl. Acad. Sci. U S A* **82**: 1291-1295.

Fisher, S. E., Francks, C., Marlow, A. J., MacPhie, I. L., Newbury, D. F., Cardon, L. R., Ishikawa-Bush, Y., Richardson, A. J., Talcott, J. B., Gayan, J., Olson, R. K., Pennington, B. F., Smith, S. D., DeFries, J. C., Stein, J. F., and Monaco, A.P., 2002, Independent genome-wide scans identify a chromosome 18 quantitative-trait locus influencing dyslexia. *Nat. Genet.* **30**: 86-91.

Fitch, R. H., Brown, C., Tallal, P., and Rosen, G., 1997, Effects of sex and MK-801 on auditory processing deficits associated with developmental microgyric lesions in rats. *Behav. Neurosci.* **111**: 404–412.

Fitch, R. H., Miller, S., and Tallal, P., 1997, Neurobiology of speech perception. *Annu. Rev. Neurosci.* **20**, 331–53.

Fitch, R. H., Tallal, P., Brown, C., Galaburda, A., and Rosen, G., 1994, Induced microgyria and auditory temporal processing in rats: a model for language impairment? *Cereb. Cortex* **4**: 260–270.

Flax, J. F., Realpe-Bonilla, T., Hirsch, L. S., Brzustowicz, L. M., Bartlett, C. W., Tallal, P., 2003, Specific language impairment in families; evidence for co-occurrence with reading impairments. *J. Speech Lang.Hear. Res.* **46**: 530-543.

Fodor, J., 1983, *The Modularity of the Mind.* MIT Press, Cambridge, MA.

Galaburda, A. M., 1994, Developmental dyslexia and animal studies: At the interface between cognition and neurology. *Cognition* **50**: 133-149.

Galaburda, A. M., and Kemper, T. L., 1979, Cytoarchitectonic abnormalities in developmental dyslexia; a case study. *Ann. Neurol.* **6**: 94–100.

Galaburda, A. M., Menard, M. T., Rosen, G. D., and Livingstone, M. S., 1994, Evidence for aberrant auditory anatomy in developmental dyslexia. *Proc. Natl. Acad. Sci.U S A* **91**: 8010–8013.

Galaburda, A. M., Sherman, G. F., Rosen, G. D., Aboitiz, F., and Geschwind, N., 1985, Developmental dyslexia: Four consecutive cases with cortical anomalies. *Ann. Neurol.* **18**: 222–233.

Giraud, A. L., and Price, C. J., 2001, The constraints functional neuroimaging places on classical models of auditory word processing. *J. Cogn. Neurosci.* **13**: 754– 765.

Habib, M., 2000, The neurological basis of developmental dyslexia: an overview and working hypothesis. *Brain* **123**: 2373-99.

Hagman, J., Wood, F., Buchsbaum, M., Flowers, L., Katz, W., and Tallal, P., 1992, Cerebral brain metabolism in adult dyslexics assessed with positron emission tomography during performance of an auditory task. *Arch. Neurol.* **49**: 734-739.

Hari, R., and Kiesla, P., 1996, Deficit of temporal auditory processing in dyslexic adults. *Neurosci. Lett.* **205**: 138–140.

Hari, R., and Renvall, H., 2001, Impaired processing of rapid stimulus sequences in dyslexia. *Trends Cogn. Sci.* **5**: 525-532.

Hari, R., Vlata, M., and Uutela, K., 1999, Prolonged attentional dwell time in dyslexic adults. *Neurosci. Lett.* **271**: 202-204.

Hebb, D. O., 1949, *The Organization of Behavior: A Neuropsychological Theory*. Wiley Press, New York.

Herman, A., Galaburda, A., Fitch, R. H., Carter, A. R., and Rosen, G., 1997, Cerebral microgyria, thalamic cell size and auditory temporal processing in male and female rats. *Cereb. Cortex* **7**: 453–464.

Hoffman, H. S., and Ison, J. R., 1980, Reflex modification in the domain of startle: I. Some empirical findings and their implications for how the nervous system processes sensory input. *Psychol. Rev.* **87**: 175–89.

Humphreys, P., Rosen, G. D., Press, D. M., Sherman, G. F., and Galaburda, A. M., 1991, Freezing lesions of the newborn rat brain: A model for cerebrocortical microgyria. *J. Neuropathol. Exp. Neurol.* **50**: 145–160.

Ison, J. R., and Hoffman, H. S., 1983, Reflex modification in the domain of startle: II. The anomalous history of a robust and ubiquitous phenomenon. *Psychol. Bull.* **94**: 3–17.

Ison, J. R., 1982, Temporal acuity in auditory function in the rat: reflex inhibition by brief gaps in noise. *J. Comp. Physiol. Psychol.* **96**: 945–54.

Ison, J. R., and Pinckney, L. A., 1983, Reflex inhibition in humans: sensitivity to brief silent periods in white noise. *Percept. Psychophys.* **34**: 84–8.

Ison, J. R., O'Connor, K., Bowen, G. P., and Bocirnea, A., 1991, Temporal resolution of gaps in noise by the rat is lost with functional decortication. *Behav. Neurosci.* **105**: 33–40.

Jernigan, T., Hesselink, J. R., Sowell, E., and Tallal, P., 1991. Cerebral structure on magnetic resonance imaging in language- and learning-impaired children. *Arch. Neurol.* **48**: 539-545.

Kaas, J. H., and Hackett, T. A., 1999. "What" and "where" processing in auditory cortex. *Nat. Neurosci.* **2**: 1045 – 7.

Kaas, J. H., and Hackett, T. A., 2000, Subdivisions of auditory cortex and processing stream in primates. *Proc. Natl. Acad. Sci.U S A* **97**: 11793 – 9.

Karni, A., and Sagi, D., 1991, Where practice makes perfect in texture discrimination: Evidence for primary visual cortical plasticity. *Proc. Natl. Acad. Sci.U S A* **88**: 4966-4970.

Karni, A., Meyer, G., Jezzard, P., Adams, M. M., Turner, R., and Ungerleider, L. G., 1995, Functional MRI evidence for adult motor cortex plasticity during motor skill learning. *Nature* **377**: 155-158.

Kilgard, M. P., and Merzenich, M. M., 1998, Plasticity of temporal information processing in the primary auditory cortex. *Nat. Neurosci.* **1**: 727-731.

Kimura, D., and Harshman, R., 1984, Sex differences in brain organization for verbal and non-verbal functions. In *Progress in Brain Research*, vol. 61 (G.J. DeVries *et al.*, eds.), Elsevier Science Publishers, Amsterdam.

Kimura, D., 1967, Functional asymmetry of the brain in dichotic listening. *Cortex* **3**: 163-178.

Kraus, N., McGee, T., Littman, T., Nicol, T., and King, C., 1994, Nonprimary auditory thalamic representation of acoustic change. *J. Neurophysiol.* **72**: 1270–7.

Kraus, N., McGee, T. J., Carrell, T. D., Zecker, S. G., Nicol, T. G., and Koch, D. B., 1996, Auditory neurophysiologic responses and discrimination deficits in children with learning problems. *Science* **273**: 971–973.

Kuhl, P. K., Williams, K. A., Lacerda, F., Stevens, K. N., and Lindblom, B., 1992, Linguistic experience alters phonetic perception in infants by 6 months of age. *Science* **255**: 606-608

Larsen, J. P., Hoien, T., Lundberg, I., and Odegaard, H., 1990, MRI evaluation of the size and symmetry of the plamun temporale in adolescents with developmental dyslexia. *Brain Lang.* **39**: 289-301.

Leitner, D. S., Hammond, G. R., Springer, C. P., Ingham, K. M., Mekilo, A. M., Bodison, P. R., Aranda, M. T.,and Shawaryn, M. A., 1993, Parameters affecting gap detection in the rat. *Percept. Psychophys.* **54**: 395-405.

Leonard, C. M., Voeller, K. K. S., Lombardino, L. J., Morris, M. K., Hynd, G. W., Alexander, A. W., Andersen, H. G., Garofalakis, M., Honeyman, J. C., Mao, J., Agee, F., and Staab, E., 1993, Anomalous cerebral stucture in dyslexia revealed with magnetic resonance imaging. *Arch. Neurol.* **50**: 461-469.

Leonard, L. B., 1998, *Children With Specific Language Impairment.* MIT Press, Cambridge, MA.

Leppänen, P. H. T., and Lyytinen, H., 1997, Auditory event-related potentials in the study of developmental language-related disorders. *Audiol. Neuro-Otol.* **2**: 308-340.

Leppänen, P. H. T., Pihko, E., Eklund, K. M., and Lyytinen, H., 1999, Cortical responses of infants with and without a genetic risk for dyslexia: II. Group effects. *NeuroReport*, 969-973.

Liberman, A. M., Cooper, F. S., Shankweiler, D. P., Kennedy, M. S., 1967, Psychological Review Perception of the speech code. *Psychol. Rev.* **74**: 431-461.

Livingstone, M. S., Rosen, G. D., Drislane, F. W., and Galaburda, A. M., 1991, Physiological and anatomical evidence for a magnocellular defect in developmental dyslexia. *Proc. Natl. Acad. Sci.U S A* **88**: 7943–7947.

McAnally, K. I., and Stein, J. F., 1996, Auditory temporal coding in dyslexia. *Proc. R. Soc. Lond.* **263**: 961-965.

McAnally, K. I., and Stein, J. F., 1997, Scalp potentials evoked by amplitude modulated tones in dyslexia. *J. Speech Lang.Hear. Res.* **40**: 939–945.

McArthur, G. M., Hogben, J. H., Edwards, V. T., Heath. S. M., and Mengler, E. D., 2000, On the "specifics" of specific reading disability and specific language impairment. *J. Child Phychol. Psychiatry* **41**: 869-874.

McCrosky, R., and Kidder, H., 1980, Auditory fusion among learning disabled, reading disabled and normal children. *J. Learn. Disabil.* **13**: 69–76.

McGlone, J., 1980, Sex differences in human brain asymmetry: a critical review. *Behav. Brain Sci.* **3**: 215–263.

Merzenich, M. M., and Jenkins, W. M., 1995, Cortical plasticity, learning, and learning dysfunction. In *Maturational Windows and Adult Cortical Plasticity* (B. Julesz and I. Kovacs, eds.), Addison-Wesley, Vol. **23**, pp. 247-271.

Merzenich, M. M., Tallal, P., Peterson, B., Miller, S. L., and Jenkins, W. M., 1999, Some neurological principles relevant to the origins of - and the cortical plasticity based remediation of - language learning impairments. In *Neuroplasticity: Building a Bridge from the Laboratory to the Clinic* (J. Grafman, ed.), Elsevier, Amsterdam, pp. 169-187.

Merzenich, M., Jenkins, W. M., Johnston, P., Schreiner, C., Miller, S. L., and Tallal, P., 1996, Temporal processing deficits of language-learning impaired children ameliorated by training. *Science* **271**: 77-81.

Mody, M., Studdert-Kennedy, M., and Brady, S., 1997, Speech perception in deficits in poor readers: Auditory processing or phonological coding? *J. Exp. Child Psychol.* **64**: 199-231.

Molfese, D. L., and Molfese, V. J., 1997, Discrimination of language skills at five years of age using event-related potentials recorded at birth. *Dev. Neuropsychol.* **13**: 135-156.

Nagarajan, S. S., Wang, X., Merzenich, M. M., Schreiner, C. E., Johnston, P., Jenkins, W. M., Miller, S., and Tallal, P., 1998, Speech modifications algorithms used for training language learning-impaired children. *IEEE Trans. Rehabil. Eng.* **6**: 257-268.

Neville, H., Coffey, S., Holcomb, P., and Tallal, P., 1993, The neurobiology of sensory and language processing in language-impaired children. *J. Cogn. Neurosci.* **5**: 235-253.

Nicolson, R., Fawcett, A. J., and Dean, P., 2001, Dyslexia, development and the cerebellum. *Trends Neurosci.* **24**: 515 – 516.

Nicolson, R., Lenane, M., Singaracharlu, S., Malaspina, D., Giedd, J. N., Hamburger, S. D., Gochman, P., Bedwell, J., Thaker, G. K., Fernandez, T., Wudarsky, M., Hommer, D. W., and Rapoport, J. L., 2000, Premorbid speech and language impairments in childhood-onset schizophrenia: Association with risk factors. *Am. J. Psychiatry* **157**: 794-800.

Paulesu, E., Frith, U., Snowling, M., Gallagher, A., Morton, J., Frackowiak, R. S. J., and Frith, C.D., 1996, Is developmental dyslexia a disconnection syndrome. *Brain* **119**: 143-157.

Pihko, E., Leppänen, P. H. T., Eklund, K. M., Cheour, M., Guttorm, T. K., and Lyytinen, H., 1999 Cortical responses of infants with and without a genetic risk for dyslexia: I. Age effects. *NeuroReport* **10**: 901-905.

Pinker, S., 1997, Words and rules in the human brain. *Nature* **387**: 547-548.

Poldrack, R. A., Protopapas, A., Nagarajan, S., Tallal, P., Merzenich, M. M., Temple, E., and Gabrieli, J. D. E., 1998, Auditory processing of temporally compressed speech: An fMRI study. Paper presented at the *Cognitive Neuroscience Society Meeting*, March, 1998, San Francisco, CA.

Ramus, F., 2003, Developmental dyslexia: specific phonological deficit or general sensorimotor function? *Curr. Opin. Neurobiol.* **13**: 212 – 218.

Ramus, F., Rosen, S., Dakin, S. C., Day, B. L., Castellote, J. M., White, S., and Frith, U., 2003, Theories of developmental dyslexia: insights from a multiple case study of dyslexic adults. *Brain* **126**: 841 – 865.

Rapin, I., 1978, Consequences of congenital hearing loss – a long term view. *J. Otolaryngol.* **7**: 473 – 483.

Rauschecker, J. P., and Tian, B., 2000, Mechanisms and streams for processing of "what" and "where" in auditory cortex. *Proc. Natl. Acad. Sci.U S A* **97**: 11800 – 6.

Recanzone, G. H., Schreiner, C. E., and Merzenich, M. M., 1993, Plasticity in the frequency representation of primary auditory cortex following discrimination training in adult owl monkeys. *J. Neurosci.* **13**: 87–104.

Reed, M. A., 1989, Speech perception and the discrimination of brief auditory cues in reading disabled children. *J.Exp. Child Psychol.***48**: 270-292.

Robin, D., Tomblin, J.B., and Kearney, A., 1989, Auditory temporal pattern learning in children with severe speech and language impairment. *Brain Lang.* **36**: 604–613.

Romanski, L. M., Tian, B., Fritz, J., Mishkin, M., Goldman-Rakic, P. S., and Rauschecker, J. P., 1999, Dual streams of auditory afferents target multiple domains in the primate prefrontal cortex. *Nat. Neurosci.***2**: 1131 –1136.

Rosen, G. D., Press, D. M., Sherman, G. F., and Galaburda, A. M., 1992, The development of induced cerebrocortical microgyria in the rat. *J. Neuropathol. Exp. Neurol.* **51**: 601–611.

Rumsey, J. M., Andreason, P., Zametkin, A. J., Aquino, T., King, A. C., Hamburger, S. D., Pikus, A., Rapoport, J. L., and Cohen, R. M., 1992, Failure to activate the left temporoparietal cortex in dyslexia. *Arch. Neurol.* **49**: 527-534.

Rumsey, J. M., Nace, K., Donohue, B., Wise, D., Maisog, J. M., and Andreason, P., 1997, A positron emission tomographic study of impaired word recognition and phonological processing in dyslexic men. *Arch. Neurol.* **54**: 562-573.

Schwartz, J., and Tallal, P., 1980, Rate of acoustic change may underlie hemispheric specialization for speech perception. *Science* **207**: 1380-1381.

Scott, S. K., and Johnsrude, I. S., 2003, The neuroanatomical and functional organization of speech perception. *Trends Neurosci.s* **26**: 100 – 107.

Scott, S. K., Blank, S. C., Rosen, S., and Wise, R. J. S., 2000, Identification of a pathway for intelligible speech in the left temporal lobe. *Brain* **123**: 2400 – 6.

Shaywitz, B. A., Shaywitz, S. E., Pugh, K. R., Constable, R. T., Skudlarski, P., Fulbright, R. K., Bronen, R. A., Fletcher, J. M., Shankwelier, D. P., Katz, L., and Gore, J. C. , 1995, Sex differences in the functional organization of the brain for language. *Nature* **373**: 607–609.

Shaywitz, S. E., Shaywitz, B. A., Pugh, K. R., Fulbright, R. K., Constable, R. T., Mencl, W. E., Shankweiler, D. P., Liberman, A. M., Skudlarski, P., Fletcher, J. M., Katz, L., Marchione, K. E., Lacadie, C., Gatenby, C., and Gore, J. C., 1998, Functional disruption in the organization of the brain for reading in dyslexia. *Proc. Natl. Acad. Sci.U S A* **95**: 2636-2641.

Snow, C. E., Burns, S. M., and Griffin, P., 1998, *Preventing Reading Difficulties in Young Children.* National Academy Press, Washington, DC.

Stein, J., 2001, The magnocellular theory of developmental dyslexia. *Dyslexia* **7**: 12–36.

Stein, J., and Talcott, J., 1999, Impaired neuronal timing in developmental dyslexia-The magnocellular hypothesis. *Dyslexia* **5**: 59-77.

Steinschneider, M., Volkov, I. O., Noh, M. D., Garell, P. C., and Howard II, M.A., 1999, Temporal encoding of the voice onset time phonetic parameter by field potentials recorded directly from human auditory cortex. *J. Neurophysiol.* **82**: 2346 – 2357.

Talcott, J. B., Witton, C., McClean, M., Hansen, P. C., Rees, A., Green, G. G. R., and Stein, J. F., 2000, Dynamic sensory sensitivity and children's word decoding skills. *Proc. Natl. Acad. Sci.U S A* **97**: 2952-2957.

Tallal, P., and Benasich, A. A., 2002, Developmental language learning impairments. *Dev. Psychopathol.* **14**: 559-579.

Tallal, P., 1980, Auditory temporal perception, phonics and reading disabilities in children. *Brain Lang.* **9**: 182-198.

Tallal, P., 2003, Language learning disabilities: integrating research approaches. *Curr. Dir. Psychol. Sci.* **12**: 206-211.

Tallal, P., and Piercy, M., 1973a, Defects of non-verbal auditory perception in children with developmental aphasia. *Nature* **241**: 468-469.

Tallal, P., and Piercy, M., 1973b, Developmental aphasia: Impaired rate of non-verbal processing as a function of sensory modality. *Neuropsychologia* **11**: 389-398.

Tallal, P., and Piercy, M., 1974, Developmental aphasia: rate of auditory processing and selective impairment of consonant perception. *Neuropsychologia* **12**: 83-93.

Tallal, P., and Piercy, M., 1975, Developmental aphasia: the perception of brief vowels and extended stop consonants. *Neuropsychologia* **13**: 69-74.

Tallal, P., Merzenich, M. M., Miller, S., and Jenkins, W., 1998, Language learning impairments: Integrating basic science, technology, and remediation. *Exp. Brain Res.* **123**: 210-219.

Tallal, P., Miller, S. L., Bedi, G., Byma, G., Wang, X., Nagarajan, S. S., Schreiner, C., Jenkins, W. M., and Merzenich, M. M., 1996, Language comprehension in language-learning impaired children improved with acoustically modified speech. *Science* **271**: 81-84.

Tallal, P., Stark, R. E., and Mellits, D., 1985, The relationship between auditory temporal analysis and receptive language development: Evidence from studies of developmental language disorder. *Neuropsychologia* **23**: 527-534.

Temple, E., Deutsch, G. K., Poldrack, R. A., Miller, S. L., Tallal, P., Merzenich, M. M. and Gabrieli, J., 2003, Neural deficits in children with dyslexia ameliorated by behavioral remediation: evidence from fMRI. *Proc. Natl. Acad. Sci.U S A* **100**: 2860 – 2865.

Temple, E., Poldrack, R. A., Protopapas, A., Nagarajan, S., Salz, T., Tallal, P., Merzenich, M. M., and Gabrieli, J. D. E., 2000, Disruption of the neural response to rapid acoustic stimuli in dyslexia: Evidence from fMRI, *Proc. Natl. Acad. Sci.U S A* **97**: 13907-12.

Thomas, M., and Karmiloff-Smith, A., 2003, Are developmental disorders like cases of adult brain damage? Implications from connectionist modeling. *Behavi. Brain Sci.***25**: 727-788 .

Thomas, J., Clark, M., Benasich, A. A., and Fitch, R. H., 2001, Developmental changes in gap detection thresholds in the rat. Paper presented at *Cognitive Neuroscience Society Meeting*, March, 2001, San Francisco, CA.

Tomblin, J. B., Records, N. L., Buckwalter, P., Zhang, X., Smith, E., and O'Brien, M., 1997, Prevalence of specific language impairment in kindergarten children. *J. Speech Hear. Res.* **39**: 1315-1320.

Trehub, S. E., and Henderson, J. L., 1996, Temporal resolution in infancy and subsequent language development. *J. Speech Hear. Res.* **39**: 1315–20.

Vellutino, F. R., and Scanlon, D., 1987, Phonological coding, phonological awareness, and reading ability: Evidence from a longitudinal and experimental study. *Merrill Palmer Q.* **33**: 312-363.

Wise, R. J. S., 2003, Language systems in normal and aphasic human subjects: functional imaging studies and inferences from animal studies. *Br. Med. Bull.* **65**: 95–119.

Wise, R. J. S., Scott, S. K., Blank, S. C., Mummery, C. J., Murphy, K., and Warburton, E.A., 2001, Separate neural subsystems within "Wernicke's area". *Brain*, **124**: 83–95.

Witton, C., Talcott, J., Hansen, P., Richardson, A., Griffiths, T., Rees, A., Stein, J., and Green, G., 1998, Sensitivity to dynamic auditory and visual stimuli predicts nonword reading ability in both dyslexic and normal readers. *Curr. Biol.* **8**: 791–797.

Wolf, M., 1986, Rapid alternating stimulus naming in developmental dyslexias. *Brain Lang.* **27**: 360-379.

Wright, B. A., Bowen, R. W., and Zecker, S. G., 2000, Nonlinguistic perceptual deficits associated with reading and language disorders. *Curr. Opin. Neurobiol.* **10**: 482-486.

Wright, B. A., Lombardino, L. J., King, W. M., Puranik, C. S., Leonard, C. M., and Merzenich, M. M., 1997, Deficits in auditory temporal and spectral resolution in language-impaired children. *Nature* **387**: 176-178.

Zatorre, R. J., Evans, A. C., Meyer, E., and Gjedde, A., 1992, Lateralization of phonetic and pitch discrimination in speech processing. *Science* **256**.

Zatorre, R.J., 2003, Sound analysis in auditory cortex. *Trends Neurosci.* **26**: 229–230.

Neuroplastic Adaptations of the Auditory System in Musicians and Nonmusicians

LARRY E. ROBERTS*, DANIEL J. BOSNYAK*, ANTOINE SHAHIN[#], and LAUREL J. TRAINOR*
*Department of Psychology; and [#]Unit for Medical Physics and Applied Radiation Sciences, McMaster University, Hamilton, Ontario, Canada

1. INTRODUCTION

The human ear has been adapted by evolutionary processes to respond to sound frequencies that are present in the environment and convey information relevant to survival and reproductive fitness. However, the specific features of most sounds that we hear on a second-by-second basis (for example, the harmonic structure, loudness, and temporal shape of a particular voice, language, or musical note) and the meaning attached to these sounds are unique for each individual and cannot be anticipated by a genetic code. The evolutionary response to this limitation on natural selection has been the development of mechanisms that represent the detailed features of sensory input (sensory maps) and update those representations on a millisecond time scale (neural plasticity). We describe two experiments which used auditory evoked potentials (AEPs) to study these processes in the human brain.

2. EXPERIMENT 1

Alteration of the tuning properties of auditory neurons by aversive conditioning in the adult guinea pig has been documented in primary (A1) and secondary (A2) auditory cortex as well as in the medial, dorsal, and ventral divisions of the auditory thalamus (Edeline 1999). When brain regions are contrasted with the same training procedure, tone-evoked plasticity is expressed more commonly by neurons in A1 (95%) than by neurons in A2 (62%; Diamond and Weinberger, 1984). Neural plasticity of the magnitude seen in these and other animal studies suggests that cortical reorganization induced by behavioral training in humans should be expressed in AEPs which reflect the activity of populations of neurons in the brain. We therefore trained 8 nonmusician subjects to detect small increments in pitch from a carrier frequency of 2.0 kHz using 40-Hz amplitude modulated tones (Bosnyak *et al.*, 2004). This stimulus procedure allowed us to separate transient components of the AEP which have been localized to spatially differentiable generators distributed in the belt and parabelt regions of A2 from the 40-Hz "steady-state" response (SSR) whose sources localize to Heschl's gyrus in A1 (Schneider *et al.*, 2002). In this way the effects of training on distributed auditory cortical representations could be examined.

Subjects were trained for 15 daily sessions to discriminate between an S1 stimulus of 2.0 kHz and an S2 stimulus of either the same or a slightly higher frequency (each stimulus 1s duration separated by 0.5 sec). On each of 480 trials in a training session subjects stated whether the S1 and S2 were the same or different and were informed of the correctness of their decision. Two "test" sessions were also administered, one before the training series and the second after its completion, which evaluated discrimination ability without feedback for the trained stimulus set (2.0 kHz S1) and for two untrained control sets 200 Hz lower (1.8 kHz S1) or 200 Hz higher (2.2 kHz S1) than the trained stimuli. All stimuli were 40-Hz amplitude-modulated pure tones. The EEG was measured continuously (64 channels, sampled dc to 100 Hz @ 500 Hz using a Cz reference, re-referenced offline to an average reference) on the two test sessions and on the 3rd and 13th sessions of discrimination training. Subjects gained familiarity with 40 Hz AM tones in a preliminary session administered before the first test session in which auditory thresholds and initial discrimination ability at 2.0 kHz were assessed without feedback using staircase procedures.

Behavioral performance (hit rate corrected for false alarms) is summarized in Figure 1a (upper panel) where it can be seen that discrimination on the trained stimulus set improved rapidly in the early sessions and more gradually thereafter. Improvement was confirmed by comparing the before/after test sessions ($p<0.001$) and sessions 3 and 13 of training ($p<0.05$). Subjects also improved on the untrained stimulus sets ($p<.02$ in each case, lower panel, Figure 1a) although to a lesser degree than on the trained stimuli. These results were corroborated by d′ values and psychophysical functions plotted for each subject and stimulus set.

We analyzed the EEG for the 2.0 kHz S1 which the subjects experienced most frequently over training and was not interrupted by behavioral responses. Two transient AEPs were augmented by discrimination training

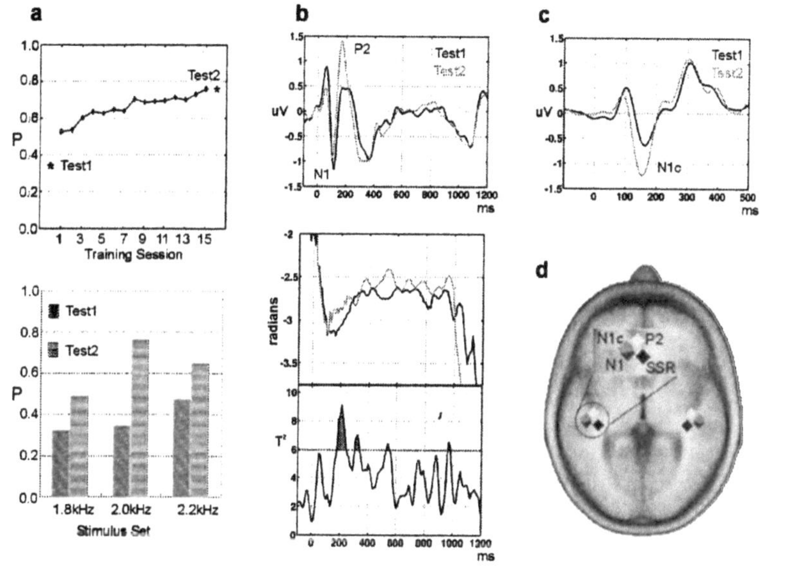

Figure 1. a. Behavioral performance [P=P(H)-P(FA)/1-p(FA)] on trained stimulus set (2.0 kHz S1, upper) and before/after differences on trained and control sets (lower). b. Transient AEP evoked by the trained 2.0 kHz S1 in before/after test sessions (upper), phase of the SSR in the two test sessions (middle), and bivariate T^2 evaluating before/after SSR differences (lower; horizontal line p<0.01, Monte Carlo determination), at electrode Fz. c. N1c before and after training (T_8 electrode, right hemisphere). d. Source localizations.

when before/after test sessions were compared, the P2 (latency 170 ms, Figure 1b, upper panel, p<0.001) and, in the right hemisphere only, the N1c (latency 155 ms, Figure 1c, p = 0.007). P2 and N1c responses evoked by untrained S1 control stimuli were also larger after training than before, but before/after differences did not reach significance. The amplitude of the N1 (Figure 1b, upper panel) did not change with training although N1 latency decreased by 9 ms (p<.001).

The SSR was analyzed by sliding a bivariate T^2 statistic sensitive to phase and amplitude in a window 100 ms wide across the EEG in 10 ms time steps, correcting for phase shifts induced by moving the window at each step. T^2 statistics comparing before/after SSR differences revealed a training-induced modification of the SSR commencing near the P2 (Figure 1b, lower panel) which was caused by a shortening of SSR phase (Figure 1b, middle panel) with no consistent effect on SSR amplitude (amplitude not shown;

animated phase and amplitude dynamics can be viewed at www.psychology.mcmaster.ca/hnplab). The phase effect was more pronounced for the trained 2.0 kHz S1 than for the untrained control S1 stimuli ($p<0.05$), although some generalization to the untrained 2.2 kHz S1 was observed (these results not shown). No 40-Hz activity corresponding to the phase effect was detected in a control experiment (n=10 subjects) in which P2 transient responses were evoked by unmodulated 2.0 kHz tones. This finding indicates that modification of SSR phase by discrimination training was not caused by a 40 Hz component of the P2 transient response but was a separate brain event.

Regional sources were modeled for each AEP (six determinations for each AEP, based on the three stimulus sets before and after training) and are averaged in the axial plane in Figure 1d. Sources for the SSR localized medially with respect to those of the N1, N1c, and P2 ($p<0.03$ or better) and posterior to the P2 ($p<0.0002$). These results are consistent with studies which have localized SSR generators by source modeling (Schneider *et al.*, 2002) and by intracortical measurements (Godey *et al.*, 2001; Liégeois-Chauvel *et al.*, 1993) to Heschl's gyrus. Differentiation of SSR sources from those of the N1, N1c, and P2 is also in agreement with previous findings reviewed by Shahin *et al.* (2003) which have localized N1, N1c, and P2 sources to the region of A2 including P2 sites anterior to the auditory core. P2 sources may reflect activation centered in anterior auditory belt regions which receive reciprocal connections from one another and from parabelt zones that project reciprocally to prefrontal cortex (Kass and Hackett, 1998). The N1 and N1c sources shown in Figure 1d localized laterally to those of the P2 ($p<0.0005$) and may reflect activation of posterior and lateral parabelt regions which have dense connections with caudal and rostral parts of the superior temporal gyrus. A note of caution is that source analysis differentiates only centers of activation and cannot resolve overlapping generators of similar orientation or determine their spatial extent.

Enhancement of the N1c and P2 in our data suggests that the number of A2 neurons depolarizing synchronously was increased by training on the discrimination task. This finding is consistent with animal data indicating that plasticity is a general property of A2 neurons. The expression of the N1c in the right hemisphere in our subjects is consistent with evidence for specialization of auditory neurons in this hemisphere for processing of spectral information (Zatorre and Belin, 2001). However, contrary to animal studies (Edeline, 1999; Recanzone *et al.*, 1993), our SSR results do not point to an expansion of the tonotopic representation for the trained frequencies in A1. Rather, the temporal properties of the response were modified such that SSR phase appeared to plateau more quickly after training than before training began. It is possible that competitive interactions induced by processing of multiple S2 frequencies during discrimination may have preserved a segregated representation in A1, such that only temporal properties of the representation were affected (Kilgard *et al.*, 2001). Temporal effects obtained for the SSR and for the N1 and N1c invite the hypothesis that acoustic properties of the S1 stimulus may have been represented more rapidly after training compared to before training.

3. EXPERIMENT 2

The results of Experiment 1 showing that P2 and N1c responses are neuroplastic implies that these responses should be augmented when evoked by musical tones in highly skilled musicians who have processed musical stimuli extensively in the context of musical practice. Experiment 2 (Shahin *et al.*, 2003) evaluated this prediction.

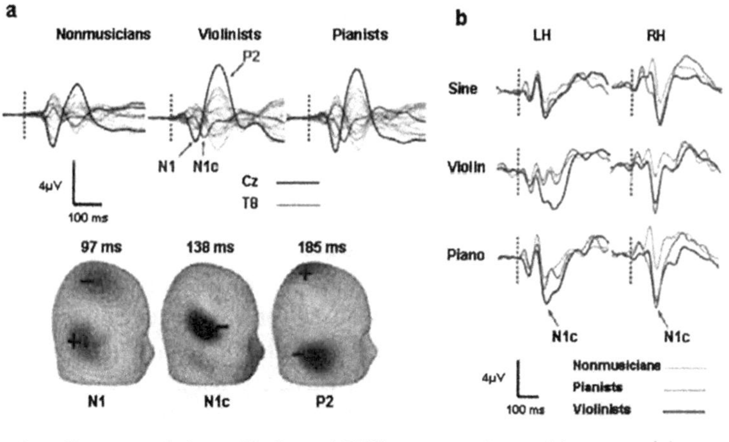

Figure 2. a. Upper panel shows 32 channel EEG response observed in nonmusician, violinist, and pianist groups averaged over sine, violin, and piano tones. N1 and P2 are identified in electrode C_z and N1c in T_8. Lower panel shows latency and scalp current density for N1, N1c, and P2. *b.* N1c evoked by each tone in the left (T_7) and right (T_8) hemispheres.

We studied 11 highly skilled violinists (age 24.3 + 2.2 years) who were members of Canada's National Academy Orchestra and 9 skilled pianists (aged 23± 2.5 years) who had at least Grade 10 certification from Canada's Royal Conservatory of Music. Nonmusician controls (n=14) were age matched university students who did not play a musical instrument and reported listening passively to music for less than 1 hour/day. Subjects were presented with violin and piano tones (A3 and C3, American notation) and pure tones of the same fundamental frequency. Each tone (500 ms duration) was presented 120 times in a randomized order (free field, ISI 2s) while subjects read a newspaper. The EEG was recorded from 32 channels (10-20 system, DC to 100 Hz) using methods described previously.

As predicted, we found an enhanced P2 in our two musician groups compared to nonmusician controls (Figure 2a, main effect of group $p<0.0001$, C_z electrode). A main effect of stimulus was also found ($p<0.0001$, not shown) which was attributable to a larger P2 occurring to piano and violin tones than to sine tones, but the interaction of stimulus with group was not significant. In addition an enhanced N1c was detected in our two musician groups compared to controls (group main effect $p<0.025$, electrodes T_7 and T_8) and is shown for each stimulus, group, and hemisphere in Figure 2b. Compared to control subjects both musician groups showed larger N1c responses to all stimuli in the right hemisphere ($p<0.04$ or better), particularly for violin and piano tones; group comparisons were not significant for any stimulus in the left hemisphere. No effects of group on N1 amplitude or latency were found. However, N1c latency was shorter for violin and piano tones compared to sine tones in the right hemisphere (tone by hemisphere interaction $p<0.025$). Regional sources fitted to the P2 localized medially to those of the N1 ($p<0.02$) and N1c ($p<0.0001$) in the region of auditory cortex, as was found in Experiment 1.

4. GENERAL DISCUSSION

In Experiment 2 the predicted enhancements of P2 and N1c in musicians were obtained for musical tones as well as for sine tones which have the quality of pitch. At present it is not clear whether musical skill is associated with augmented brain responses for sounds in general or only for sounds processed during musical practice (Pantev *et al.*, 2001; Shahin *et al.*, 2003). In our study enhancement of the P2 and N1c in musicians was not specific to the instrument of practice, perhaps because several violinists reported piano as a secondary instrument. Laboratory training results which show P2 and N1c responses to be neuroplastic (Experiment 1; cf. Tremblay *et al.* 2001 and Atienza *et al.* 2002 for the P2) indicate that intrinsic genetic and/or prenatal factors need not be invoked to explain augmentation of these responses in musicians, although such factors could play a contributing role.

Recently, the N19-P30 source waveform underlying the 40-Hz SSR has been reported to be augmented by 102% in professional musicians compared to nonmusicians when extracted by deconvolution from SSRs near 39 Hz (Schneider *et al.*, 2002). The SSR source waveform also correlated highly ($r = .87$) with the volume of gray matter in the anteromedial portion of Heschl's gyrus well as with musical aptitude ($r = .71$). Our results taken with those of Schneider *et al.* (2002) therefore suggest a dissociation of transient P2 and

N1c AEPs from the SSR, with the neuroplastic P2 and N1c expressing as amplitude enhancements in training studies and in musicians but SSR amplitude enhancement in musicians only where it could be an anatomical marker for musical skill. These findings call for study of the principles and mechanisms that govern cortical reorganization induced by experience over the life span and point to the tractability of their investigation in humans. Training procedures other than the one we studied in Experiment 1 may modify SSR amplitude and its anatomical substrate depending on the type of training that is given, its duration, and when it is delivered in the course of brain development.

ACKNOWLEDGEMENTS

This research was supported by grants from NSERC (operating) and CIHR (Operating and NET) of Canada.

REFERENCES

Atienza, M., Cantero, J. L., and Dominguez-Marin, E., 2002, The time course of neural changes underlying auditory perceptual learning. *Learn. Mem.* **9**: 138-150.

Bosnyak, D. J., Eaton, R. A., and Roberts, L. E., 2004, Distributed auditory cortical representations are modified when nonmusicians are trained at pitch discrimination with 40-Hz amplitude modulated tones. *Cereb. Cortex*, in press.

Diamond, D. M., and Weinberger, N. M., 1984, Physiological plasticity of single neurons in auditory cortex of the cat during acquisition of the pupillary conditioned response: II. Secondary field (AII). *Behav. Neurosci.* **98**: 189-210.

Godey, B., Schwartz, D., de Graaf, J. B., Chauvel, P., and Liégeois-Chauvel, C., 2001, Neuromagnetic source localization of auditory evoked fields and intracerebral evoked potentials: A comparison of data in the same patients. *Clin. Neurophysiol.* **112**: 1850-1859.

Edeline, J., 1999, Learning-induced physiological plasticity in the thalamo-cortical sensory systems: A critical evaluation of receptive field plasticity, map changes and their potential mechanisms. *Prog. Neurobiol.* **57**: 165-224.

Kaas, J. H., and Hackett, T. A., 1998, Subdivisions of auditory cortex and levels of processing in primates. *Audiol. Neurootol.* **3**: 73-85.

Kilgard, M. P., Pandya, P. K., Vazquez, J., Gehi, A., Schreiner, C. E., and Merzenich, M. M., 2001, Sensory input directs spatial and temporal plasticity in primary auditory cortex. *J. Neurophysiol.* **86**: 326-338.

Liégeois-Chauvel, C., Musolino, A., Badier, J. M., Marquis, P., and Chauvel, P., 1993, Evoked potentials recorded from the auditory cortex in man: evaluation and topography of the middle latency components. *Electroenceph. Clin. Neurophysiol.* **92**: 204-214.

Pantev, C., Roberts, L. E., Schulz, M., Engelien, A., and Ross, B., 2001, Timbre-specific enhancement of auditory cortical representations in musicians. *NeuroReport* **12**: 169-174.

Recanzone, G. H., Schreiner, C. E., and Merzenich, M. M., 1993, Plasticity in the frequency representation of primary auditory cortex following discrimination training in adult owl monkeys. *J. Neurosci.* **13**: 87-103.

Schneider, P., Scherg, M., Dosch, H. G., Specht, H. J., Gutschalk, A., and Rupp, A., 2002, Morphology of Heschl's gyrus reflects enhanced activation in the auditory cortex of musicians. *Nat. Neurosci.* **5**: 688-694.

Shahin, A., Bosnyak, D. J., Trainor, D. J., and Roberts, L. E., 2003, Enhancement of Neuroplastic P2 and N1c Auditory Evoked Potentials in Musicians. *J. Neurosci.* **23**: 5545-5552.

Tremblay, K., Kraus, N., McGee, T., Ponton, C., and Otis, B., 2001, Central auditory plasticity: Changes in the N1-P2 complex after speech-sound training. *Ear Hear.* **22**: 79-90.

Zatorre, R. J., and Berlin, P., 2001, Spectral and temporal processing in human auditory cortex. *Cerebral Cortex* **11**: 946-953.

Functional and Structural Characteristics of Auditory Cortex in the Blind

ALEXANDER A. STEVENS
Department of Psychiatry, Oregon Health & Science University, Portland, Oregon, USA

1. INTRODUCTION

While the acoustic environment is similar for the sighted and blind, the reliance on auditory perception is much greater in the blind because events beyond their fingertips reach them primarily through sound. There is increasing evidence that extensive use of a particular sensory modality leads to functional and structural changes in cortical fields. For instance, early-onset blindness has been linked to superior auditory abilities in attention, memory and speech processing (Niemeyer and Starlinger, 1981; Muchnik *et al.*, 1991; Roder *et al.*, 2000, 2001). Recent neuroimaging studies have established that occipital areas normally involved in visual processing are responsive during auditory and language processing tasks, indicating substantial functional reorganization in posterior cortex (Arno *et al.*, 2001; Burton *et al.*, 2002). Less attention has been paid to auditory cortical areas in the temporal lobes despite the fact that many auditory functions that are enhanced in the blind no doubt rely on these structures. This raises the possibility that blindness alters functional and structural properties in auditory cortices as well.

Recently, evidence for alterations in auditory cortices in the blind was provided using MEG, indicating that tonotopic maps on Heschl's gyrus may be expanded (Elbert *et al.*, 2002). Changes in both function and structure have been detected in Heschl's Gyrus (HG) of professional musicians suggesting that neurophysiological and cortical hypertrophy are associated with enhanced auditory abilities (Schneider *et al.*, 2001). These findings raise the question whether functional alterations and structural changes in HG reflect developmental and environmental influences or rather that the functional and structural changes underlie enhanced auditory abilities. We set out to explore the issue of auditory functional and structural changes in the auditory cortex in the blind. We reasoned that, if changes are due to heavy reliance on auditory perceptual functions then we may be able to detect changes in metabolic activity in auditory cortical areas using functional magnetic resonance imaging (fMRI). Similarly, if auditory perceptual enhancement is associated with cortical hypertrophy in HG then we may be able to detect such changes using structural MRI.

2. APPROACH

Given the importance of hearing to the blind for interpreting the environment, we hypothesized that the blind, particularly congenitally blind (CB), may have functional alterations in the auditory cortical fields. In addition, we hypothesized that if increases in cortical volume results from use-dependent plasticity, then the CB subjects may also show increased grey and white matter volume in HG. We tested these two hypotheses using fMRI and structural MRI. Nineteen healthy CB (N=7), adventitiously blind (AB N=5), and sighted control (SC N=7) subjects received fMRI and structural MRI scans, and structural scans were obtained for two additional CB and two additional AB subjects for the volumetric analyses of HG.

3. FUNCTIONAL MRI OF AUDITORY STIMULATION

The fMRI study compared the location and extent of signal in auditory cortical areas along the superior temporal plane in blind and sighted subjects in order to detect changes in auditory functional organization associated with visual deprivation. We used an EPI-BOLD sequence with "sparse sampling" in order to reduce the stimulation of auditory cortex by scanner noise (Robson *et al.*, 1998). Each acquisition of brain image volumes was separated by an 8.3 sec interval during which the stimuli were presented. This is important due to the possibility that blind and sighted individuals may differ in their sensitivity to the scanner noise, causing different degrees of attenuation of stimulus-based signal in auditory cortical fields.

Pure Tones (PT) and Frequency Modulated (FM) tone stimulation conditions were compared to a quiet baseline condition. PT stimuli were pure sinusoids produced at frequencies from 0.4 kHz to 0.48 kHz (LO) for the low frequency set and 4.0 kHz – 4.8 kHz (HI) for the high frequency set. FM stimuli were created in the same LO and HI frequency ranges as the PT stimuli but each tone was frequency modulated at 10% of its centre frequency. Stimuli were 750 ms in duration (50 ms rise/fall time) and presented at 75-78 dB SPL. Eleven tones were presented between each volume acquisition. Each condition was presented 12 times per scan and subjects underwent four scans. Subjects were instructed to press a button whenever they detected a change from a PT to an FM condition. Only the analyses of the combined PT and FM stimulation conditions compared to baseline will be presented here.

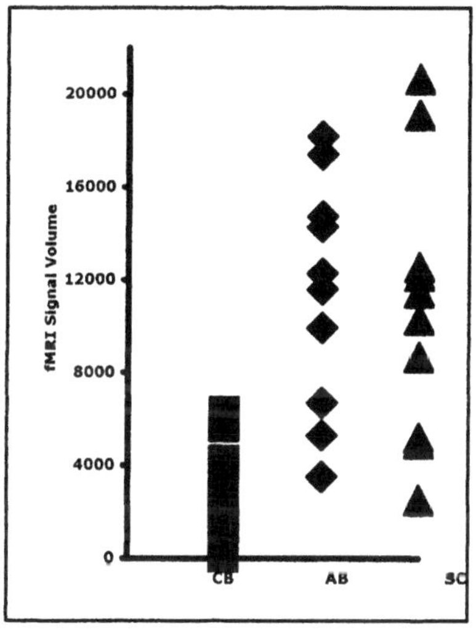

Figure 1. Total fMRI signal volume in the temporal lobes by group.

The data were analyzed using Brain Voyager 4.9 (Brain Innovation). Following correction for head-motion and normalization of intensity within scans, the data were spatially smoothed using a 4 mm (FWHM) Gaussian kernel. A general linear model was used to generate statistical parametric maps for each subject with thresholds for active pixels set at ($P < 0.01$) corrected for multiple comparisons. The number of active voxels within each temporal lobe were subjected to a repeated measures analysis of variance with Group as the between subjects factor and hemisphere a within subject factor. Analysis of the combined PT and FM stimulation conditions revealed two major findings. First, the CB group had significantly less activity in the temporal lobes than both the AB and SC groups (main effect of group: $F(2,13) = 6.19$, $P = 0.02$) confirmed with a Scheffe' test (Figure 1). This significantly lower level of activity in the CB group was present bilaterally as indicated by no significant main effect of hemisphere (P = 0.97) nor a significant interaction of group and hemisphere (P = 0.83). Second, calculations of the center of mass of activity indicated that the CB group had a significantly more posterior center of mass of activity in the left temporal lobe than the SC, with the AB group falling in between (Table 1). This result suggests that substantial functional reorganization occurred in auditory cortices in the CB group and to a lesser degree in the AB group.

Table 1. Coordinates of the centre-of-mass of activation in temporal lobes.

	Left Hemisphere			Right Hemisphere		
Group	X	y	z	x	y	z
AB	-47	-28	6	49	-24	7
CB	-48	-31*	5	52	-25	5
SC	-45	-23	7	50	-20	2

* Significantly more posterior than SC y-coordinate (P < .05).

Table 2. Heschl's Gyrus volumes (in mm^3) by group.

	White Matter		Grey Matter	
Group	Left*	Right	Left	Right
AB	740	862	2348	1912
CB	774	1016	1983	1882
SC	733	799	2288	1840

Left greater than right (P < .04)

4. STRUCTURAL MRI

The structural MRI study investigated whether there were measurable changes in HG as a result of blindness. The structural scans were acquired with a voxel size of 0.94 mm^3. Each HG was traced by hand using landmarks described by Penhune *et al.* (1996). The grey matter and white matter within HG was segmented using a partially automated algorithm implemented in Brain Voyager. Gray and White matter volumes were analyzed using a within and between subjects repeated measures design. There were no reliable differences detected between the CB, AB, and SC groups on measures of HG grey matter or white matter volumes (Table 2). There was a main effect of hemisphere indicating that the left HG had significantly greater WM than the right HG [F(1,20) = 4.41, P = .048) and marginally greater grey matter (P = 0.07). Similar results were also obtained when the analyses were restricted to the medial two-thirds of HG, where primary auditory cortex is normally localized in humans.

5. DISCUSSION

The results of the present functional MRI and structural MRI studies suggest that there are complex functional changes in auditory cortical areas in the congenitally blind individuals, and to a lesser degree in the adventitiously blind. The reduced signal to auditory stimulation in the CB group relative to both AB and SC groups in the superior temporal plane under low demand conditions indicates that functional alterations of cortical fields occur in the CB. The reduced fMRI signal volume in the temporal lobes of the CB group during auditory stimulation raises several possibilities. One, the decreased metabolic activity could reflect greater efficiency in processing of auditory stimuli thereby reducing the overall neural activity. Second, a change in the functional maps reflecting a posterior shift in the centre of mass of fMRI signal activity in the left hemisphere of CB subjects, suggests that a reorganization of auditory cortical fields occurs in congenital blindness. There was no evidence of greater grey matter or white matter in HG of the CB group. These results suggest that in the case of blindness gross structural changes are not present and therefore cortical hypertrophy is unlikely to underlie the enhanced auditory perceptual abilities documented in CB individuals. However, the changes in the organization of other posterior cortical regions due to visual deprivation clearly suggest that experience dependent changes occur. The lack of hypertrophy may be a result of expansion of auditory fields into more posterior temporal lobe areas due to the lack of visual stimulation that would otherwise influence the organization of those areas. Therefore, auditory cortical fields may have expanded so that they no longer reside within the normal anatomical boundaries. It is also possible that neuro-developmental factors such as retinopathy of prematurity, a cause of blindness in some of our subjects, may mask hypertrophy, due to loss of cortical tissue and ventricular enlargement present in many cases of premature birth. Further investigations of the functional and anatomical correlates of auditory enhancements in the blind may further clarify the role of experience on brain organization.

REFERENCES

Arno, P., De Volder, A. G., Vanlierde, A., Wanet-Defalque, M. C., Streel, E., Robert, A., Sanabria-Bohorquez, S., and Veraart, C., 2001, Occipital activation by pattern recognition in the early blind using auditory substitution for vision. *Neuroimage* **13**: 632-645.

Burton, H., Snyder, A. Z., Diamond, J. B., and Raichle, M. E., 2002, Adaptive changes in early and late blind: a FMRI study of verb generation to heard nouns. *J. Neurophysiol.* **88**: 3359-3371.

Elbert, T., Sterr, A., Rockstroh, B., Pantev, C., Muller, M. M., and Taub, E., 2002, Expansion of the tonotopic area in the auditory cortex of the blind. *J. Neurosci.* **22**: 9941-9944.

Muchnik, C., Efrati, M., Nemeth, E., Malin, M., *et al.*, 1991, Central auditory skills in blind and sighted subjects. *Scand. Audiol.* **20**: 19-23.

Niemeyer, W., and Starlinger, I., 1981, Do the blind hear better? Investigations on auditory processing in congenital or early acquired blindness. II. Central functions. *Audiology* **20**: 510-515.

Penhune, V. B., Zatorre, R. J., MacDonald, J. D., and Evans, A. C., 1996, Interhemispheric anatomical differences in human primary auditory cortex: probabilistic mapping and volume measurement from magnetic resonance scans. *Cereb. Cortex* **6**: 661-672.

Roder, B., Rosler, F., and Neville, H. J., 2000, Event-related potentials during auditory language processing in congenitally blind and sighted people. *Neuropsychologia* **38**: 1482-1502.

Roder, B., Rosler, F., and Neville, H. J., 2001, Auditory memory in congenitally blind adults: a behavioral-electrophysiological investigation. *Brain Res. Cogn. Brain Res.* **11**: 289-303.

Schneider, P., Scherg, M., Dosch, H. G., Specht, H. J., Gutschalk, A., and Rupp, A., 2002, Morphology of Heschl's gyrus reflects enhanced activation in the auditory cortex of musicians. *Nat. Neurosci.* **5**: 688-694.

Participants

Ahissar, Merav
Hebrew University
Dept. of Psychology
Mount Scopus,
Jerusalem, 91905
ISRAEL

Altschuler, Richard
University of Michigan
Kresge Hearing Research Institute
1301 East Ann,
Ann Arbor, 48109-0506
USA

Amitay, Sygal
MRC Institute of Hearing
Research, University Park
Nottingham, NG7 2RD
UNITED KINGDOM

Ari-Even Roth, Daphne
Tel-Aviv University, Department
of Communication Disorders,
Sackler Faculty of Medicine
6, Rav-ashi Str., Tel-Aviv, 69395
ISRAEL

Balasubramanian, Venu
Seton Hall University
Dept. of Speech-Language
Pathology & Audiology
400 South Orange Ave
South Orange, 7079
USA

Banai, Karen
Hebrew University of Jerusalem
Dept. of Neurobiology
P.O.B 24429 Mt. Scopus
Jerusalem, 91243
ISRAEL

Benson, Christina
Duke University Medical Center
Dept. of Neurobiology
Bryan Research Building
Research Drive,
Durham, NC 27612
USA

Ben-Yehudah, Gal
Hebrew University
Dept. of Neurobiology
Givat Ram, Jerusalem, 91904
ISRAEL

Berrebi, Albert
West Virginia Univ. School of
Medicine,
Sensory Neuroscience Research
Ctr. & Dept. of Otolaryngology
Hsc PO Box 9303
Morgantown, 26506-9303
USA

Bizley, Jennifer
University of Oxford
University Laboratory of
Physiology, Parks Road
Oxford, OX1 3PT
UNITED KINGDOM

Campbell, Robert
Oxford University
Laboratory of Physiology
Parks Road
Oxford, OX1 3PT
UNITED KINGDOM

Caspary, Donald
Southern Illinois University
School of Medicine, Dept.of
Pharmacology
801 N. Rutledge
Springfield, 62702
USA

Castellano, Orlando J.
University of Salamanca
Institute of Neuroscience of
„Castilla y León“ (INCyL)
Avda. Alfonso X, El Sabio, S/n
Salamanca, 37007
SPAIN

Collet, Lionel
CNRS UMR 5020 Neurosciences
et Systemes sensoriels,
Audiologie, Hospices Civils de
Lyon, 50 avenue Tony Garnier
Lyon, 69366
FRANCE

Cramer, Karina
University of California, Irvine
Dept. Neurobiology & Behavior
2205 McGaugh Hall
Irvine, 92697-4550
USA

Denham, Susan
University of Plymouth
Institute of Theoretical
Neuroscience
Drake Circus
Plymouth, PL4 8AA
UNITED KINGDOM

Doubell, Timothy
University of Oxford
University Laboratory of
Physiology, Parks Road
Oxford, OX26EG
UNITED KINGDOM

Druga, Rastislav
Department of Functional Anatomy, 2nd Medical Faculty, Charles University
José Martího 31, 162 52 Prague 6
CZECH REPUBLIC

Eggermont, Jos J.
University of Calgary
Dept. of Psychology
2500 University Drive N.W.
Calgary, T2N
CANADA

Ehret, Günter
University of Ulm
Dept. of Neurobiology
Albert-Einstein-Allee 11
Ulm, D-89069
GERMANY

Eichhammer, Peter
University of Regensburg
Dept. of Psychiatry
Universitaetsstrasse 84
Regensburg, 93053
GERMANY

Fujioka, Takako
The Rotman Research Institute
MEG Unit, 3560 Bathurst Street
Toronto, M6A 2E1
CANADA

Griffin, Sarah
University College London
Dept. of Physiology
Gower Street
London, WC1E 6BT
UNITED KINGDOM

Groh, Daniel
2nd Medical Faculty, Charles University, ENT Clinic
V Úvalu 84, 150 00 Prague 5
CZECH REPUBLIC

Grothe, Benedikt
Max-Planck-Institute of Neurobiology
Dept. of Auditory Processing
Am Klopferspitz 18a
Martinsried, D-82152
GERMANY

Grube, Manon
University of Leipzig
Dept. of Neurobiology
Talstr. 33, Leipzig, 4103
GERMANY

Guillemot, Frederick
Ecole Polytechnique de Montreal
Genie Informatique
Edouard Montpetit
Montreal, H3C 3J7
CANADA

Guillemot, Jean-Paul
Université du Québec a Montréal
Dept. of Kinesiology
St-Denis, Montréal, H3C 3P8
CANADA

Harris, Julie
University of Washington
Depts. Oto-HNS and Neurobiology & Behavior
Box 357923, Seattle, 98195-7923
USA

Herdman, Anthony
University of Toronto
Rotman Research Institute
3560 Bathurst Street
Toronto, M6A 2E1
CANADA

Hernández, Olga
Institute for Neuroscience of
Castilla y León, University of
Salamanca, Campus Miguel De
Unamuno, Salamanca, 37007
SPAIN

Hildesheimer, Minka
Tel-Aviv University, Department
of Communication Disorders
The Chaim Sheba Medical Center
Tel-hashomer, Ramat-gan, 52621
ISRAEL

Hromádka, Tomáš
Cold Spring Harbor Laboratory
Watson School of Biological
Sciences, 1 Bungtown Road
Cold Spring Harbor, 11724
USA

Hubka, Peter
Comenius University,
Inst. of Pathological Physiology
Sasinkova 4, Bratislava 811 08
SLOVAK REPUBLIC

Cheung, Steven
Univ. of California San Francisco,
Department of Otolaryngology-
Head and Neck Surgery
Box 0342 A730 400 Parnassus Av.
San Francisco, 94143-0342
USA

Jilek, Milan
Institute of Experimental
Medicine, Academy of Science
Dept.of Auditory Neuroscience
Videnska 1083, 142 20 Prague
CZECH REPUBLIC

Karmos, George
Institute for Psychology of the
Hungarian Academy of Science
Department of Psychophysiology
Szondi U.83-85 P.O.Box 398
Budapest, H-1394
HUNGARY

Kelly, Jack B.
Carleton Univ., Psychology Dept.
329 Life Science Building
1125 Colonel By Drive
Ottawa, K1S 5B6
CANADA

King, Andrew
University of Oxford
Laboratory of Physiology
Parks Road, Oxford, OX1 3PT
UNITED KINGDOM

Kishon-Rabin, Liat
Tel-Aviv University
Communication Disorders Dept
Sackler Faculty of Medicine
POB 39040, Tel-Aviv, 69978
ISRAEL

Klinke, Rainer
Goethe-Universitaet Frankfurt
Physiologisches Institut II
Theodor Stern Kai 7
Frankfurt, D 60590
GERMANY

Kraus, Nina
Northwestern University
Dept. of Communication Sciences
and Neurobiology & Physiology
2299 N, Campus Drive
Evanston, 60208 IL
USA

Kudo, Motoi
Shiga Univ. of Medical Sciences
Department of Anatomy 2
Seta Tsukinowa-cho
Otsu, 520-2192
JAPAN

Langguth, Berthold
University of Regensburg
Dept. for Psychiatry and
Psychotherapy
Universitaetsstr. 84,
Regensburg, 93049
GERMANY

Langner, Gerald
Darmstadt Univ. of Technology
Institute of Zoology
Schnittspahsntr. 3,
Darmstadt, 64287
GERMANY

Lippé, Sarah
Univ. of Montreal Ste-Justine`s
Hospital, Dept. of Psychology
C. P. 6128, Succursale Centre-ville
Montreal, H3C 3J7
CANADA

Lonsbury-Martin, Brenda
University of Colorado
Health Sciences Center
Dept. of Otolaryngology (B205)
4200 East Ninth Ave
Denver, 80262-0001
USA

Malmierca, Manuel S.
Univ. Salamanca
Institute for Neuroscience
Campus Miguel De Unamuno
Salamanca 37007
SPAIN

Martiat, Benoit
IFAC Bastogne Belgium
Dept. Otolaryngology
Strainchamps, 31 E
Fauvillers, 6637
BELGIUM

Martin, Glen
University of Colorado
Health Sciences Center
Dept. of Otolaryngology (B205)
4200 East Ninth Avenue
Denver, 80262-0001
USA

Maureira, Carlos
Centro de Estudios Científicos
Dept. of Biophysics and Molecular
Physiology
Arturo Prat 514, Valdivia, 1469
CHILE

Merchan, Miguel
University of Salamanca
Institute of Neuroscience
of Castilla y León
Campus Miguel de Unamuno
Salamanca, 37007
SPAIN

Merzenich, Michael M.
University of California, Keck
Center for Integrative
Neurosciences
HSE-828
San Francisco, CA 94143-0732
USA

Middlebrooks, John
University of Michigan
Kresge Hearing Research Institute
1301 E. Ann St.
Ann Arbor, 48109-0506
USA

Milenkovic, Ivan
Zoological Institute, University of
Leipzig, Dept. of Neurobiology
Talstr. 33, Leipzig, 4103
GERMANY

Moore, David
MRC Institute of Hearing
Research, University Park
Nottingham, NG7 2RD
UNITED KINGDOM

Neville, Helen
University of Oregon
Dept. of Psychology
1227 University of Oregon
Eugene, 97403-1227, Oregon
USA

Nikitin, Nikolay
I.P.Pavlov Institute of Physiology,
Russian Academy of Science
Laboratory of Hearing Physiology
Nab.Makarova, 6
Saint Petersburg, 199034
RUSSIA

Ogbo, Fidel C. Nwabueze-
Institute of Experimental
Medicine, Academy of Sciences
Dept. of Auditory Neuroscience
Videnska 1083, 142 20 Prague
CZECH REPUBLIC

Ohl, Frank
Leibniz Institute for Neurobiology
Dept. of Acoustics, Learning &
Speech
Brenneckestr. 6,
Magdeburg, D-39118
GERMANY

Ouda, Ladislav
Institute of Experimental
Medicine, Academy of Sciences
Dept. of Auditory Neuroscience
Videnska 1083, 142 20 Prague
CZECH REPUBLIC

Palmer, Alan
MRC Institute of Hearing
Research,
Dept. of Neurophysiology
University Park
Nottingham, NG7 2RD
UNITED KINGDOM

Peretz, Isabelle
University of Montreal
Dept. of Psychology
C.p. 6128, Succ. Centre-ville
Quebec, H3C 3J7
CANADA

Popelář, Jiří
Institute of Experimental
Medicine, Academy of Sciences
Dept. of Auditory Neuroscience
Videnska 1083, 142 20 Prague
CZECH REPUBLIC

Portfors, Christine
Washington State University
School of Biological Sciences
14204 NE Salmon Creek Avenue
Vancouver, 98663
USA

Putter-Katz, Hanna
Hebrew University of Jerusalem,
Sheba Medical Center, Dept. of
Psychology, Speech and Hearing
Center
Mount Scopus, Jerusalem, 92950
ISRAEL

Radionova, Elena
I.P.Pavlov Institute of Physiology
Russian Academy of Sciences
Laboratory of Hearing Physiology
Nab. Makarova, 6
Saint Petersburg, 199034
RUSSIA

Rajan, Ramesh
Monash University,
Dept. of Physiology
Wellington Road
Clayton, VIC 3800
AUSTRALIA

Renvall, Hanna
Brain Research Unit, Low
Temperature Laboratory
Helsinky University of
Technology
Otakaari 3A, FIN-02015
FINLAND

Riquimaroux, Hiroshi
Doshisha University
Knowledge Engineering &
Computer Sciences
Tatara Miyakotani
Kyotanabe, 610-0321
JAPAN

Roberts, Larry E.
McMaster University
Department of Psychology
Hamilton, L8S 4K1
CANADA

Robitaille, Nicolas
Université de Montréal
Dept. de Psychologie
C.p. 6128, Succursale Centre-ville
Montréal, H3C 3J7
CANADA

Ross, Bernhard
Muenster University Hospital
Institute of Experimental
Audiology
Kardinal-von-Galen Ring 10
Muenster, D-48129
GERMANY

Rubel, Edwin
University of Washington
Virginia Merrill Bloedel Hearing
Research Center and
Otolaryngology-HNS, Box 357923
Seattle, 98195
USA

Rübsamen, Rudolf
University of Leipzig
Faculty of Bioscience, Pharmacy
and Psychology
Talstrasse 33, Leipzig, 4103
GERMANY

Rybalko, Natalia
Institute of Experimental Medicine
Academy of Sciences
Dept. of Auditory Neuroscience
Videnska 1083, 142 20 Prague
CZECH REPUBLIC

Saint-Amour, Dave
Université de Montréal
Dept. de Psychologie
C.P. 6128 Succ. Centre Ville
Montréal, H3C 3J7, Québec
CANADA

Saldana, Enrique
University of Salamanca
Institute of Neuroscience of
Castilla y León (INCyL)
Avda. Alfonso X, El Sabio, S/n
Salamanca, 37007
SPAIN

Schlosshauer, Burkhard
NMI Natural and Medical Sciences
Institute, Dept. of Neurobiology
Markwiesenstr. 55
Reutlingen, D-72770
GERMANY

Schnupp, Jan W.
Oxford University
Dept. of Physiology
Parks Road, Oxford, OX1 3PT
UNITED KINGDOM

Schreiner, Christoph
Univ. of California, San Francisco
Dept. of Otolaryngology
513 Parnassus Avenue
San Francisco, CA 94143-0732
USA

Sorensen, Staci
University of Washington
Dept. of Neurobiology and
Behavior, Box 357923
Seattle, 98195-7923
USA

Spirou, George
West Virginia University, Sensory
Neuroscience Research Center
PO Box 9303 HSC
One Medical Center Drive
Morgantown, 26506-9303
West Virginia
USA

Stevens, Alexander
Oregon Health & Science
University
Dept. of Psychiatry
3181 Sw Sam Jackson Park Rd.
Mail Code Cr139
Portland, 97239-3098
USA

Suga, Nobuo
Washington University
Department of Biology
St. Louis, MO 63130
USA

Supin, Alexander J.
Severtsov Institute of Ecology and
Evolution,
Russian Academy of Sciences
33 Leninsky Prosp.,
119071 Moscow
RUSSIA

Sur, Mriganka
MIT, Department of Brain and
Cognitive Sciences
45 Carleton St, E25-235
Cambridge, MA 02139
USA

Šuta, Daniel
Institute of Experimental Medicine
Academy of Sciences
Dept. of Auditory Neuroscience
Videnska 1083, 142 20 Prague
CZECH REPUBLIC

Swirszcz, Krystyna
Warsaw University
Faculty of Psychology
Stawki 5/7, 00-183 Warsaw
POLAND

Syka, Josef
Institute of Experimental Medicine
Academy of Sciences
Dept. of Auditory Neuroscience
Videnska 1083, 142 00 Prague 4
CZECH REPUBLIC

Tallal, Paula
Rutgers University
Center for Molecular and
Behavioral Neuroscience
197 University Avenue
Newark 7102
USA

Tureček, Rostislav
Institute of Experimental Medicine
Academy of Sciences
Dept. of Auditory Neuroscience
Videnska 1083, 142 20 Prague
CZECH REPUBLIC

Tzounopoulos, Thanos
Oregon Health & Sciences University,
Oregon Hearing Research Center
& Vollum Institute, L335A
3181 Sw Sam Jackson Park Rd
Portland, 97239
USA

Ungan, Pekcan
Hacettepe Univ. Medical Faculty
Department of Biophysics
Hacettepe, Ankara, 6100
TURKEY

Vinuela, Antonio
University of Salamanca
Institute of Neuroscience of
„Castilla y León“ (INCyL)
Avda. Alfonso X, El Sabio, S/n
Salamanca, 37007
SPAIN

Wang, Xiaoqin
Johns Hopkins University
Dept. of Biomedical Engineering
720 Rutland Avenue, Ross 424
Baltimore, 21205
USA

Winer, Jeffery
Univ. of California at Berkeley
Dept. of Molecular and Cell Biology
Room 289 Life Sciences Addition
Berkeley, 94720-3200
USA

Wu, Shu Hui
Carleton University
Psychology Department
1125 Colonel By Drive
Ottawa, K1S 5B6
CANADA

Young, Eric
Johns Hopkins University
Dept of Biomedical Engineering
720 Rutland Ave.
Baltimore, MD 21205
USA

Zatorre, Robert
Montreal Neurological Institute, McGill University, Dept. of Neuropsychology/Cognitive Neuroscience
Montreal, H3A 2B4
CANADA

International Symposium on Plasticity of the Central Auditory System and Processing of Complex Acoustic Signals
held July 7 – 10, 2003
Prague, Czech Republic

Index

MIX
Papier aus verantwortungsvollen Quellen
Paper from responsible sources
FSC® C105338

If you have any concerns about our products,
you can contact us on
ProductSafety@springernature.com

In case Publisher is established outside the EU,
the EU authorized representative is:
Springer Nature Customer Service Center GmbH
Europaplatz 3, 69115 Heidelberg, Germany

Printed by Libri Plureos GmbH
in Hamburg, Germany